Environmental Science, Engineering and Technology Series

Check Dams, Morphological Adjustments and Erosion Control in Torrential Streams

ENVIRONMENTAL SCIENCE, ENGINEERING AND TECHNOLOGY SERIES

Nitrous Oxide Emissions Research Progress
Adam I. Sheldon and Edward P. Barnhart (Editors)
2009. ISBN: 978-1-60692-267-5

Fundamentals and Applications of Biosorption Isotherms, Kinetics and Thermodynamics
Yu Liu and Jianlong Wang (Editors)
2009. ISBN: 978-1-60741-169-7

Environmental Effects of Off-Highway Vehicles
Douglas S. Ouren, Christopher Haas, Cynthia P. Melcher, Susan C. Stewart, Phadrea D. Ponds, Natalie R. Sexton, Lucy Burris, Tammy Fancher and Zachary H. Bowen
2009. ISBN: 978-1-60692-936-0

Agricultural Runoff, Coastal Engineering and Flooding
Christopher A. Hudspeth and Timothy E. Reeve (Editors)
2009. ISBN: 978-1-60741-097-3

Agricultural Runoff, Coastal Engineering and Flooding
Christopher A. Hudspeth and Timothy E. Reeve (Editors)
2009. ISBN: 978-1-60876-608-6
(Online Book)

Conservation of Natural Resources
Nikolas J. Kudrow (Editor)
2009. ISBN: 978-1-60741-178-9

Conservation of Natural Resources
Nikolas J. Kudrow (Editor)
2009. ISBN: 978-1-60876-642-6
(Online Book)

Directory of Conservation Funding Sources for Developing Countries: Conservation Biology, Education and Training, Fellowships and Scholarships
Alfred O. Owino and Joseph O. Oyugi
2009. ISBN: 978-1-60741-367-7

Forest Canopies: Forest Production, Ecosystem Health and Climate Conditions
Jason D. Creighton and Paul J. Roney (Editors)
2009. ISBN: 978-1-60741-457-5

Soil Fertility
Derek P. Lucero and Joseph E. Boggs (Editors)
2009. ISBN: 978-1-60741-466-7

The Amazon Gold Rush and Environmental Mercury Contamination
Daniel Marcos Bonotto and Ene Glória da Silveira
2009. ISBN: 978-1-60741-609-8

Process Engineering in Plant-Based Products
Hongzhang Chen
2009. ISBN: 978-1-60741-962-4

Buildings and the Environment
Jonas Nemecek and Patrik Schulz (Editors)
2009. ISBN: 978-1-60876-128-9

Tree Growth: Influences, Layers and Types
Wesley P. Karam (Editor)
2009. ISBN: 978-1-60741-784-2

Syngas: Production Methods, Post Treatment and Economics
Adorjan Kurucz and Izsak Bencik (Editors)
2009. ISBN: 978-1-60741-841-2

Syngas: Production Methods, Post Treatment and Economics
Adorjan Kurucz and Izsak Bencik (Editors)
2009. ISBN: 978-1-61668-214-9
(Online Book)

Potential of Activated Sludge Utilization
Xiaoyi Yang
2010. ISBN: 978-1-60876-019-0

Psychological Approaches to Sustainability: Current Trends in Theory, Research and Applications
Victor Corral-Verdugo, Cirilo H. Garcia-Cadena and Martha Frias-Armenta (Editors)
2010. ISBN: 978-1-60876-356-6

Carbon Capture and Storage including Coal-Fired Power Plants
Todd P. Carington (Editor)
2009. ISBN: 978-1-60741-196-3

Process Engineering in Plant-Based Products
Hongzhang Chen
2009. ISBN: 978-1-60741-962-4

Fluid Waste Disposal
Kay W. Canton (Editor)
2010. ISBN: 978-1-60741-915-0

Recent Progress on Earthquake Geology
Pierpaolo Guarnieri (Editor)
2010. ISBN: 978-1-60876-147-0

Check Dams, Morphological Adjustments and Erosion Control in Torrential Streams
Carmelo Consesa Garcia and Mario Aristide Lenzi (Editors)
2010. ISBN: 978-1-60876-146-3

ENVIRONMENTAL SCIENCE, ENGINEERING AND TECHNOLOGY SERIES

CHECK DAMS, MORPHOLOGICAL ADJUSTMENTS AND EROSION CONTROL IN TORRENTIAL STREAMS

CARMELO CONESA GARCIA
AND
MARIO ARISTIDE LENZI
EDITORS

New York

For permission to use material from this book please contact us:
Telephone 631-231-7269; Fax 631-231-8175
Web Site: http://www.novapublishers.com

LIBRARY OF CONGRESS CATALOGING-IN-PUBLICATION DATA

Check dams, morphological adjustments and erosion control in torrential streams / Carmelo Consesa Garcia and Mario Aristide Lenzi, editors.
xii, 339 p. : ill., maps (some col.) ; 27 cm.
Includes bibliographical references and index.
ISBN: 978-1-62618-856-3 (soft cover)
1. Environmental science, engineering and technology series. I. Consesa Garcia, Carmelo II. Aristide Lenzi, Mario.
TC549 .C54 2010
627/.8

Published by Nova Science Publishers, Inc. † New York

CONTENTS

PREFACE

"Remember when discoursing about water or fluids to adduce first experience and then reason"
Leonardo da Vinci (1452-1519)

"In geomorphologic systems the ability to measure may always exceed ability to forecast or explain"
L.B. Leopold and W.B. Langbein

For several decades, the check dam building has been one of the most commonly used structural measures in the Watershed and Catchments Management, Forestry Hydrological Restoration Programs and Plans in Torrential Catchments. Thanks to updated knowledge on river dynamics and sediment transport in steep streams, check-dams and hydraulic works in mountain rivers improved their planning out in the design phase. Bedload discharge and debris flow researches focused on total discharge assessment, together with main hydraulic works to control sediments in mountain torrents (closed and open check dams, grade control structures, retention systems, canalisations), become the key words of the control works in torrential streams. These works are since long time in use for mountain catchments protection and mitigation of flood hazards, but they could increase the efficiency if we clarify, during the design, their functioning objectives.

The main purpose of this book is to analyze the effectiveness of check dams and its influence on the channel stabilization and the erosion control in torrential catchments. In view of the great number of adjustable components in a torrential corrected channel (structure types, geometry of cross-sections, bed material, channel pattern, bed slope, bed roughness, etc.), response to sediment discharge changes (excess of solids -overfeeding- or sediment deficit) induced by check dams is very rapid and complex. The magnitude and type of channel adjustment change in accordance with the morphological organization of the fluvial system, even if the supply of sediments is kept stable after the refill of drop structures. According to it, a significant part of the book is focused to acquire a greater comprehension of the bed stability variations, on the other hand necessary to recover an equilibrium slope stable. Other important aims are to establish criteria of optimization for a suitable location of the check dams, capable of providing relative stability to the system, and analyze the morphological effects upstream and downstream associated with different sedimentation, bed-scouring and local scouring processes.

Experiences and examined study cases represent different hydroclimatic regimes (from arid, semiarid, mediterranean, temperate, cold humid, and sub-tropical climate) associated with different geographical regions in Europe, North America, Asia and Australia. These purposes are contained on the 14 chapters, written by different professors, researchers and practitioners working on the worldwide. We thank each of these individuals for their efforts and their contributions to each chapter.

This has been a very enjoyable book to edit; and we hope that you gain equal pleasure and understanding from reading it. We would like to dedicate this book to ours under-grade and graduate students, who have been a continuing source of intellectual exchange, inspiration, and surprise.

Carmelo CONESA-GARCIA
Mario Aristide LENZI

ABOUT THE EDITORS

Carmelo CONESA-GARCIA is Professor (from 1990) of *Applied Physical Geography* and *Fluvial Morphology* at the University of Murcia, Spain. Researcher and Professor invited by the 'Centre de Géomorphologie du CNRS' y CREGEPE (Centre de Recherches en Géographie Physique de l'Environnement), France (1991); Professor at the GIS Summer School, "ERASMUS Curriculum Development Programme" (1995). Professor invited by the 'UFR des Sciences de la Terre et de l'Aménagement Régional, GÉOPHEN (Géographie Physique et Environnement)', Université de Caen Basse Normandie, France (2001). Editor of 8 books. Author and co-author of more than 140 publications of which 26 are in International ISI Journals; subjects: desertification in semiarid environments, fluvial processes, morphological channel adjustments, ephemeral channels, sediment transport, bed erosion, hydrologic management and forestry restoration, environmental changes, natural hazards and floods. Collaborator in 18 national and international research projects and contracts, among them MEDALUS III (Mediterranean Desertification and Land Use), Project 4 (*Ephemeral Channels and Rivers*), European Commission (1996-98), Projects with Foreign Institutions, Agreements between the CSIC (Spain) and the CNRST (Morocco) (2002-03); Project of the Interuniversity Cooperation Agency (2004-05); Scientific Coordinator of international projects such as the financed one by the Foundation Euro-Mediterranean Institute of Water, Council of Europe (2003-05). Project scientific assessor in the Spanish Plan of I+D+I., National Program of Biodiversity, Sciences of the Earth and Global Change. Lecturer chief invited in national congresses, and participant in different scientific international conferences, e.g. International Geomorphology Conferences (Hamilton, 1993; Bologna, 1997; Zaragoza, 2005), Conferences on "Erosion and land degradation in the Mediterranean" (Aveiro, 1995; London, 1997; Cagliari, 1998), Symposiums and Workshops on Applied Hydrology (Sao Paulo, 1993; Barcelona, 2002). Member of international Editorial Boards in journals such as the Bulletin of the Spanish Association of Geography (Journal ISI) and GeoFocus (International Journal of Geographical Information Science and Technology). Chief Organizer of the XI National Congress of Quantitative Methods, GIS and Remote Sensing in Spain, 2004.

Mario Aristide LENZI is Full Professor (from 2005) of "Fluvial Morphology and River Restoration" and Scientific Coordinator (from 2001) of PhD in "Catchments Management and Geomatic" (Doctoral School on Land, Environment, Resources and Health), at the University of Padova, School of Agriculture, Italy. Head of the

Degree-Course Board “Land protection and maintenance” (from 2003) and Academician of the “Accademia Italiana di Scienze Forestali” (from 2004), he is also Lecturer of “Mountain fluvial morphology and stream restoration” at the European Erasmus Mundus MSc “SUFONAMA” (Sustainable Forestry and Nature Management), Course offered by a consortium of five European universities: The University of Copenhagen, Centre for Forest, Landscape and Planning, Copenhagen, Denmark (Coordinator); Bangor University, School of Environment and Natural Resources, Bangor, UK; University of Göttingen, Faculty of Forest Sciences and Ecology, Göttingen, Germany; Swedish University of Agricultural Sciences, Alnarp, Sweden; University of Padova, Faculty of Agriculture, Padova, Italy. Author and co-author of more than 130 publications of which 35 are in International ISI Journals; subjects: sediment transport, local scouring, debris flow, mountain rivers, catchments management, large wood debris in streams, stream and fluvial hydrodynamics and morphodynamics, river and stream restoration. Scientific Coordinator of the University of Padova research team in: 1) International Program for Western Europe "Flow hydraulics along step-pool channels", supported by the Colorado State University and N.S.F. of USA; 2) European Research projects: EPIC FORCE: “Evidence-based policy for integrated control of forested river catchments in extreme rainfall and snowmelt”, 2005-08; DAMOCLES: “Debris-fall hazard assessment in mountain catchments for local end users”, 2000-03; EROSLOPE II "Dynamic of sediments and water in Alpine catchments: processes and prediction", 1996-99. He has focused his research on mountain rivers and catchments management. His extensive field work in Italy, Spain, Europe, South America and North America has provided him the kind of invaluable, first-hand experience that makes this book a work of authority.

In: Check Dams, Morphological Adjustments…
Editors: C. Conesa and M. A. Lenzi
ISBN: 978-1-62618-856-3

Chapter 1

TRADITIONAL USES OF CHECK DAMS: A GLOBAL AND HISTORICAL INTRODUCTION

William E. Doolittle
Department of Geography and the Environment,
The University of Texas at Austin, USA

ABSTRACT

This chapter uses examples from various parts of the world and from ancient, historic, and modern times to provide a brief overview of the varied uses of check dams. It begins with a few personal reflections and ideas about check dams as human constructs, literally and philosophically. Discussion then turns to their settings, construction materials and how check dams might have originated, and functions, both obvious and nuanced. It concludes with a discussion of some new concepts stemming from the recent literature on environmental degradation.

INTRODUCTION

While watching my grandchildren play outdoors recently, I reflected back on my own childhood, recalling memories of losing battle after battle with nature. My childhood was probably typical. I built sand castles on the beach only to watch them wash away under the waves of rising tides. I played in the dirt behind my house, building roads with and for my toy trucks, only to watch slopes all too steep collapse under their own weight and bury my newly constructed roads as well as my trucks. While playing in the rain with a stick I scratched a "ditch," a groove really, in order to divert runoff from a rivulet where it flowed naturally to where I wanted it to flow. But alas, the water cut its way back in a self-selected course. As a teenager, I stood with pride looking at the lawn I just mowed so neatly, only to experience a few days later the frustration of seeing it had grown again to unacceptable heights. Success after success was followed always by failure after failure. These experiences though provided a valuable education. Unwittingly, I learned something about coastal processes, mass wasting, sedimentation, hydrology, fluvial geomorphology, and vegetation succession. These proved to be profound lessons in my understanding of the resiliency of nature, the fragility of human constructions, and the foolishness of the concept of permanence, lessons apparently not learned by many planners, architects, civil engineers, and policy-makers. In addition to being a time of physical growth and intellectual development (as well as fun, of course), my childhood became for me a personal microcosm and a metaphor of humans in relationship to the biophysical environment—an accumulation of knowledge that provided me insights to, and wisdom about, deliberate actions with both intended and unintended consequences.

These basic childhood activities were for me, and are for others, self-learned. No one has to teach a toddler how to scoop and pile sand. No one has to teach a young child how to move dirt. No one has to teach a youth how to scratch the earth with a stick; chimpanzees even do it! The same holds true for vegetation removal. Things that come naturally to children today doubtless came naturally to children millennia ago, a sort of behavioral uniformitarianism. It is logical to assume, therefore and by extension, that ancient peoples, individuals and cultures, learned from nature, and from their interactions with it.

Some features on the landscape that were deliberately made by humans look as though they came about naturally, while some that formed naturally appear as though made by people (Figure 1). And, therein lays the mystique of check dams, at least for those of us whose work is principally in archaeological and historical contexts (Doolittle, Neely, and Pool 1993). Viewed in terms of environmental change, there may well be no deliberate human activity that has a greater impact given its minimal input than the construction of check dams. The simple act of dropping an obstacle across a small water course affects land both upstream and downstream.

Figure 1. Were these rocks deposited naturally, or is this a check dam built by people, near Quetta, Pakistan? If it is a check dam, when was it constructed, years ago, decades ago, centuries ago, or millennia ago? Photograph by Diana Davis.

The act itself might take only seconds, but the impact can last for years, even decades, and create countless ripples throughout thc total ecosystem in which the feature was built. Were this action not simple enough, check dams can also be “constructed,” if that word applies, inadvertently during the course of other activities. For example, trees felled during the course of swidden cultivation have been known to fall across small stream beds thereby becoming check dams even though they were neither planned nor built as such (Kunstadter 1978:328).

Fascinating; check dams are simply fascinating in so many different ways, as the following chapters in this volume attest. To begin with, however, the remainder of this chapter will introduce, briefly, some basic aspects of check dams, citing a few examples from around the world, in prehistoric, historic, and present-day contexts.

“Check” and “dam.” The words themselves remind me of something once said by the late Bernard Q. Nietschmann (1974). “When two amorphous terms…are joined we end up with something where the whole is less than the sum of its parts.” In this particular case it has to be because of “dam.” Mere mention of the term connotes images of the High Aswan *Dam*, Hoover *Dam*, or the Three Gorges *Dam*. Comparing any “check” dam to one of these structures is not unlike comparing a paper airplane to the Airbus A380. In the latter case, both have wings and fly. In the former case, both are structures paralleling contours that serve to impede runoff. Similarities end there, however, beginning in terms of locations.

SETTINGS

Unlike their larger counterparts, check dams have been built traditionally in two general types of settings or situations, across channel bottoms and on hill slopes. Channel bottoms are the most common places, and, of course, “channel bottom” itself is a rather loose term. The huge Sayano-Shushenskaya Dam in Russia was built across the bottom of the Yenisei River channel; the small(er) Alcántara Dam in Spain lies across the bottom of the Tagus River channel; and the even smaller yet Derwent Dam in England is situated in the bottom of the River Derwent’s channel. Typically, however, check dams cross the bottoms of even far smaller channels. And, these channels are typically dry and formed by erosion—*wadis*, *arroyos*, gullies, gulches, ravines.

Regardless what name they go by, the channels in which check dams were or are built have dry bottoms most of the year. Characteristic of arid lands, these channels carry water only after extreme rainfall events upstream, such as summertime convective thunderstorms. Often referred to as “flash floods,” stream flows of this nature typically involve huge amounts of rapidly moving water (Reid, et al. 1994). They carry an abundance of sediment, both in suspension and in bed load (Garcia, Laronne, and Sala 2000). They also carry a great deal of organic matter such as leaves, branches, and dead animals. Their onset is sudden, with high velocities and volumes. They end as rapidly and as dramatically as they begin. In those places of higher elevation and/or higher latitude, springtime snowmelt can also provide temporary flows. These, however, begin slowly, increase in volume and velocity at a moderate rate, and then gradually decrease to zero. Erosion and sediment transport are markedly less in these cases than in those where thunderstorms are involved.

Stream channels in arid lands vary in size from a few meters in width and less than a meter deep, to a kilometer or more in width and several meters deep. For the most part, however, check dams tend to be built in those that are shallow and not excessively wide—something on the order of less than a meter deep to less than 10 meters wide.

Given that check dams are found across channel bottoms that are wide and shallow, thereby resulting in features that are long and low, it should come as no surprise that check dams are also found along hill slopes. In such settings, it is sometimes difficult to distinguish between check dams and low, long terraces (Humphreys and Brookfield 1991). Indeed in one ancient site in southern Arizona, USA, check dams constructed across the bottom of a shallow drainage—that is concave in cross-sectional shape—become long and low terraces as one moves upstream/upslope onto the hill slope—that is convex in cross-sectional shape (Doolittle and Neely 2004:9-10). Depending on the materials with which they are made, check dams on hill slopes can be easily confused with lynchets (Macnab 1965), and terracettes (Higgins 1982).

Materials

Dead trees falling across small streams (Marston 1982) or on slopes paralleling contours (Doolittle 2000:259) might well have been nature's prototypical check dams that humans mimicked. As such, wood is one material long used in check dam construction. Wooden check dams typically involve logs lying horizontally on the surface. One example of this comes from northern Thailand (Kunstadter 1978:332). Cases have been documented, however, in which trees and branches approximately 10 cm in diameter were cut into short sections and driven into the ground, touching each other, thereby forming a small, less than 20 cm high, palisade-like row across small drainages. Perhaps the best examples of this come from Melanesia (Barrau 1958:41-42) and Haiti (Figure 2). Much more common, in fact nearly ubiquitous, is the construction of stake-and-brush check dams. These features involve stakes 2-3 cm in diameter driven into the ground 1-2 m apart, perpendicular to the flow. The stakes protrude above the ground approximately 1/2 m, and small branches, twigs, and brush are then woven between them. The Food and Agriculture Organization (FAO) of the United Nations has long advocated the construction of such features in dry land agricultural development projects around the world (Geyik 1986:28-33). Indeed, stake-and-brush check dams are particularly useful in the establishment of new agricultural fields when disposal of cleared brush is as important as the construction of other field features (Doolittle 1984:128).

One material that is never in short supply for the construction of check dams is dirt—the earth itself. Earthen check dams constructed across channel bottoms are typically trapezoidal in cross-sectional shape, 3-4 times wider at the base than they are high, at least according to FAO guidelines (Geyik 1986:28-29). By virtue of being subjected to highly erosive stream flows, there is no archaeological or historic remains of earthen check dams across channel bottoms. In contrast, there is much evidence of earthen check dams having been built on hill slopes. Better known as "tie-ridging" or "ridge-and-bed structures," some of the best examples are from southern México and highland Guatemala (Wilken 1987:136).

Nearly as ever-present as earth, rocks have been, without question, the most commonly used check dam building material. Rocks have many advantages over wood and earth.

Figure 2. A palisade-like check dam in Haiti. Photograph by Charles Palmer.

They do not deteriorate like wood, and they do not erode like earth. Rocks have size and weight, and check dams made with them have gravity working to their advantage. Simple one-row check dams typically involve rocks that are approximately the size of footballs (soccer balls), laid adjacent to each other, and perpendicular to the stream/sheet flow. The same holds true for check dams that are two rows wide at the base and two courses high. Larger check dams frequently involve a mixture of rock sizes, with smaller rocks filling the chinks between larger rocks. Irregularly shaped rocks tend to be preferred over water-worn cobbles in these cases as the sharp edges interlock, thereby increasing the structural integrity of the check dams. The archaeological literature on the American Southwest and the Middle East is replete with thousands of examples of rock check dams. As with wood and earth, rocks used in check dam construction are usually found close at hand, they are rarely brought in from a distant source.

Rocks are undoubtedly one of nature's great building materials. Humans, however, always seem to want to improve on nature, and such is also the case with check dams. In recent years, engineers have attempted to improve the permanence of check dams by encasing rocks in heavy gauge wire cages or baskets a meter or so square, and connecting these baskets together to make a larger structure. Known formally as "gabions," these features are always built across the channels of recently dissected streams, and have been dubbed as "gully plugs" (Geyik 1986:40). Rarely do they survive very long (Figure 3).

In addition to the use of steel, synthetic materials are increasingly being used in check dams. Drive by any construction site today in a country with strict building and environmental codes and you will have a hard time missing the sediment traps enclosing the area in which work is taking place. They are everywhere; alongside highways being widened, at the bases of hills on which buildings are being erected, in parks where playground equipment is being installed. Invariably they are made of synthetic materials, steel rods driven into the earth a few meters apart and extending upward 1/3 to 1/2 m. Nylon or plastic sheeting is strung between and attached to the rods, thereby forming a moderately flexible and vertical barrier that catches sediment.

Figure 3. A breached gabion in the southwestern United States. It that failed shortly after construction, despite the best intentions of its designer/builders.

Unlike check dams built with other materials, these are intentionally temporary. The others have short lives as well, but are not built to be torn down later once their job is finished. These synthetic and non-biodegradable materials are not reused and they are not recyclable. They end up in landfills, essentially becoming what they once were intended to keep in place.

FUNCTIONS

Check dams serve one or more of three functions. They can control water, conserve soil, and improve land. Although these tasks may appear at first glance to be relatively straightforward, there are actually a number of fascinating subtleties and nuances that make them quite complex. Be they located in channel bottoms or on gentle hill slopes, check dams, regardless of the materials of which they are made, impede or literally "check" the flow of water. Obstacles placed across the direction of flow dissipate the energy of flowing water. Once slowed, flowing water is more easily managed than rapidly flowing water. But, why slow the flow one might ask? Unlike their larger counterparts, true "dams," check dams are not intended to impound water in reservoirs. This is particularly true for those built on hill slopes where true dams are never built. Check dams built across gentle hill slopes are typically low but vary in length. In addition to slowing the overland flow these features spread the sheet flow or runoff uniformly across the surface so that some areas are not subjected to heavy flows while other areas receive minimal amounts of water. In this respect check dams can be envisioned much like the bumpers in a pinball machine. The indigenous

Zuni people of North American continue to use check dams in this manner, as they have for centuries (Cushing 1920:249-256; Norton, Sandor, and White 2007).

Taking this notion one step future, check dams on hillsides can also serve to divert water away from places down slope where it is not wanted or to places down slope where it is wanted. Examples of this can be found from Nabatean times in the Negev Desert (Evenari, Shanan, and Tadmor 1982) and those of the latter variety are associated with what has been called "water harvesting" (United Nations Environment Programme1983). Check dams built across stream channels can divert water into canals for the irrigation of crops at other locales, or other purposes. Again, there are plenty of examples from the Negev (Yair 2001). In this latter context, they are perhaps better described as "weirs," structures that are either permeable or easily topped, and that slow all of the flow but divert only part of it. Check dams that double as weirs are usually much larger than typical check dams.

Small check dams, often built in series or multiples, slow the velocity of flowing water principally to prevent, reduce, or halt soil erosion, and in some cases as attempts to restored degraded lands. Rapidly flowing water tends to be erosive. Accordingly, a series of check dams built on a gentle hill slope can prevent the formation of rills and rivulets, thereby mitigating the hazard of sheet erosion. Similarly a set of check dams built in a small drainage with a parabolic cross-sectional shape in a formerly grassy area that was cleared in order to plant crops might prevent the normal runoff from cutting a straight-sided gully. And, if such a gully is in its infant stages of formation, check dams can be constructed in attempts to stop the erosion. Given that rapidly flowing water tends to carry in suspension the sediment eroded from upstream or upslope, decelerated flow normally results in the deposition of sediment behind each check dam. In at least one case from the American Southwest, prehistoric farmers built series of check dams in drainages to trap coarse sediment eroding off upland surfaces and kept it from burying crops cultivated in a shallow basin downstream (Doolittle 1985). Doubtless, this was done in many other parts of the world as well.

Check dams can make land less susceptible to erosion than it might normally be, and they can help restore eroded land to its former condition. Additionally, check dams can improve land in at least two ways, by simultaneously leveling the land surface and increasing soil depth.

By virtue of its setting, its impeding the flow of water, and its trapping of sediment, a single check dam results in the transformation of the physical environment in its immediate proximity. Usually within a brief period after its construction, a check dam will begin collect sediment on its upstream side. Sediment will be continuously collected until its surface elevation equals that of the top of the check dam. If a check dam is 30 cm high, sediment collected behind it will be a maximum of 30 cm deep. The depth of this sediment decreases as one moves upslope or upstream of the check dam until the point at which the natural surface elevation is equal to that of the check dam's top. This, then, is the optimum point at which the next check dam upstream in the series should be constructed. Check dams everywhere in the world, throughout history, and regardless of building material, have resulted in level land and deep soils. What people have done with these new lands varies. In some places such as the Negev, they were farmed, and therefore what began as a check dam became a terrace. In other places, these newly created (albeit inadvertently) surfaces, remained uncultivated; features constructed as check dams remained check dams. However, if the velocity of flowing water and the amount of sediment in suspension increased, then check dams might well be breached (Figure 4), resulting in increased erosion both upslope or upstream, and in the immediate vicinity of the check dams, with subsequent heavy amounts of sedimentation further downstream or down slope. In sum, check dams carry with them the seeds of their own destruction. They are victims of their own success. First they are buried; then they are breached. The problem that gave rise to the construction of check dams reappears shortly after it was mitigated by the check dams.

Figure 4. Ancient and abandoned check dams in the American Southwest. Victims of their own success.

CONCLUSION

Check dams are perhaps humanity's most primordial form of geomorphic transformation. Dropping an obstacle such a rock or a branch into or across a rivulet can have a small but profound impact. Such actions can be, and indeed are often thought of as, intentional. They can, however, also be inadvertent. Just as humans can learn from nature, our ancestors probably learned from the unintended consequences of their accidents. For example, the finding of sediment trapped behind a piece of firewood dropped on the way home the previous day, before a night time thunderstorm, might have been a real "eye-opener," a discovery that gave rise to later attempts to replicate sedimentation intentionally, with other materials in other settings. The first check dam discoverer/experimenter did not know anything about hydrology, fluvial geomorphology, soil science, or agricultural engineering. But she or he doubtless discovered, in short order, the ability of a system to return to its previous state after a small disturbance. Ecologists thinking in terms of equilibrium used to call this "stability" (Holling 1973). The capacity of a system to absorb such small changes without significantly altering other aspects of the system is now better-known as "resilience" (Johnson and Lewis 2007:5).

Humans have long transformed their environments by building various features. The process of altering the natural world and replacing what was present with a new system state that is productive in human terms for the foreseeable future has been labeled "creative destruction" (Johnson and Lewis 2007:28). Check dams certainly fit this description. Resource systems can contain critical flaws internal to the system that undermine its viability. Such flaws are often insidious, masked by short-term superficial signs of success, leaving the system vulnerable to catastrophic collapses, and are labeled "destructive creation" (Johnson and Lewis 2007:42). Check dams fit this description as well. They conform, in other words, to concept of"self-organized criticality," perpetually out of balance, but poised in a critical state in which something will happen (Bak 1996).

The construction of check dams is simultaneously creative and destructive. They are the answer to one or more problems, but in the process they also create new problems to be answered. Indeed, they are as much a problem as they are an answer. Check dams are fascinating features geomorpologically and intellectually. They were one of humanity's earliest constructions, and they will doubtless be built as long as there are humans.

REFERENCES

Bak, P. (1996). *How nature works: the science of self-organized criticality*. New York, USA: Copernicus.

Barrau, J. (1958). *Subsistence agriculture in Melanesia.* Honolulu, USA: Bernice P. Bishop Museum.

Cushing, F. H. (1920). *Zuni breadstuff.* New York, USA: Heye Foundation and Museum of the American Indian.

Doolittle, W. E. (1984). Agricultural change as an incremental process. *Annals of the Association of American Gegraphers* 74, 124-137.

Doolittle, W. E. (1985). The use of check dams for protecting downstream agricultural lands in the prehistoric Southwest: A contextual analysis. *Journal of anthropological research* 41, 279-305.

Doolittle, W. E. (2000). *Cultivated landscapes of native North America*. Oxford, UK: Oxford University Press.

Doolittle, W. E., and Neely, J. A. (2004). A checkered landscape. In W. E. Doolittle and J. A. Neely (Eds.), *The Safford Valley grids: Prehistoric cultivation in the southern Arizona desert* (pp. 1-17). Tucson, USA: University of Arizona Press.

Doolittle, W. E., Neely, J. A., Pool, M. D. (1993). *A method for distinguishing between prehistoric, and recent water and soil control features*. Kiva 59, 7-25.

Evenari, M., Shanan, L., and Tadmor, N. (1982). *The Negev; The challenge of the desert* (second edition). Cambridge, USA: Harvard University Press.

Garcia, C., Laronne, J. B. and Sala, M. (2000). Continuous monitoring of bedload flux in a mountain gravel-bed river. *Geomorphology* 34, 23-31.

Geyik, M. P. (1986). FAO watershed management field manual: Gully control. Rome: Food and Agriculture Organization of the United Nations, *Conservation guide* 13/2.

Higgins, C. G. (1982). Grazing-step terracettes and their significance. *Zeitschritf der Geomorphogie* 26, 459-472.

Holling, C. S. (1973). Resilience and stability of ecological systems. *Annual review of ecology and systematics*, 4, 1-23.

Humphreys, G. S., and Brookfield, H. C. (1991). The use of unstable steeplands in the mountains of Papua New Guinea. *Mountain research and development* 11, 295-318.

Kunstadter, P. (1978). Subsistence agricultural economies of Lua' and Karen hill farmers, Mae Sariang District, Northwestern Thailand. In P. Kunstadter, E. C. Chapman, and S. Sabhasri (Eds.), *Farmers in the forest: Economic development and marginal agriculture in northern Thailand* (pp.74-133). Honolulu, USA: The East-West Center and the University of Hawaii Press.

Johnson, D. L., and Lewis, L. A. (2007). *Land degradation: Creation and destruction* (second edition). Lanham, USA: Roman and Littlefield Publishers.

Marston, R.A. (1982). The geomorphic significance of log stops in forest streams. *Annals of the Association of American Geographers* 72, 99-108.

Macnab, J. W. (1965). British strip lynchets. *Antiquity* 39, 279-288.

Neitschmann, B. Q. (1974). *Cultural ecology: Something old, something new, something borrowed, something blue*. Unpublished paper presented at the annual meeting of the Association of American Geographers. Seattle, USA.

Norton, J. B., Sandor, J. A., and White, C. S. (2007). Runoff and sediments from hillslope soils within a native American agroecosystem. *Soil Science Society of America Journal* 71, 476-483.

Reid, I., Powell, M., Laronne, J. B., Garcia, C. (1994). Flash floods in desert rivers: studying the unexpected. *Transactions, American Geophysical Union* 75, 452.

United Nations Environment Programme (1983). Rain and stormwater harvesting in rural areas. *Water Resources Series Vol. 5*. Dublin, Ireland: Tycooly International Publishing Limited for the United Nations Environment Programme.

Wilken, G. C. (1987). *Good farmers: Traditional agricultural resource management in Mexico and Central America*. Berkeley, USA: University of California Press.

Yair, A. (2001). Water-harvesting efficiency in arid and semiarid areas. In S-W. Breckle, M Vest, and W. Wucherer (Eds.), *Sustainable land use in deserts* (pp. 289-302). Berlin, Germany: Springer-Verlag.

In: Check Dams, Morphological Adjustments…
Editors: C. Conesa and M. A. Lenzi

ISBN: 978-1-62618-856-3

Chapter 2

CRITERIA FOR OPTIMIZING CHECK DAM LOCATION AND MAINTENANCE REQUIREMENTS

Ali Morad Hassanli[1] and Simon Beecham[2]

[1]Shiraz University, Iran,
[2]University of South Australia
Centre for Water Management and Reuse, Australia

ABSTRACT

Check dams generally consist of a vertical barrier constructed on ditches, small streams, channels and gullies that have often been formed by the erosive activity of water. These structures are commonly constructed using stone, gravel-filled sand bags, masonry, logs, woven-wire and brush fills. A check dam serves many purposes such as reducing runoff velocity, reducing erosive activities, reducing the original channel gradient, improving bed sediment moisture in adjoining areas, sediment retention and allowing percolation to recharge aquifers. A check dam interferes with flows in the upstream and downstream channels and dissipates the energy of flowing water. Therefore if it is not designed, installed and located properly it may either bypass concentrated runoff or be damaged by excess discharges. On the other hand, since each check dam is expected to control a limited drainage area, selecting an appropriate space between a series of check dams along a channel is important. Check dam design is influenced by various conditions that can occur in practice. For example, large drainage areas produce increased runoff volumes and flow rates that must be compatible with the check dam design criteria. Also a large horizontal distance between check dams will cover a larger drained area and will lead to higher runoff energies. In addition, the design height of check dams will influence the distance between check dams, the number of check dams and the total cost of the project. In general, the design variables for check dams such as height, distance between dams, gradient between the crest of a check dam and the foot of the next upstream check dam, shape of the check dam, construction materials, spillway design, length of structure across the channel and also economic considerations lead us to develop optimised design criteria. In order to optimise the design of check dams it is important to understand the relationships between the various design variables. In this chapter, we will describe a detailed methodology for developing relationships between check dam heights, distance between check dams, stream bed slope and compensation slope above dead level. The optimisation of check dam design goes beyond solely the hydrological performance of the system and the geological conditions. In order to fully optimise the location of check dams it is necessary to consider all of these design criteria.

2.1. Introduction

The work presented in this chapter is based largely on the authors' experience on arid drainage basins, although much of the theory applies to check dams in other catchment types. Check dams are generally used in concentrated flow areas, such as gullies, vegetated ditches, streams and swales. They can either be permanent or temporary barriers to prevent erosion and promote sedimentation by slowing flow velocity and/or filtering concentrated flows. Check dams are also used in flood control projects, particularly in some regions where construction of large dams is not possible. Certain types of check dam are designed to control sediment by allowing it to settle out above the check dam or to settle in the ponds upstream where the flow velocity is reduced. The impounded runoff is usually allowed to percolate through the porous rock filter.

One of the functions of check dams or trap dams is reducing sediment inflow into the reservoirs. They are also used to decrease the gradient of channels and to reduce the flow velocity. Check dams can be constructed from a variety of materials. Choice of material depends on the situation and the purpose of check dam construction. Under high-flow conditions, water flows over and/or through the structure. Under low-flow conditions, water ponds behind the structure and then seeps slowly through the check dam, or it infiltrates through the upstream stream bed or it evaporates. In a catchment, depending on the application, various types of dams may be constructed. These can include productive dams for creating farmlands, flood control dams for preventing flood water and intercepting sediments, water-storage dams for irrigation, rock check dams for stabilizing vegetation or reducing bed gradient and gully check dams for controlling gully development. In Figure 1 four different types of check dams are shown.

Rock check dams may be keyed into the channel floor, typically by a depth of 15cm (Minneapolis Metropolitan Council, 2003) in order to be more stable and to reduce the likelihood of slippage. When rock check dams are to be temporary structures, using a geotextile fabric beneath the check dam will facilitate easy removal when they are no longer required. Check dams must be designed and constructed correctly and maintained properly, in order to prevent material from being washed out.

For small reservoirs, upstream check dams can be used for gully control and they can be built with cheap, local material, if available. These dams are usually less than 2 metres high and their maintenance costs are not considerable (Durgunoglu and Singh, 1993). In gully control, all structural measures generally protect bed sediment against erosion and conserve water. However, earth plugs and check dams in particular, also make more water available to the bed sediment's underground storage by retaining surface run-off. As a result, well and spring water in the immediate vicinity is increased during the first few years and permanently regulated after a vegetative cover is established.

Figure 1. Four different types of check dam: (a) Loose stone, (b) Woody vegetation, (c) Masonry and (d) Log.

Cultivated lands on both sides and at the base of the gully are protected from flood and sediment damage, because structural measures such as diversion ditches, earth plugs and check dams, prevent minor floods by diverting and holding the surface run-off (Geyik, 1986).

2.2. Factors Influencing Check Dam Location

Check dams as a traditional water harvesting system have made a significant contribution to the sustainability and profitability of agricultural catchments in a number of regions such as Kumbadaje panchayat in the state of Kerala in India (Balooni et al., 2008). Check dams on gullies intercept large quantities of sediment, and productive flat land has been created above the dams. Thus, the benefit of check dams may be measured by the area of land created above it (Jiongxin, 2004). Boulder check dams have been employed for torrent management particularly for high-gradient stream stabilization. The design criteria employed for boulder check dams in Northern Italy were taken from the step-pool morphology approach (Lenzi, 2002).

Nyssen et al. (2000) assessed the performance of 400 loose rock check dams in northern Ethiopia. They identified several factors that influenced check dam collapse but no clear threshold values could be established for single factors. They reported that the frequent collapse of the dams is strongly associated with drainage area (A) and slope (S), with the product of these factors (S x A) being a measure of runoff energy. This product is a proxy for stream power. They concluded that effective dams have a spillway, an apron, a concave plan form (looking down slope) and are built at vertical intervals and with heights that result in a negative gradient for the line connecting the spillway and the foot of the upstream check dam (Figure 2). To decrease the risk of failure, horizontal distances between check dams have to be carefully calculated as a function of bed slope and drainage area. Lenzi and Comiti (2003) describe the bed profile and gradient size distribution adjustment in steep channels in a mountain river with check dams. They concluded that the drop height, flow depth and step spacing affect scouring dynamics in a complex manner.

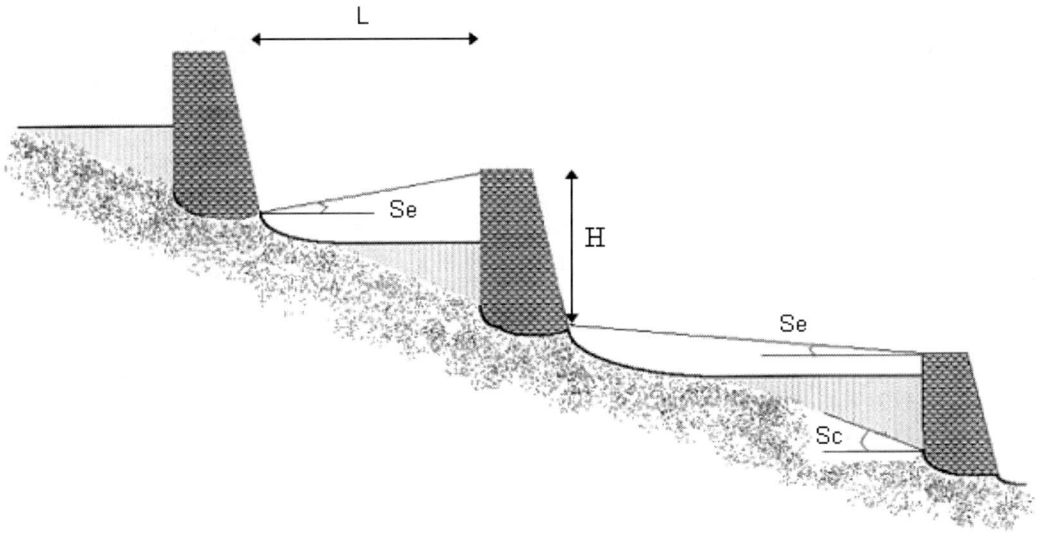

Figure 2. A typical check dam configuration of height (H), horizontal distance (L), slope of channel (Sc), slope of line connecting spillway and foot of upstream check dam (Se) (after Nyssen et al., 2004).

2.3. Environmental Impacts of Check Dams

Any intervention such as construction of a sequence of check dams along a river or stream in a complex ecosystem will affect the ecological stability and balance of its biotic communities. To minimize the ecological disturbance, the construction of engineering structures and other measures should be as compatible as possible with the stream's natural tendency to reach a stable configuration over a long period. Such criteria should be applied with care and be accompanied by traditional interventions (Lenzi, 2002).

Xu et al. (2004) claim that the implementation of check dams in gullies as a widespread engineering measure is one of the most effective ways to conserve bed sediment and water and is the primary method for controlling the sediment pouring into the Yellow River. The Yellow River has the highest sediment load of any river in the world. It typically delivers over 1.6 billion tonnes of sediment annually. This corresponds to an average sediment concentration of 35 kg per cubic metre, one quarter of which is trapped inland through structures such as check dams. Control structures such as check dams have enabled the Yellow River catchment to support 12% of China's huge population while only generating

2% of its total runoff. Over 100,000 check dams have been built on the Loess Plateau in China in the last 50 years.

Heede and Mufich (1973) claim that effective check dam height is one of the most important design criteria in gully control because it influences the spacing of dams, materials and costs of treatments. They also maintain that there is one optimum dam height at which total treatment costs are lowest and volume requirements for rock smallest. They describe how check dams not only act as flood-retarding structures but that they decrease original sediment yield sometimes by 50% or more. They add that the main criteria determining gully control are the magnitude and gradient of gully, the bed sediments in which the gully developed, the vegetation in and alongside the gully, the peak flow generated by the design storm and perhaps most important, the availability of funds. They developed functional relationships to facilitate the design of effective treatments with minimum cost of installation. They presented cost estimation for various types of check dams with varying heights and developed the CAGCOM computer program to assist with the design calculations.

Hassanli et al. (2008) evaluated the influence of porous (loose) check dam location on the retention of fine sediments. They concluded that check dams in the far downstream sections of a stream are more effective in fine sediment retention than those located in the upstream sections. They also added that using broken and angular rocks instead of rounded rocks in porous check dam construction improves the effectiveness of the check dams for the retention of fine sediments. Their study showed that erosion of the banks underneath check dams is the primary cause of dam collapse.

Tulu (1999) developed an approximate equation for optimum check dam spacing for economical gully stabilization. Tulu examined the geometry of the gully bed, bed sediment type and the allowable channel slope above the dead level (compensation slope or the stable gradient of a channel after a long term check dam functioning, say 10 years). His equation estimated check dam spacing and hence the number of check dams required for gully erosion control.

The values obtained for different bed sediment types agreed with the recommendations of Ayres and Scoates (1939). Esmaeli et al. (2008) developed three empirical models to estimate the slope of deposited sediments (compensation slope) behind porous check dams to calculate the optimum space between the dams. They concluded that the factors significantly affecting the slope of deposited sediments are stream bed slope, check dam height, stream width, and D_{50} of sediments. The compensation slope obtained from these models was used to estimate the optimum distance between check dams.

Zhou et al. (2004) claim that some problems such as failure of check dams can be resolved through improved design and construction. They explained how a group of check dams along a channel constitutes a check dam system. To secure the relative stability of a check dam or a check dam system many elements such as hydrological, geographical and geological conditions of the catchment should be taken into account.,

The check dam system in gullies becomes relatively stable when the ratio of the dam farmland area to that of the controlled catchment is between 1:25 and 1:15. When the impounded water depth is less than 0.8m and the storage time is shorter than 3 to 7 days, a dam designed for enduring the largest rainstorm in 100 years is found to be relatively stable (Zhou et al., 2004).

2.4. Design Criteria for Check Dam Structures

2.4.1. Distance between Check Dams and Effective Check Dam Height

An understanding of the required distance between a series of check dams is necessary for planning and optimum design of check dams and effective erosion control. The design of a gully control by check dams depends on several criteria whose combination changes between gullies, since no two gullies are identical (Heede and Mufich, 1973). To decrease the risk of failure of check dams to an acceptable level, the horizontal distance between check dams has to be carefully calculated. Nyssen et al. (2004) discusses how the horizontal distance is a function of drainage area and the slope of the bed sediment surface. For a given channel bed slope (S_c) the ideal spacing between check dams (L) would be a function of height (H) as shown in Equation 1:

$$L = H/S_c \tag{1}$$

This means the steeper the channel gradient, the closer will be the check dams. The above equation is based on the rule that distance between two check dams is equal to distance between the toe of the upstream check dam and head of the downstream dam. In practice, this ideal spacing is not recommended because too many structures would be required particularly for steep channels, as shown in Figure 3. In the long term, even the most erosive bed sediments would become stable on some gradient steeper than the dead level (Tulu, 1999). This is called the "compensation slope" (Figure 4). The compensation slope is the gradient between the top of the lower check dam and the bottom of the upper one. This is the final gradient of the gully channel. It is formed when material carried by flowing water fills the check dams to spillway level. FAO (Geyik, 1986) reported that field experience has demonstrated that the compensation slope of gullies is often not more than three percent.

Figure 3. A series of check dams in a steep stream (Hassanli et al., 2007).

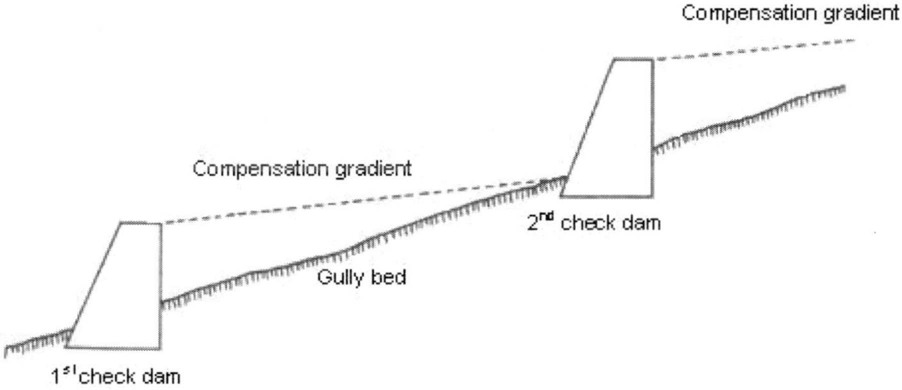

Figure 4. Compensation slope between two check dams (after Geyik, 1986).

On the other hand, when the distance between two consecutive check dams is too small (less than the distance between the toe of the upper check dam and the top of the lower one) not only might it be unjustifiable economically but the upstream check dam is likely to become buried, as shown in Figure 5. Table 1 shows the maximum allowable slopes as recommended by Ayres and Scoates (1939). The effective dam height is the elevation of the crest of the spillway above the original stream or gully bottom. The effective check dam height affects the spacing and thus the number of required dams, the total cost and the expected sediment deposits behind the check dams.

Table 1. Maximum Allowable Slopes (after Ayres and Scoates, 1939)

Sediment type	Slope (%)
Coarse sand with small gravel	2
Fine sand and silt loam	1
Silts and clays	0.5

Figure 5. Example of a buried check dam because of incorrect spacing (Esmaeli et al., 2008).

Equation 2 was developed by Heede and Mufich (1973) for spacing between check dams in gullies. This equation is based on the original bed slope and the check dam height.

$$L = \frac{H_e}{KS_c \, Cos\beta} \tag{2}$$

where:

L: distance between check dams (m)
He: effective height of check dam (m)
S_c: original bed slope as a ratio (tan β)
β: the angle corresponding to the channel gradient
K: an empirical coefficient for local conditions.

Equation (2) is based on the assumption that the gradient of the sediment deposits above a check dam (dead level) is (1 - K)S_c.

K changes with two groups of gradients as follows:
$K = 0.3$ for $S_c \leq 0.2$
$K = 0.5$ for $S_c > 0.2$

Heede and Mufich (1973) assert that the spacing between check dams is affected by the stabilized sediment deposits. Hence each structure is installed at the upstream toe of the expected deposit wedge formed behind the downstream dam. Spacing decreases with increasing channel gradient and increases with effective check dam height. It is not economical to construct a check dam in such a way that the bottom of the upper dam and the top of the lower dam are at the same level. Even the most erosive bed sediments will remain stable on some gradient steeper than the dead level. Tulu (1999) assumed a stable channel slope (equilibrium or compensation slope) to be an allowable channel slope above the dead level for each bed sediment type. Based on this assumption he developed an equation for check dam spacing for a constant slope, S_e, above the dead level as given in Equation 3:

$$L = \frac{100 H_e}{S_c - S_e} \quad \text{for } S_c - S_e > 0 \tag{3}$$

In this equation the check dam spacing is affected by the check dam effective height, original channel gradient and the compensation slope (S_e) above the dead level (S_d). Esmaeli (2006) examined Equation 3 and concluded that for small slopes, particularly for those less than 7%, it estimates the distance between check dams very accurately. Figure 6 shows the original gradient (S_c), the dead level (S_d) and the compensation or equilibrium slope (S_e) between two consecutive check dams located a distance L apart. In Equation 3, H_e is known and S_c can be measured. Only S_e is unknown. In fact, S_e is the gradient of deposited sediments behind each check dam which has formed and stabilized over a long period of time, often over 10 years or more (Heede and Mufich, 1973). Tulu has recommended a number of values for S_e that are related to bed sediment type and channel bed slope as listed in Table 2.

Table 2. Values of S_e in Equation 3 for different bed sediment types and channel bed gradients (after Tulu, 1999)

Bed sediment type	Channel slope (%)	S_e (%)
Coarse sand and bed sediment containing considerable amounts of gravel	> 2	2.0
	≤ 2	0.0
Fine sand and silt loam	≥ 1	1.0
	< 1	0.0
Light silt and clays	≥ 0.5	0.5
	< 0.5	0.0

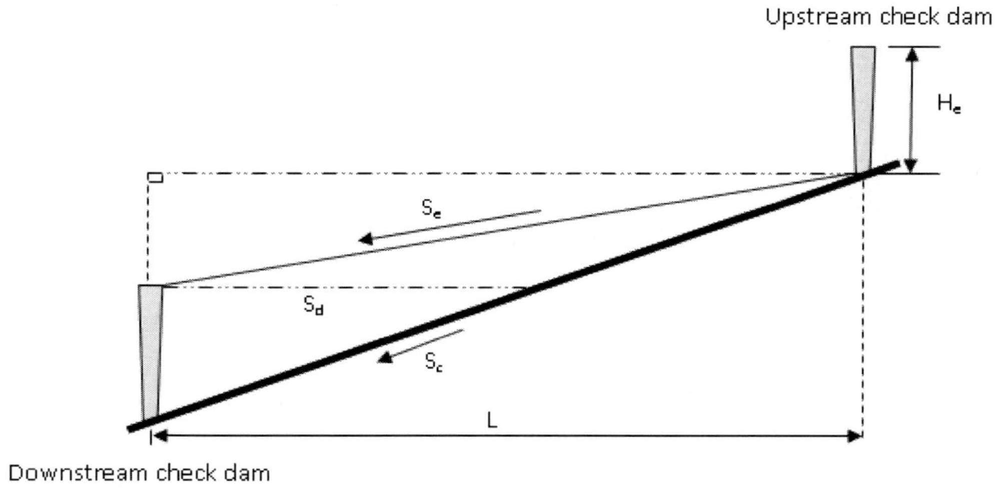

Figure 6. Relationship between channel bed slope (Sc), compensation slope (Se) and dead level (Sd) between two consecutive check dams.

The issue of estimating suitable S_e values was further investigated by Esmaeli et al. (2008). The models they developed for compensation slope may be used to calculate the distance L between check dams using Equation 3. The simplest empirical model developed by Esmaeli et al. (2008) for compensation slope needs only three variables that are simple to measure. These are shown in Equation 4.

$$\ln(S_e) = -2.27 + 1.38 \ln(S_c) - 0.56 \ln(H_e) + 0.25 \ln(B) \qquad (4)$$

where:

S_e: compensation or equilibrium slope (%)

S_c: stream bed gradient (%)

H_e: effective height of check dam (m)

B: stream width where the check dam is constructed (m).

Equation 4 is used to estimate the compensation slope and hence the optimum distance between check dams. Esmaeli (2003) reported that the models developed in his research for Southern Iran (in the Droodzan catchment) give a better estimation for distance between check dams compared to the models developed by Heede and Mufich (1973) and Tulu (1999). Figure 7 shows a comparison between the three models for S_e = 1% in the case of Tulu's model, and for H_e = 1m in the case of the model developed by Esmaeli et al. (2008). It can be seen that the three models are in close agreement. While the model of Esmaeli et al. (2008), as presented in Equation 4, estimates S_e using only three variables (S_c, H_e and B), in practice only two of these (H_e and B) are design variables since the channel bed slope (S_c) is usually measured. Figure 8 shows the influence of channel stream width (B) on dam spacing, when using Equation 4. Figure 9 shows the influence of effective height (H_e) on dam spacing, when using Equation 4.

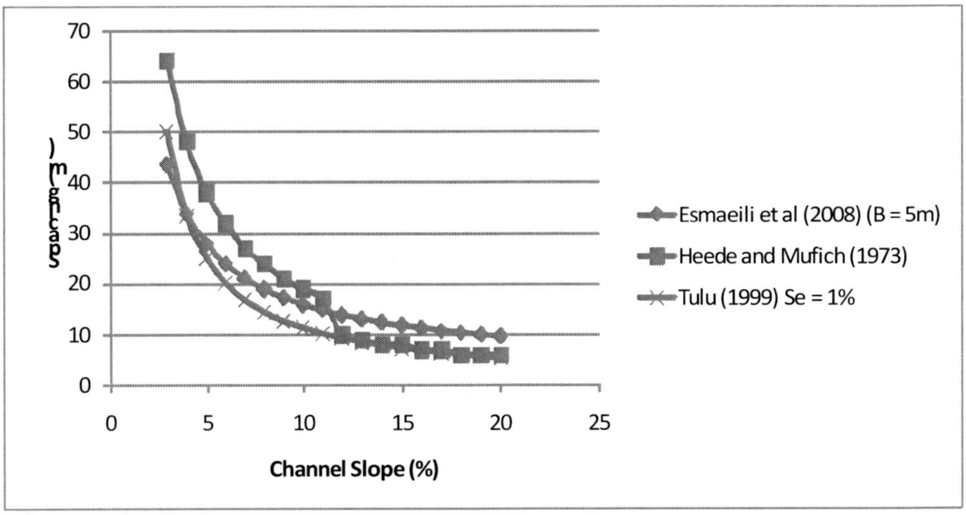

Figure 7. A comparison of three models for estimating dam spacing.

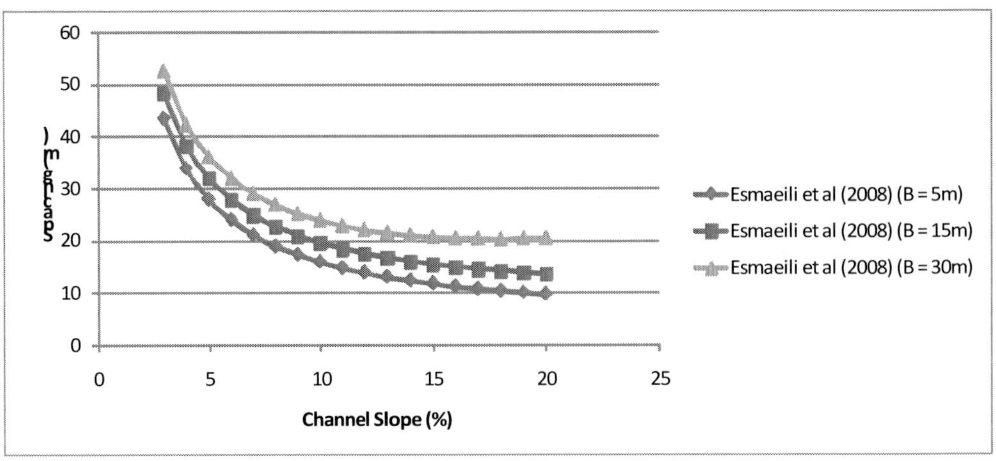

Figure 8. Influence of channel bed width on dam spacing.

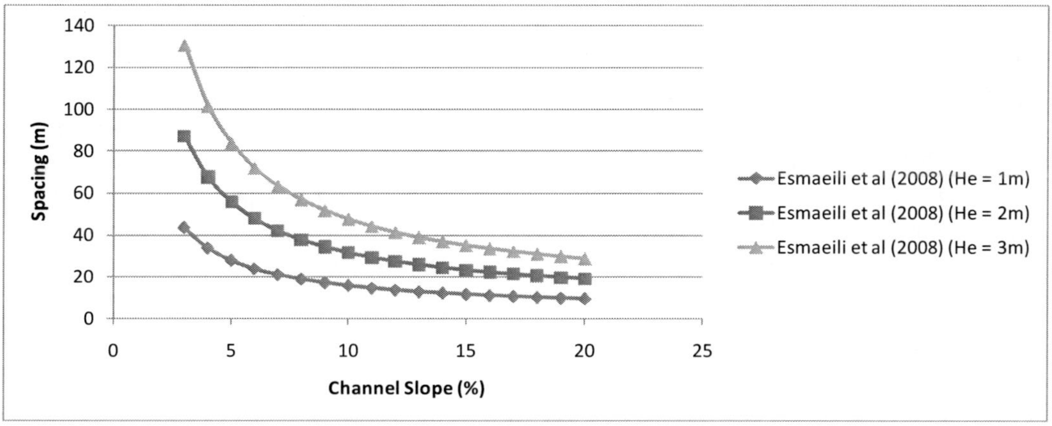

Figure 9. Influence of effective height (He) on dam spacing.

Nyssen et al. (2004) presented an equation to calculate the horizontal distance between check dams of height 1 m. In this equation (Equation 5) the horizontal distance (L) is a function of the slope of the bed sediment surface (S) and the gradient of the line connecting the spillway crest and the foot of the nearby upstream check dam (S_e). They developed a monograph for optimal horizontal spacing of loose rock check dams of height 1m as a function of drainage area (A) and steepness of the bed sediment surface slope. They discussed how for their study area (Northern Ethiopia), this optimal spacing (L) would reduce the risk of collapse of loose rock check dams to a probability of 20% in the first two years after construction.

$$L = \frac{1}{(S - S_e)} \tag{5}$$

where $S_e = 0.21 - 0.01A$

2.4.2. Drainage Area

For a long channel, the main criteria for selecting structural control measures are based on the drained catchment area, the gradient and the length of the channel. There are reports showing that the frequent collapse of loose check dams is strongly associated with drainage area and slope. Nyssen et al. (2004) reported that the product of drainage area and mean channel slope is a measure of runoff energy. As drainage area increases, both the volume and flow rate of runoff increase.

2.4.3. Height

When the spacing between downstream and upstream check dams is extended the foundation depth of the upstream check dam needs to increase. However, when the spacing is shortened, the foundation depth decreases. As the foundation depth is increased, the total height of the check dam (effective height plus foundation depth) should not exceed the permissible maximum total height (Geyik, 1986). Increasing check dam height reduces the required number of check dams more in steep rather than in low-gradient gullies (Heede and Mufich, 1973). Also high dams produce a larger magnitude of total sediment deposits than low dams. Heede and Mufich add the relationship between spacing and effective dam height is linear. The height of check dams will be limited by the effective upstream drainage area, the depth of the channel, the project objectives and the type of check dam.

2.4.4. Number of Check Dams Needed Along a Channel

The number of check dams needed to provide the desirable slope depends on the project objectives and economic considerations. The height of a check dam will influence the number of check dams and the total cost of the project. Rather than a high number of low height check dams, it is more desirable to have a fewer number of high check dams. For gabion check dams this height can be less than 5 metres. The number of check dams in a channel is a function of the original gradient, the compensation slope, the height of the check dams and the spacing between them as shown in Equations 6 and 7:

$$N = \frac{(S_c - S_e)L}{H_e} \tag{6}$$

or:

$$N = \frac{(h_1 - h_2)}{H_e} \tag{7}$$

where:

N: the number of check dams required

S_c and S_e: original bed slope and compensation slope (%), respectively

L: horizontal length between the first and the last check dam in the channel (m)

H_e: the average effective height of check dams, excluding the foundation depth

h_1: total vertical distance calculated from the average gully channel gradient and the horizontal distance between the first and last check dam in that portion of the gully bed

h_2: total vertical distance calculated from the compensation slope and horizontal distance between the first and the last check dam in that portion of the gully bed.

2.4.5. Maximum Discharge and Spillway Dimensions

Since the drainage area associated with each check dam along a stream or gully mainly is generally not large, the peak discharge from the drainage area may be calculated using standard run-off equations. The Rational Method is the simplest method to determine peak discharge from a drainage basin. It is the most common method used for check dam design.

Use of the rational formula is possible only when rainfall intensities for the area for the design rainfall event are available:

$$Q_{max} = \frac{CIA}{3.6} \qquad (8)$$

where:

Q_{max}: maximum discharge of the catchment at the check dam site (m^3/s)

C: runoff coefficient (typical values of C are given in Chow et al., 1988).

I: Rainfall intensity (mm/hr) is based on the time of concentration. I is typically found from Intensity/Frequency/Duration curves for rainfall events in the geographical region of interest. For small drainage areas, rainfall intensity is calculated according to the maximum rainfall intensity for that area for a given frequency and storm duration.

A: Catchment area above the proposed check dam (km^2).

If there is no rainfall intensity data for the area, either the Kresnik discharge formula or the general continuity equation together with the Manning velocity formula could be used. The Kresnik equation (Geyik, 1986) is given in Equation 9:

$$Q_{max} = \frac{32KA}{A^{0.5} + 0.5} \qquad (9)$$

Q_{max}: maximum discharge of the catchment at the check dam location (m^3/s)

K: Coefficient, which varies between 0.6 and 2.0 depending on land use type

A: Catchment area above the check dam (km^2).

The cross-sectional area or continuity equation is given in Equation 10:

$$Q = A_w V_m \qquad (10)$$

Q: maximum discharge of the catchment at the check dam site (m^3/s).

A_w: Wetted cross-sectional area of the steam or gully bed considering the highest flood water level at the proposed check dam site (m^2)

V_m: Mean velocity of the flowing water at the cross-section (m/s).

Since measuring the velocity of the highest flood is often difficult in seasonal streams, the Manning formula could be employed as presented in Equation 11:

$$V = \frac{1}{nR^{2/3} S_c^{1/2}} \qquad (11)$$

V: Flow velocity at the check dam site (m/s)

n: Manning roughness coefficient Which depends on the roughness of the channel

R: Hydraulic radius for highest flood water level (m)
S_c: Channel bed slope (%).

The main Kresnik formula (Equation 9) gives better results for torrents in catchment areas greater than 300 ha. A simplified version of the Kresnik formula as shown in Equation 12 gives more suitable results for catchment areas of less than 20 ha (Geyik, 1986).

$$Q_{\max} = 25A^{1/2} \tag{12}$$

Q_{max}: maximum discharge (m^3/s) and A: Catchment area (km^2).

2.4.6. Spillway Design

In the design of check dams, the dimension of the spillway is of critical importance. If the spillway is too small, flows may overtop and destroy the structure. Clogging by debris in small spillways may lead to similar destructive overflows. Generally the form of cross section of most gullies is trapezoidal, rectangular, V-shaped or concave. In trapezoidal channels the effective length of the spillway equals the bottom width of the channel where possible. Spillway dimensions are calculated from the broad-crested weir formula (Equation 13).

$$Q = CWD^{3/2} \tag{13}$$

Q: discharge of the gully or channel catchment at the proposed check dam site (m^3/s)
C: discharge coefficient which is 3.0 for loose rock, boulder log and brushwood check dams and 1.8 for gabion and cement masonry check dams
W: effective length of spillway (m)
D: depth of water over spillway (m).

The length of the foundation must always be longer than the length of the spillway to prevent scouring and undermining by falling water. Also the crest of spillways should be level.

2.5. The Effect of Porous Check Dams on Fine Sediment Retention

The use of check dams and catchment management can reduce the inflow of sediments into a reservoir. However, a significant amount of sediment may still enter the reservoir, either because of their inefficient use or because of their inability to trap fine and suspended sediments. Hassanli et al. (2008) claim that when loose (porous) check dams are placed in the far downstream sections of a channel, their performance in the retention of fine sediments is much better than for those located in the upstream sections (Figure 10). Their research showed that using broken and angular rocks in loose stone check dams compared to using rounded rocks improved the effectiveness of check dams for fine sediment retention.

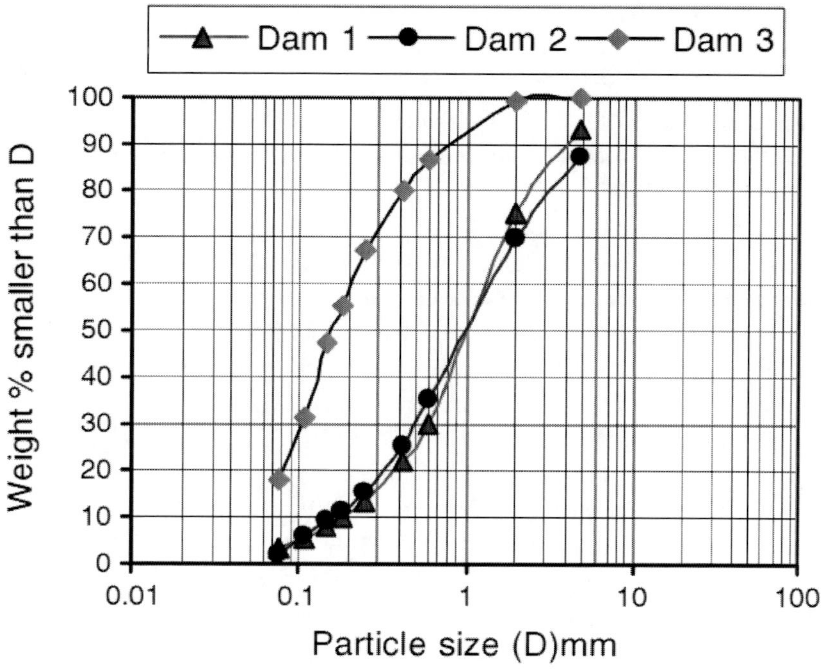

Figure 10. Sediments trapped by dam 3 are finer than the trapped sediments for the two other check dams. Check dams 1 to 3 were located in the far upstream, midway and far downstream reaches of the channel, respectively (Hassanli et al., 2008).

2.6. Stability of Check Dams

2.6.1. Gravity or Bulk Dams

Gravity check dams are mainly constructed from loose stones, boulders, gabions or cement masonry. Their downstream face slopes backward while their upstream face is generally vertical. A gravity check dam's stability against overturning, collapsing or sliding is usually calculated using empirical formulas. The widths of a bulk dam's crest and base are calculated according to their height. The weight of the dam and the water flowing through its spillway section act as vertical forces against it. The pressure on its upstream face, from material filled to the spillway level and from water in front of check dam act as the horizontal force. For stability, the weight of gravity dams on stony ground or new alluviums must not exceed 5 kg/cm^2. On the other hand, the dam weight must work against the overturning forces imposed by water and sediments. Loose stone and boulders are usually used on relatively small dams. (Geyik, 1986).

2.6.2. Temporary Check Dams

Temporary check dams are small temporary structures constructed across a swale or drainage ditch. Their purpose is to reduce the velocity of concentrated stormwater flows, thereby reducing erosion of the swale or ditch. This practice is limited to use in small open channels that drain about 4 ha or less. It should not be used in a stream, including perennially flowing streams or intermittent stream channels (USDA SCS, 2003). According to FAO (Geyik, 1986) the following criteria should be taken into account when specifying temporary check dams:

- The drainage area of the ditch or swale being protected should not exceed 4 ha.
- The maximum height of the check dam should be 65 cm.
- The centre of the check dam should be at least 15 cm lower than the outer edges.
- The maximum spacing between the dams should be such that the toe of the upstream dam is at the same elevation as the top of the downstream dam.
- Check dams should be installed before runoff is directed to the swale or drainage ditch.
- Log check dams should be constructed with logs of diameter 10 to 15 cm. The logs should be embedded into the bed sediment by at least 50 cm.
- All bales should be either wire-bound or string-tied.
- The barrier should be entrenched and backfilled. A trench should be excavated the width of a bale and the length of the proposed barrier to a minimum depth of 10 cm.
- After the bales are staked and chinked, the excavated bed sediment should be backfilled against the barrier.
- The gaps between bales should be filled by wedging with hay to prevent water from escaping between the bales.

2.7. Maintenance Considerations

2.7.1. Erosion Control

One of the main functions of a check dam is prevention of gully formation which is much easier than controlling it. Once gullies have begun to form, they must be controlled as soon as possible. If the initial formation of a gully is not controlled, it starts to extend and become deeper, wider and larger. Prevention is also more economical because structural measures are considerably more expensive than preventive measures. Therefore, in erosion control, emphasis should be given to proper land management practice rather than structural measures. In addition to proper land-management practices to decrease the cost of structural gully-control measures, specific slope-treatment measures, such as retention and infiltration ditches, terraces, wattles and using tree branches for filling depressions should be carried out in the eroded area between the branch gullies (Geyik, 1986).

Maintenance for structural measures must be continued for at least two years after construction. The areas must be inspected at least once a year. The plants and grass established upstream of the check dams need to be managed and protected. It is necessary to maintain dams as soon as collapsing starts. For this purpose, a maintenance team should be active during any rainy season. Labour investment in early repair works is often much less than repairing dams after total collapse. Ecological gully control, possibly in association with check dams, appears to be a sustainable method of gully rehabilitation. Rapidly recovering vegetation will lead to dam stabilization and increased sediment deposition. In addition, fodder will be produced, releasing pressure on other areas (Nyssen et al., 2004). Check dams should be inspected for sediment accumulation after each significant rainfall. Regular inspections should be made to ensure that the centre of the dam is lower than the edges. Rock should be added, removed or replaced as needed to maintain the design height and cross section. All damage should be repaired immediately if significant erosion occurs. Additional protection may be required. This may include protective riprap or additional check dams. After high flow around the edges of the dam erosion, submergence and deposition above the check dams are expected. Any loose or displaced stone should be repaired according to the original specifications. Figure 11 shows a collapsed check dam due to poor maintenance. Heede and Mufich (1973) claim maintenance costs are approximately 1% of the costs of check dam installation.

Figure 11. A collapsed check dam due to poor maintenance (South of Iran).

2.7.2. Size of Gully and Its Relationship to a Torrent

In deteriorated mountain catchments, each continuous gully in a network usually has a distinct catchment and a main gully channel. A torrent catchment can comprise several gully systems, forest and rangelands, which can be in deteriorated or in good condition, hillside farming areas, low croplands and urban areas. Therefore, the catchment area of a torrent may spread over more than 1000 ha and be more than 2 km long. In order to control a torrent and avoid flooding, it is essential to stabilize all the gullies throughout the entire catchment area (Geyik, 1986).

2.7.3. Criteria for the Selection of Gully Control Measures

Catchments deteriorate because of human misuse and mismanagement of the land. This includes overgrazing, deforestation, urbanization, mining, short intensive rainstorms, prolonged rainfall of moderate intensity. These land use activities and precipitation factors also lead to high run-off rates which can cause flooding and promote gully formation. Therefore minimizing surface run-off is essential in gully control, which is one of the most important restoration methods used in catchment management. Timing is an essential element. In areas receiving seasonal rainfall, the field work for all structural and vegetative control measures should be completed during the dry and early rainy season. This is an important aspect of gully control, particularly in tropical and subtropical regions. In addition, vegetative measures cannot be implemented until structural work is complete (Geyik, 1986). In gully control, temporary structural measures such as brushwood, logs, woven-wire, loose stone, porous check dams and boulder check dams are used to facilitate the growth of permanent vegetative cover. Check dams are constructed across the gully bed to stop channel and lateral erosion. Check dams reduce the original gradient of the channel bed, the velocity of water flow and ultimately the erosive power of run-off. Temporary check dams, which may be in place for three to eight years, collect and hold bed sediment and moisture in the bottom of the channel. Tree seedlings, as well as shrub and grass cuttings planted in gullies can grow without being washed away by flowing water. Using these techniques, a permanent vegetative cover can be established in a short time (Geyik, 1986). Various portions of the main gully channel and branch gullies are stabilized by brush fills, earth plugs, brushwood, log and loose-stone check dams. The lower parts are often treated with loose-stone, boulder and porous check dams. At a stable point in the lowest section of the main gully channel, for example, on a rock outcrop, a gabion check dam or cement masonry check dam should be constructed. If there is not a stable point, a counter-dam such as a gabion or a cement masonry check dam should be constructed in front of the first check dam. The positions where the other check dams will be constructed should be determined according to the compensation slope (deposited sediment slope) and the effective height of the check dams. General standards for selecting control measures for each portion of a continuous gully are given by the FAO (Geyik, 1986). These criteria should be used for continuous gullies, gully networks, or multiple-gully systems located in deteriorated mountain catchments. In upstream catchments with very steep slopes, the gradient of the main gully channel can easily reach 100% or even more. In using these standards, there is not much difference between the gullies located in mountain catchments and those in rolling lands. In rolling lands, the highest gradient of a main gully channel may be 30 to 40%. In nearly-flat areas (agricultural and rangelands on foothills), this gradient is often much less than 30 to 40%. The remaining criteria are the length of the main gully channel's portions and the catchment area of the gully portions.

2.7.4. Criteria for Protection Measures

A very effective practice is to provide appropriate diversion structures constructed above the gully area to direct run-off away from gully heads and to discharge it either into natural channels or stable areas that are not susceptible to erosion. This practice is important to protect critical planted areas from being washed away. Small diversion ditches constructed either alone or with other structures such as check dams and earth plugs are commonly used in gully control. To prevent scouring along the diversion channel, the gradient of small diversion ditches must be less than one percent. However, if there is permanent vegetative cover in the channel, the gradient may be as high as two to three percent. Diversion ditches should be large enough to carry all the water that is discharged from the gully catchment area during periods of maximum run-off. In regions subject to particularly heavy rainfall, in addition to the above measures, a series of check dams can be constructed along the gully channels (Geyik, 1986). Where natural plant growth does not occur or if certain plant species of economic value are desired, introduced vegetation may be considered. When planting decisions are made for erosion control alone, it makes little difference whether trees, shrubs or grasses are used. Any of these, when well established, will provide good protection for the bed sediment. Consequently, the type of vegetation should be considered according to how the planted area will ultimately be utilized. In regions with heavy rainfall, filling, shaping and diversions alone may not suffice to control gully expansion. Additional gully control and slope stabilization measures, such as check dams, stone terraces, wattles and revegetation should be undertaken (Geyik,1986). To obtain satisfactory results from structural measures, a series of check dams should be constructed for each portion of the gully bed. After low check dams silt up and rot away, vegetation can control the low overfalls much more easily than on high check dams. Stabilized catchment slopes are the best assurance for the continued functioning of gully control structures. Therefore, attention must always be given to keeping the gully catchment well vegetated. If this fails, the structural gully control measures may fail as well (Geyik, 1986).

Conclusions

Catchments deteriorate because of human misuse and mismanagement of the land. This includes overgrazing, deforestation, urbanization, mining. The impact of climate change on catchment deterioration likely needs to be taken into account. Check dams as a traditional water harvesting system have made a significant contribution to the sustainability and profitability of agricultural catchments. Depending on the catchment objectives, various types of check dam may be utilised. Any intervention such as the construction of a sequence of check dams in a complex ecosystem will affect the ecological stability and balance of its biotic communities. Thus engineering structures and other measures in a check dam system design need to be made as compatible as possible as with the natural ecosystem of the catchment to secure a stable configuration over a long period. Many elements such as the hydrological and geomorphological conditions of a catchment need to be taken into account to secure the relative stability of a check dam system. To decrease the risk of failure, horizontal distances between check dams have to be carefully calculated as a function of bed slope, dam height and drainage area. The ideal spacing between check dams would simply be a function of check dam height and the channel bed slope. This relationship is based on the rule that the distance between two dams is equal to the distance between the toe of the upstream check dam and the head of the downstream one. However, this ideal spacing is not recommended because two many structures would be required particularly for steep gradients and this would not be justified economically. For optimum check dam spacing, the use of the compensation slope is recommended (Ayres and Scoates, 1939). Empirical models have been developed (Esmaeli et al., 2008) and equations for direct calculation of check dam spacing were presented (Heede and Mufich, 1973; Tulu, 1999) in this chapter. It is shown then when loose (porous) check dams are placed in the far downstream sections of a channel, their performance in the retention of fine sediments is much better than for those located in the upstream sections. Also the effectiveness of the loose stone check dams for fine sediment retention is improved when using broken and angular rocks compared to using rounded rocks. The many factors that affect the optimum distance between check dams have been discussed. These include check dam height, bed slope, compensation slope, dead level, drainage area, the number of check dams in a check dam system, maximum discharge and spillway dimensions. Other topics discussed include maintenance requirements for check dams, criteria for gully control measures and gradient stabilization. One of the main functions of a check dam is prevention of gully formation which is much easier than controlling it. If the initial formation of a gully is not controlled, it starts to extend and become deeper, wider and larger. Prevention is more economical because structural measures are considerably more expensive than preventive measures. Therefore, in erosion control, emphasis should be given to proper land management practice rather than just structural measures. It is necessary to repair dams as soon as collapsing starts. For this purpose, a permanent maintenance team needs to be active during any rainy season. Ecological gully control, possibly in association with check dams, appears to be a sustainable method of gully rehabilitation.

REFERENCES

Ayres Q.C. and Scoates, W.D. (1939). *Land Drainage and Reclamation,* 2nd edn. New York and London. McGraw-Hill Book Company.

Balooni, K., Kalro, A. H. and Kamalamma, A.G. (2008). Community initiatives in building and managing temporary check-dams across seasonal streams for water harvesting in South India, *Agricultural Water Management. doi:10.1016/j.agwat.2008.06.012.*

Chow V.T., Maidment D.R. and Mays, L.W. (1988). *Applied Hydrology*. McGraw-Hill Education, New York, p572.

Durgunoglu, A. and Singh, K.P. (1993). The economics of using sediment entrapment reduction measures in lake and reservoir design. University of Illinois, UILU-WRC-93-216. *Research Report 216.*

Esmaeli, N.A., Hassanli, A.M. and Soufi, M. (2008). A Study of the Influential Factors on the Slopes of Deposited Sediments Behind the Porous Check dams and Model Development for Prediction. *Journal of Desert, Vol. 12:* pp 113-119. Online at http://jdesert.ut.ac.ir.

Esmaeli, N.A. (2003). Models for optimum distance between the porous check dams in Droodzan watershed. Thesis submitted to the school of graduate studies in partial fulfilment of the requirements for the MSc degree. Shiraz University, 129p.

Geyik, M.P. (1986). FAO watershed management field manual. Gully control. Food and Agriculture Organization of the United Nations conservation guide 13/2, Rome.

Hassanli, A. M., Esmaeli, N, A. and Beecham, S. (2007). Evaluation of the effect of porous check dam location on fine sediment retention (a case study). *Environmental. Monitoring. Assessment. DOI 10.1007/s10661-008-0318-2.*

Heede, B.H. and Mufich J.G. (1973). Functional Relationships and a Computer Program for Structural Gully Control. *Journal of Environmental. Management. Vol.1*, pp. 321-344.

Jiongxin X. (2004). Effect of human activities on overall trend of sedimentation in the lower Yellow River, China. *Environmental Management, Vol. 33*. No.5, pp. 637–653.

Lenzi, M.A. (2002). Stream bed stabilization using boulder check dams that mimic step-pool morphology features in Northern Italy. *Geomorphology, Vol.45*, pp. 243-260.

Lenzi, M.A. and Comiti, F. (2003). Local scouring and morphological adjustments in steep channels with check-dam sequences. *Geomorphology, Vol.55,* pp. 97-109.

Minneapolis Metropolitan Council (2003). *Minnesota Urban Small Sites BMP Manual.* Metropolitan Council. Minneapolis. 2000.

Nyssen, J., Veyret-Picot, M., Poesen, J., Moeyersons, J., Mitiku, H., Deckers, J. and Govers, G. (2004). The effectiveness of loose rock check dams for gully control in Tigray, northern Ethiopia. *Bed sediment Use and Management, Vol.20*, pp.55-64.

Tulu, T. (1999). Optimum check dam spacing for gully stabilization. *International Journal of Tropical Agriculture. Vol.17* (1-4), pp. 69-75.

USDA Bed sediment Conservation Service. (2003). Temporary check dams. Produced by Cumberland County SWCD, Knox-Lincoln SWCD, Maine Bed sediment and Water Conservation Commission, US Environmental Protection Agency. http://www.maine.gov/dep/blwq/docstand/escbmps/escsectionb2.pdf.

Zhou, Xu X, , Zhang, H.W. and Zhang, Q.Y. (2004). Development of check-dam systems in gullies on the Loess plateau, China. *Environmental Science and Policy, Vol.7*, pp.79-86.

In: Check Dams, Morphological Adjustments… ISBN: 978-1-62618-856-3
Editors: C. Conesa and M. A. Lenzi

Chapter 3

MODERN STRATEGIES FOR TORRENT CONTROL: SLIT AND W-WEIR CHECK DAMS

Vito Ferro*

Dipartimento di Ingegneria e Tecnologie Agro-Forestali,
University of Palermo, Viale delle Scienze 13, 90128 Palermo, Italy

ABSTRACT

In this Chapter the hydraulic behavior of a slit check dam is firstly presented. Then, for a slit check-dam with a rectangular and a trapezium contracted cross-section, the equations available for solving technical problems are introduced. The design problem of a slit check dam is solved using the flow motion equation and the continuity equation and a relationship between the slit width and the mean flow velocity in the first river section upstream the check dam is deduced.

For an inverse trapezium slit cross-section (the wider width of the trapezium is placed on the bed), the analysis also shows both that the slit width depends on the maximum value of the mean flow velocity in the river section just upstream the check dam and that the trapping efficiency is independent of the flow discharge.

In this chapter the equations available for solving technical problems of a slit check-dam with a hyperbolic cross-section are also presented.

Finally a new check dam having a W-weir is presented. For a given channel width, a W-weir is characterized by a length of the sill greater than a linear weir. This circumstance produces the decrease of the head corresponding to a given discharge and it is useful when the small height of the river cross-section allows to place a check dam with a small height. Furthermore, the length of the weir and the increased turbulence of the stream downstream the check dam will improve flow aeration. Applying the dimensional analysis and the Π-Theorem five dimensionless groups describing the outflow process are deduced. The experimental data allow to specify the groups from which the outflow process is really dependent and to determine a power stage - discharge relationship useful for solving the design problem.

* Corresponding author: vferro@unipa.it

INTRODUCTION

The construction of transverse structures (check dams) along the thalweg is a method widely used for stabilising streams, reducing channel degradation and preventing sediment deposition downstream. The streambed upstream of the check dams holds deposited sediments and results in a decrease of the longitudinal slope. In this case the check dam has a *bed stabilization* function.

Check dams are also used to trap coarse sediment particles and to prevent debris flow effects; in this case the check dam has a *trapping* function [32].

The *open check dams* are able to establish a selective retention of the sediment particles transported by flow and to preserve the bed equilibrium of the reaches downstream the stream rehabilitation works [21], [29], [34].

This innovative processes, starting in France after 1950, aimed to search design solutions useful for different torrent control problems (selection of the transported sediments, breaking of debris flows, sediment feeding at the river mouth, ecc.) [60], [61].

The first example of open check dam is the one proposed by Poncet [59]. This Author designed a check dam having a reinforced concrete sill (1 in Figure 1) embedded into the stream banks and two reinforced concrete side shares (flanges 2 in Figure 1) which are also embedded into the stream banks for an height equal to 12 m. The central zone of the check dam was filled using a filtering system realized using larch-trunks ballasted by coarse stones. The crowning of the central zone (3 in Figure 1) was constructed using old rails disposed along the flow direction. During a flood event this check dam was damaged from the scour phenomena, occurring downstream the check dam, due to the high check dam height. For avoiding the scour damages a new larch-trunks filter was constructed below the reinforced concrete sill (4 in Figure 1).

In 1952, Genet [40] proposed a new open check dam, named "comb" check dam, having in the central zone two rows of concrete teeth, embedded into a sill. The central filtering system aimed to trap the coarse transported sediments and to divide the front of the upstream flow.

In 1955 Genet patented the new type of open check dam showed in Figure 2. In the central part of the check dam a downstream sloping or vertical grid is installed. The interdistance between two elements of the grid (train rails, I beam, ecc.) is established for obtaining the mechanical trapping of the transported material having a given size.

For open check dam having a *bed stabilization* function, the partial sediment trapping (associated to the size of the trapped sediments) determines a sediment transport downstream the check dam avoiding localized scouring phenomena. This sediment transport also allows to maintain the equilibrium of the river which could be changed if an increased sediment transport capacity due to a total sediment trapping upstream the check dam should occur.

During moderate river stages, the flow could be able to transport a quote of the sediments, deposited during the flood event, downstream the check dam and could restore a deposition zone upstream the dam. The result of this natural action of *self cleaning* can be improved by a man-made cleaning project.

For bed stabilization check dams the design has mainly to take into account the sediment selection ability of the check dam and the need to realize a reduction of the upstream sediment discharge [23].

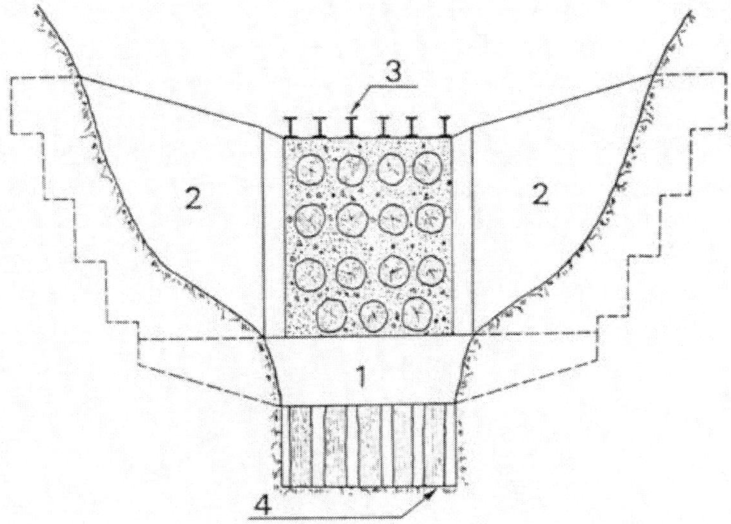

Figure 1. Scheme of the check dam proposed by Poncet.

Figure 2. View of the check dam proposed by Genet.

For *trapping* check dams the ability of the check dam to trap only the coarse transported sediments allows to maximize its stream control effect during the flood events. In fact the sediment storage volume upstream the check dam will be only available for trapping the coarse material transported by river flow [1], [2], [51]. A *trapping* check dam has to be constructed in a site morphologically suitable for trapping sediments; in other words an high sediment volume should be stored using a check dam having a low height dam.

After the prototypes suggested by Poncet and Genet, many types of open check dams were proposed and designed. Notwithstanding in these last years many theoretical and experimental investigations were carried out [1], [2], [6], [8], [9], [10], [15], [16], [19], [20], [21], [22], [30], [31], [32], [33], [34], [35], [54], [55], the hydraulic behaviour of the open check dam is not simple and it is not completely known. Among the available open check dams, the slit check dam is widely applied for its geometrically simple shape and its applicability was tested in many practical cases.

In this Chapter the hydraulic behaviour of a slit check dam is firstly presented. Then, for a slit check-dam with a rectangular and a trapezium contracted cross-section, the equations available for solving technical problems are presented. The design problem of a slit check dam is solved using the flow motion equation and the continuity equation and a relationship between the slit width and the mean flow velocity in the first river section upstream the check dam is deduced.

In this chapter the equations available for solving technical problems of a slit check-dam with a hyperbolic cross-section are also introduced. Finally a new check dam having a W-weir is presented.

HYDRAULIC BEHAVIOUR OF A SLIT CHECK DAM

The slit check dam is an obstacle to the free motion of the river flow which is obliged to cross a section, localized inside the check dam, which is more *narrow* than the river one.

The stream flow is generally supercritical and can have two different hydraulic behaviours when approaches the obstacle:

(a) the flow has a specific energy so high that it can cross the obstacle without changing its characteristics of *supercritical* flow;
(b) the flow is obliged to reduce its velocity upstream the check dam in order to cross the obstacle with the minimum specific energy value which corresponds to the *critical* condition.

The slit check dam has a slot, located in the central zone, which can be so deep to reach the sill. The scheme of Figure 3 represents a straight reach of a channel, having a width equal to B, in which is located a slit check dam having a rectangular slot with a width equal to b.

We indicate with $q_1 = Q/B$ the specific discharge, i.e. the discharge for unit of channel width, in the channel (cross-section 1 of Figure 3) and with $q_2 = Q/b$ the specific discharge of the flow crossing the narrow section (cross-section 2 of Figure 3) inside the check dam.

If the channel is steep [46], the flow in the channel upstream the cross-section 2 is uniform and supercritical. If we hypothesize that the specific energy of the flow in cross-section 2 is equal to the one e_o corresponding to the upstream uniform flow condition, the relationship between the flow depth h and the specific discharge q is the following:

$$q = h \sqrt{2\, g (e_o - h)} \qquad (1)$$

and the stage-discharge relationship for the uniform flow condition is:

$$q = h\, \chi \sqrt{R\, i} \qquad (2)$$

in which χ is the Chezy's coefficient, R is the hydraulic radius and i is the bed channel slope.

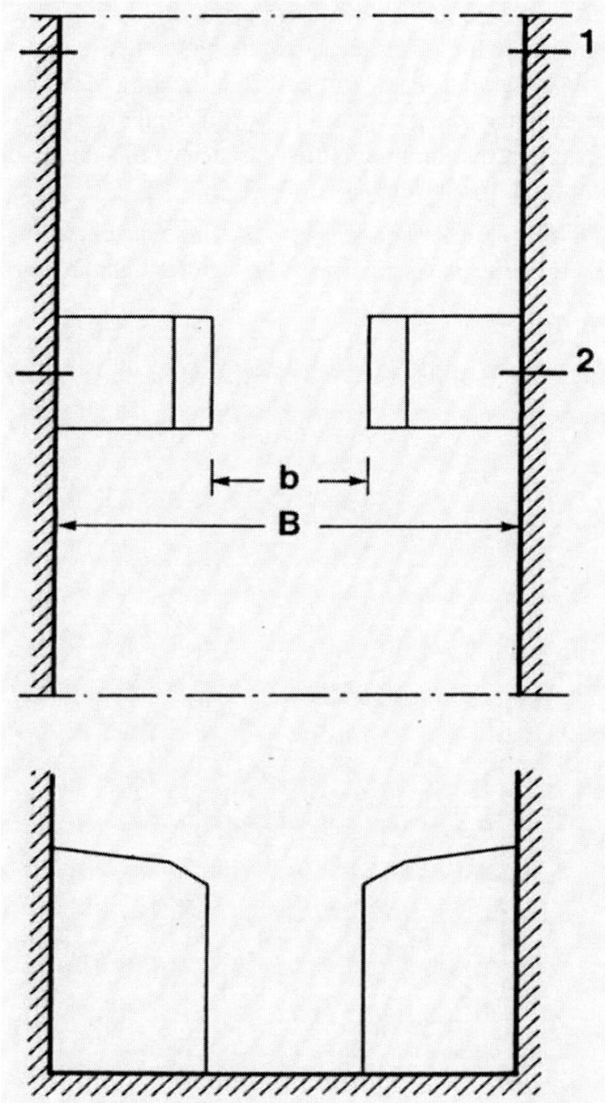

Figure 3. Scheme of a reach into which a slit check dam is located.

If the width contraction due to the presence of the check dam is *wide* the flow crossing the obstacle is supercritical (case a) [32]; in other words, the specific discharge q_2 is less than the limit value q*, which is the specific discharge of a flow crossing the obstacle with a specific energy equal to e_o in a critical condition.

Using the relationship $q(h)$ corresponding to the specific energy e_o, the water depth of the supercritical flow $h_2 > h_o$, being h_o the water depth of the upstream uniform flow, can be deduced for the known specific discharge q_2.

Therefore, in the stream reach upstream the check dam the flow motion is uniform while downstream of the obstacle the permanent flow is supercritical (backwater curve of S2 type).

This hydraulic behaviour, which is characterized by high values of the mean flow velocity both upstream and downstream of the check dam, is not useful for the stream restoration aims.

In the case b) the flow is obliged to slow down upstream of the check dam in order to get the minimum specific energy, which is greater than the value e_o of the uniform flow, allowing the flow to cross the obstacle. The flow will cross the narrow section of the obstacle as a critical flow with a flow depth $h_2 = q_2^{2/3} g^{-1/3}$ and a specific energy value equal to $3\,h_2/2$.

In this case, the width contraction due to the presence of the check dam is so *narrow* [32] that the specific discharge q_2 is greater than $q*$.

Using the relationship $q(h)$ corresponding to the specific energy $e_2 = 3h_2/2$ plotted in Figure 4, the water depth of the subcritical flow h_1 can be deduced for the known specific discharge q_1.

Therefore upstream of the check dam the shift from a uniform supercritical flow to a subcritical one (backwater curve of the S1 type) occurs by means of an hydraulic jump.

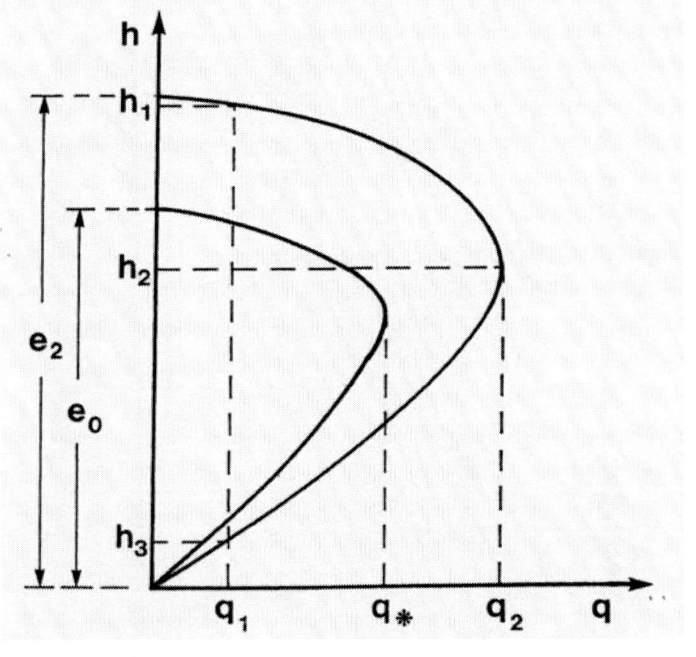

Figure 4. Relationship $q(h)$ corresponding to a specific energy e_2.

Downstream of the check dam the flow is supercritical and has to dissipate the specific energy excess equal to e_2-e_o. Using the relationship $q(h)$ corresponding to the specific energy $e_2 = 3h_2/2$ represented in Figure 4, the water depth of the subcritical flow h_3 can be deduced for the known specific discharge q_1.

The backwater curve downstream of the check dam is of S3 type and asymptotically tends to the uniform flow condition Figure 5.

A well designed slit check dam is characterized by the hydraulic behaviour of the case b). The slowing down in the flow determines the occurring of a zone in which low flow velocity values induce sediment deposition.

The study of the hydraulic behaviour of a slit check dam can be carried out, as suggested by Cola [19] and by Ferro and Ferreri [34], using an energetic approach. This approach is based on the circumstance that the flow passing through the slot becomes critical and the energy line is parallel to the bottom:

$$h_1 + \frac{V_1^2}{2g} = k + \frac{\bar{k}}{2} \tag{3}$$

in which h_1 is the water depth in the rectangular cross-section 1 (Figure 3) having a width equal to B, which is located upstream the check dam, V_1 is the mean flow velocity, k is the critical depth in the narrow cross-section 2 having a linear profile (Figure 6) and $\bar{k}$ is the corresponding mean critical water depth [32].

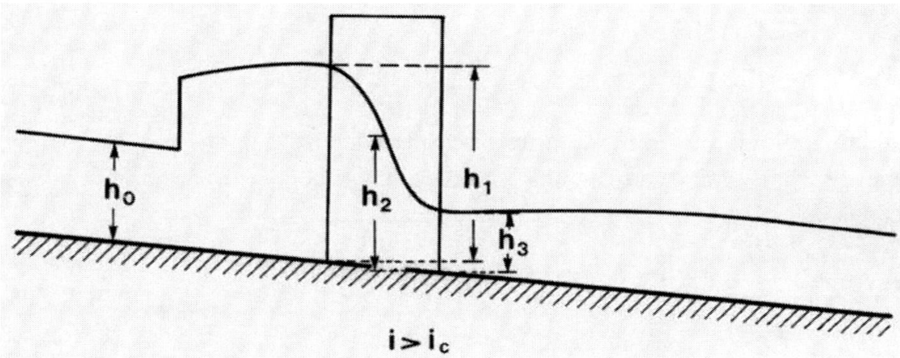

Figure 5. Backwater curves for the case b).

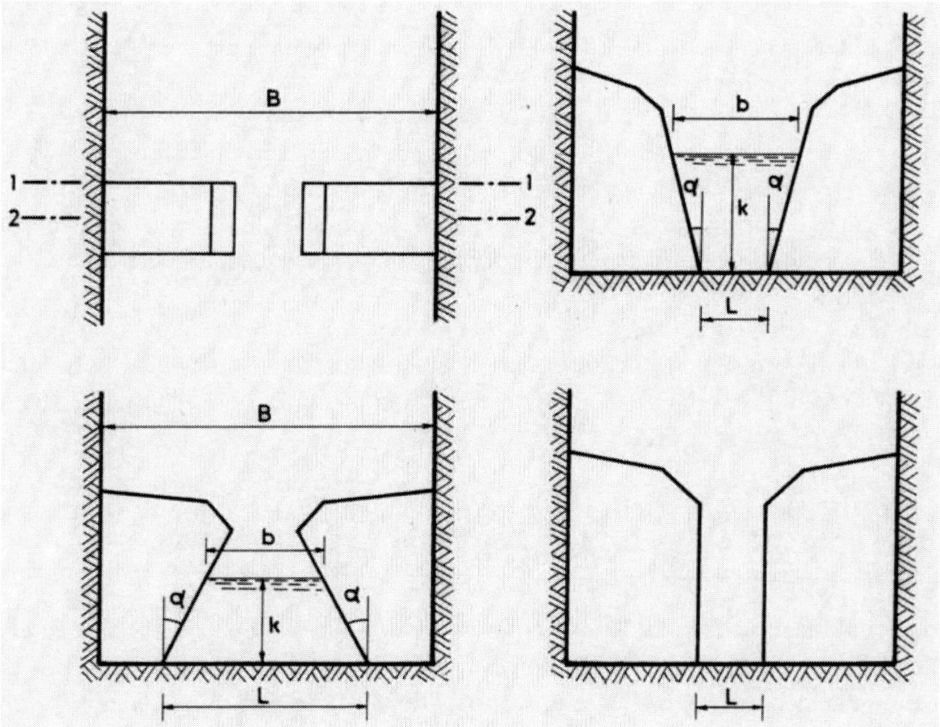

Figure 6. Scheme of slit check dams with linear profile.

The flow motion equation (3) has to be coupled with the following continuity equation:

$$B\, h_1\, V_1 = \sigma \sqrt{g\, \bar{k}} \tag{4}$$

in which $\sigma = \sigma(k)$ is the cross-section area for a water depth equal to critical depth k.

Deducing h_1 from both eq.(3) and eq.(4) and equating the two expressions the following relationship is obtained:

$$\frac{\sigma \sqrt{g\, \bar{k}}}{B\, V_1} = k + \frac{\bar{k}}{2} - \frac{V_1^2}{2\, g} \tag{5}$$

Taking into account the definition of mean critical depth [23], [32] and denoting with $b = b(k)$ the water surface width corresponding to the critical depth k into the narrow section 2, from eq.(5) we obtain

$$\sigma = \frac{B\, V_1 \left(k + \dfrac{\sigma}{2\, b} - \dfrac{V_1^2}{2\, g} \right)}{\sqrt{g\, \dfrac{\sigma}{b}}} \tag{6}$$

Eq. (6) is the general equation of the slit check dams which can be used to solve two different technician problems:

(1) for a slit check dam having a slot of given shape and dimensions, for which the function $b = b(k)$ is known, the relationship between the mean flow velocity V_1, in the channel cross-section 1 having known shape and dimensions, and the critical water depth k in the narrow cross-section of the slot can be determined (*control problem*);

(2) for a slit check dam having a slot of given shape, located in a channel having a cross-section of known shape and dimensions, the relationship $b = b(k)$ corresponding to a given mean flow velocity value V_1 can be determined (*design problem*).

THE CONTROL PROBLEM FOR A SLIT CHECK DAM WITH A LINEAR CROSS-SECTION

Consider a linear profile slot having a bottom width L and a slope of the sides equal to s ($s = \tan\alpha$, being α (Figure 6) the angle between the side and the vertical. For the rectangular cross-section the slope s is equal to zero, for the straight trapezium cross-section (having the smallest base, with a width equal to L, located on the channel bottom) the slope is greater than zero and for the upside down trapezium cross-section (having the largest base, with a width equal to L, located on the channel bottom) the slope s is less than zero.

The cross-section area and the water surface width can be calculated by the following relationships:

$$\sigma(k) = k(L + s\,k) \tag{7a}$$

$$b(k) = L + 2\,s\,k \tag{7b}$$

Introducing eqs.(7) into eq.(6) we obtain:

$$k(L + s\,k) = \frac{B\,V_1\left(k + \dfrac{k(L + s\,k)}{2\,(L + 2sk)} - \dfrac{V_1^2}{2\,g}\right)}{\sqrt{g\,\dfrac{k(L + s\,k)}{L + 2sk}}} \tag{8}$$

Eq.(8) can be simplified to obtain the following third order equation:

$$\left(\frac{B}{2\,g}\right)V_1^3 - B\,k\left(\frac{3\,L + 5\,s\,k}{2\,L + 4\,s\,k}\right)V_1 + \sqrt{\frac{g\,k^3\,(L + s\,k)^3}{L + 2\,s\,k}} = 0 \tag{9}$$

Eq. (9) is an incomplete third order equation which can be rewritten in the following form:

$$V_1^3 + p\,V_1 + q = 0 \tag{10}$$

where

$$p = -\frac{g\,k\,(3\,L + 5\,s\,k)}{(L + 2\,s\,k)} \tag{11}$$

$$q = \frac{2\,g^{3/2}\,k^{3/2}\,(L + s\,k)^{3/2}}{B\,(L + 2\,s\,k)^{1/2}} \tag{12}$$

Taking into account that the quotient Q_{pq} having the following expression:

$$Q_{pq} = \left(\frac{p}{3}\right)^3 + \left(\frac{q}{2}\right)^2 \tag{13}$$

is always negative, for any value of the variables *B*, *L*, *s* and *k*, then three different real solutions of eq.(10) exist. Taking into account that h_1 is a water depth of a backwater curve of the S1 type, the only solution having a physical meaning is the one being always positive and corresponding to a water depth h_1 greater than the critical depth:

$$V_1 = -2\sqrt{\frac{g\,k\,(3\,L+5\,s\,k)}{3\,(L+2\,s\,k)}}\;\cos\left(\frac{\alpha+\pi}{3}\right) \tag{14}$$

where the angle α has the following expression:

$$\alpha = \arccos\left\{\frac{-\dfrac{2\,g^{3/2}\,k^{3/2}\,(L+s\,k)^{3/2}}{B\,(L+2\,s\,k)^{1/2}}}{2\sqrt{\left[\dfrac{g\,k\,(3\,L+5\,s\,k)}{3\,(L+2\,s\,k)}\right]^3}}\right\} \tag{15}$$

As an example, for given values of B, L and s, the velocity values V_1 calculated by eqs. (14) and (15) and corresponding to different critical water depth k are plotted in Figure 7.

Figure 7 shows that the shape of the curve plotting the relationship V_1 - k for the straight trapezium cross-section is appreciably different from the one corresponding to the upside down trapezium cross-section. For a cross-section characterized by $s \geq 0$ the relationship V_1 monotonically increases with k while for the upside down trapezium cross-section(s < 0) the relationship $V_1 - k$ shows a maximum value [22], [23], [35]. In this last case, the mean flow velocity in the cross-section located upstream the check dam assumes a maximum value for a given critical water depth value k_p, i.e. for a given discharge Q_p.

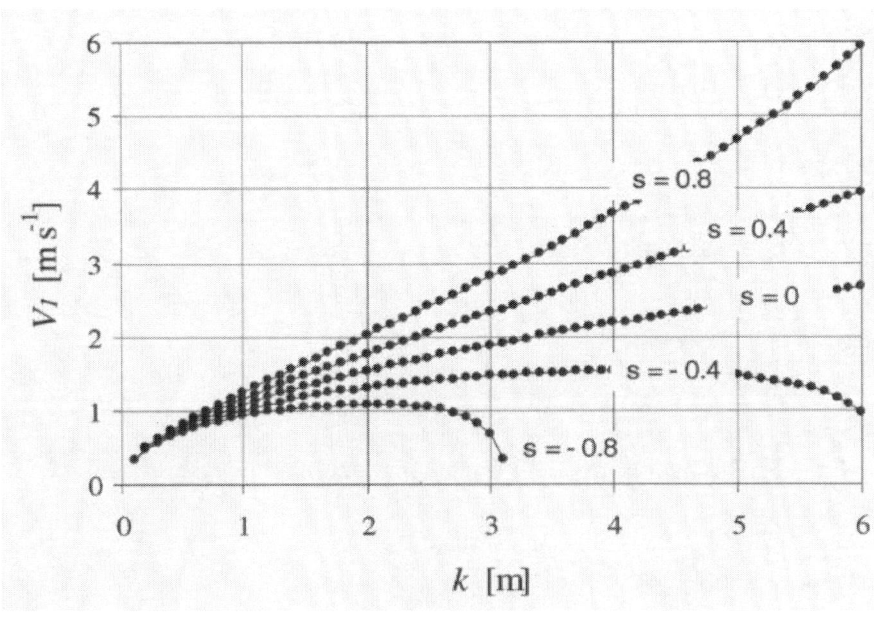

Figure 7. Values of mean flow velocity upstream the check dam calculated by eqs. (14) and (15).

In other words, if the check dam having a narrow upside down trapezium cross-section is designed using a discharge equal to Q_p, corresponding to the maximum value of the mean flow velocity V_1, then for the other conditions (when the discharge Q is less than Q_p or $Q > Q_p$) the velocity values in the cross-section upstream the check dam are always less than the maximum value V_1 corresponding to the design discharge. Using this design criterion the sediment deposits located in the area upstream the check dam will not be removed even if discharge values greater than Q_p occur. In other words, using this design criterion and this type of slot (upside down trapezium cross-section) the *bed stabilization* function is prevailing.

Therefore, the use of a check dam with a narrow upside down trapezium cross-section is not adequate when the aim of the check dam is the *trapping* function because the sediment self-clearing effect in the area upstream the dam is negligible. The trapping efficiency of the area upstream the dam should be time limited and a man-made clearing action should be carried out to maintain the storage volume useful to trap the sediments.

Neglecting the kinetic depth of the flow in the cross-section upstream the check dam, from eq.(16) the following equation is obtained:

$$V_1 = \frac{g^{1/2}\,\sigma^{3/2}}{B\,b^{1/2}\left(k + \dfrac{\sigma}{2\,b}\right)} \tag{16}$$

This hypothesis can be applied because both the V_1 values are low (subcritical flow, backwater curve of the S1 type) and neglecting $V_1^2/2g$ is partially balanced from the hypothesis that for the flow passing through the slot the energy line is parallel to the bottom.

For a slit check dam with a linear cross-section and using eqs.(7), eq.(16) becomes:

$$V_1 = \frac{2\sqrt{g\,k}\,(L + s\,k)^{3/2}\,(L + 2\,s\,k)^{1/2}}{B\,(3\,L + 5\,s\,k)} \tag{17}$$

Ferro and Bonsangue [25] demonstrated that the used hypothesis (neglecting of the kinetic depth) determines negligible errors in calculating V_1. In particular, the numerical analysis, carried out for different B, L and s values, showed that using of eq.(17) determines an error which is independent of s and is always less than 4% when the ratio B/L is greater than or equal to 2. Eq.(16) allows to solve the *control problem* for a slit check dam having a slot of known shape and sizes. For this slit check dam the relationship $b = b(k)$ is known and the flow velocity V_1, being known the channel width B, can be calculated for each value of the critical flow depth in the narrow section.

THE DESIGN PROBLEM FOR A SLIT CHECK DAM WITH A LINEAR CROSS-SECTION

For check dam having a rectangular slot, from eq.(17) with $s = 0$ the following relationship is obtained:

$$V_1 = \frac{2\,L\,\sqrt{g\,k}}{3\,B} \tag{18}$$

From eq.(18) the following expression of the contraction ratio $r = L/B$ is obtained:

$$r = \left(\frac{3}{2}\right)^{3/2} \frac{V_1^{3/2}}{g^{1/2}\,q_1^{1/2}} \tag{19}$$

where q_1 (m^3 s^{-1} m^{-1}) is the channel specific discharge equal to the ratio between the discharge Q and the channel width B.

The rough solution (19) was compared with the exacted one obtained by Ferro and Ferreri [32], [34]:

$$r = \frac{3^{3/2}\,g\,q_1}{\left(\frac{2\,g\,q_1}{V_1} + V_1^2\right)^{3/2}} \tag{20}$$

Using of both the exact solution (20) and the rough one (19) needs a relationship for evaluating the mean flow velocity V_1. This value can be set equal to V_c, being the mean flow velocity for which the sediment particles having a diameter equal to or greater than a given value d are stopped. For evaluating the critical mean flow velocity V_c are not available relationships applicable for a permanent flow motion condition [32]. For this reason the following empirical relationship linking the critical mean flow velocity V_c and the sediment particle diameter d can be used [32]:

$$V_c = 0.1215 + 0.2024\,\sqrt{d} \tag{21}$$

in which the diameter d is espressed as mm and V_c as m s^{-1}.

Di Stefano and Ferro [23] demonstrated that, for any value of the diameter d, the two solutions give contraction ratio r values which are practically coincident when the r value is less than 0.5. In other words, when the contraction is high the solution (20) has to be applied because this last equation has no applicability limits.

For the straight trapezium cross-section the solution of the design problem, i.e. the evaluation of the bottom width L of the slot having a known slope of the sides equal to s, can be obtained using the following equation deduced by eq. (17):

$$\frac{B\,V_1}{2\,\sqrt{g}} = \frac{k^{1/2}\,(L + s\,k)^{3/2}\,(L + 2\,s\,k)^{1/2}}{B\,(3\,L + 5\,s\,k)} \tag{22}$$

For given values of both the slope s of the trapezium slot and the diameter d of the particles to stop upstream the check dam (i.e. for a known value of the mean flow velocity V_1 equal to V_c), the design solutions are the pairs (k, L) verifying eq.(22).

For a given design case, i.e. for known values of B, s and d, the relationship $L(k)$ has to determined solving eq.(22) by trials, searching the k value corresponding to a given L value, and jointing to it the following stage-discharge relationship for the critical flow:

$$Q = g^{1/2} \frac{\left(L\,k + s\,k^2\right)^{3/2}}{\left(L + 2\,s\,k\right)^{1/2}} \tag{23}$$

in which L is the unknown width of the slot.

In other words, for solving the design case the relationship $L(k)$ and the stage-discharge relationship (23) have to be solved contemporaneously to determine the curve $L(Q)$.

For the upside down trapezium cross-section, the design case has to be solved for the hydraulic condition, i.e. for the critical water depth k, corresponding to the maximum value of the mean flow velocity V_1.

Deriving eq. (17) and setting the derivative equal to zero, the following critical water depth k^* corresponding to the maximum value $V_{1,max}$ (Figure 8) of the mean flow velocity in the channel cross-section upstream the check dam is obtained:

$$k^* = -\frac{L}{3\,s} \tag{24}$$

with $s < 0$.

Substituting eq.(24) into eq.(17) we obtain:

$$V_{1,\max} = \frac{2\sqrt{-g\,\frac{L}{3\,s}}\left(L - s\,\frac{L}{3\,s}\right)^{3/2}\left(L - 2\,s\,\frac{L}{3\,s}\right)^{1/2}}{B\left(3\,L - 5\,s\,\frac{L}{3\,s}\right)} \tag{25}$$

From eq.(25) the following expression of $V_{1,max}$ is obtained:

$$V_{1,\max} = \sqrt{-\frac{2\,g\,L^3}{27\,s\,B^2}} \tag{26}$$

Eq.(26) can be simply solved respect to the width L located on the channel bottom obtaining the following design relationship:

$$L = \left(-\frac{27\,B^2\,s\,V_{1,\max}^2}{2\,g}\right)^{1/3} \tag{27}$$

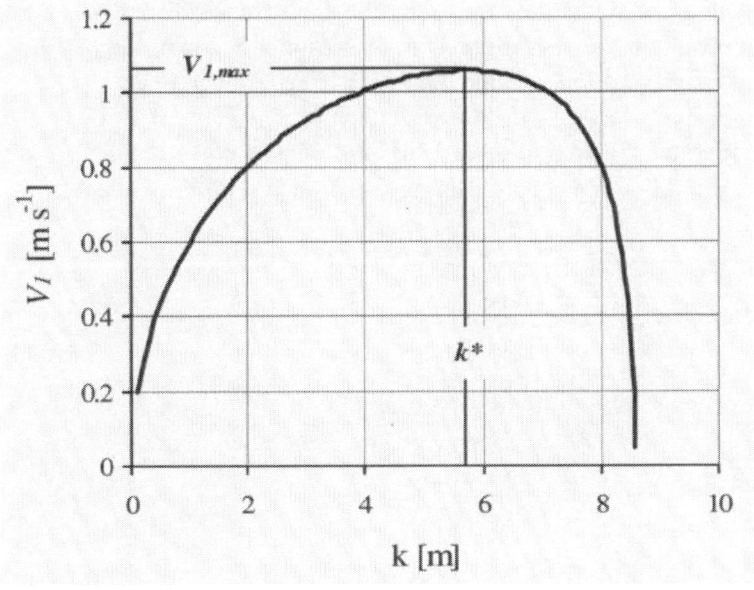

Figure 8. Values of mean flow velocity upstream the check dam calculated by eqs. (17).

For a rectangular cross-section channel having a width B, eq.(27) is useful to calculate the bottom width L of the upside down trapezium cross-section, having a know slope $s < 0$, of a check dam able to establish a mean flow velocity in the cross-section upstream to it equal to $V_{1,max}$.

In conclusion, the hydraulic design of a upside down trapezium cross-section check dam can be carried out using eq. (27), only knowing the diameter of the sediment particles to stop within the channel. Then the critical water depth k^* corresponding to L can be calculated by eq. (24). For a different value of the flow discharge Q the corresponding critical water depth k will be different to k^* but, in any case, the mean flow velocity V_1 will be less than $V_{1,max}$. In other words a variation of the flow discharge determines a reduction of the mean flow velocity upstream the check dam respect to the value $V_{1,max}$ established for the design condition and does not reduce the trap efficiency of the check dam.

SLIT CHECK DAM WITH AN HYPERBOLIC CROSS-SECTION

For studying the slit check dam having a generic shape of the slot eq. (16) has to be solved and the functions $\sigma = \sigma(k)$ and $b = b(k)$ have to be determined.

For solving eq. (16) it is useful to introduce the *shape factor* of the slot which is defined as the ratio between the mean critical depth $\overline{k}$ and the critical depth k. The factor α is a no-dimensional parameter varying with k and characterizing the geometry of the slot. For a linear cross-section of the slot the parameter α is defined by the following equation [22]:

$$\alpha = \frac{\overline{k}}{k} = \frac{L\,k + s\,k^2}{(L + 2\,s\,k)\,k} = \frac{L + s\,k}{L + 2\,s\,k} \tag{28}$$

where L and s are the bottom width and slope of the sides of the trapezium slot.

From eq. (28) for s = 0 (rectangular slot) $\alpha = 1$ is obtained. For $L = 0$ (triangular section) eq. (28) gives $\alpha = 0.5$. The parameter α is always less than 1 for a straight trapezium cross-section while α is always greater than 1 for the upside down trapezium cross-section.

Using the definition of the parameter α eq.(6) becomes

$$\sigma(k) = \frac{B V_1 \left(2g\, k + g\, \alpha\, k - V_1^2\right)}{2\, g^{3/2}\, \alpha^{1/2}\, k^{1/2}} \tag{29}$$

and can be rewritten as follows:

$$\sigma(k) = \frac{B V_1}{2\, g^{3/2}\, \alpha^{1/2}} \left(2g\, k^{1/2} + g\, \alpha\, k^{1/2} - V_1^2\, k^{-1/2}\right) \tag{30}$$

Considering that the variability of the parameter α with the critical water depth is negligible, the surface width $b(k)$ can be obtained deriving eq. (30):

$$b(k) = \frac{d\,\sigma(k)}{dk} = \frac{B V_1}{4\, g^{3/2}\, \alpha^{1/2}\, k^{3/2}} \left(2g\, k + g\, \alpha\, k + V_1^2\right) \tag{31}$$

The shape factor α assumes the following expression:

$$\alpha = \frac{\bar{k}}{k} = \frac{\sigma(k)}{b(k)\, k} = \frac{4\, k^{3/2} \left(2g\, k + g\, \alpha\, k - V_1^2\right)}{2\, k^{1/2}\, k \left(2g\, k + g\, \alpha\, k + V_1^2\right)} = \frac{4g\, k + 2g\, \alpha\, k - 2V_1^2}{2g\, k + g\, \alpha\, k + V_1^2} \tag{32}$$

Solving eq.(32) respect the parameter α the following simple relationship is obtained:

$$\alpha = 2 - \frac{V_1^2}{g\, k} \tag{33}$$

Substituting eq.(33) into eq.(29) we obtain:

$$\sigma(k) = \frac{B V_1}{g} \left(2g\, k - V_1^2\right)^{1/2} \tag{34}$$

Substituting eq.(33) into eq.(31) we obtain:

$$b(k) = \frac{B V_1}{\left(2g\, k - V_1^2\right)^{1/2}} \tag{35}$$

For given values of both the mean flow velocity V_1 in the cross-section upstream the check dam and channel width B, the curve representing the relationship $b(k)$ deduced by eq.(35) has an hyperbolic shape.

In other words, the slot which allows to establish a constant value of the velocity V_1, i.e. independent of the water depth in the slot, has an *hyperbolic shape* (Figure 9).

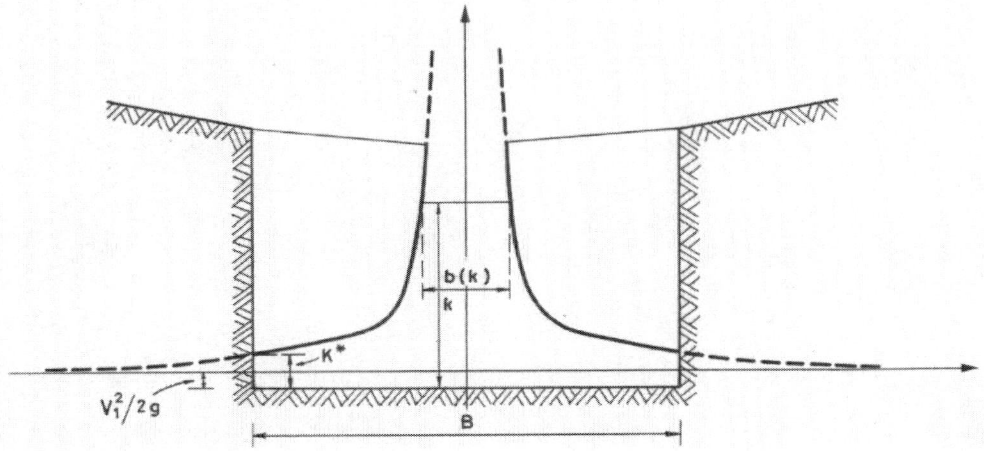

Figure 9. Scheme of slit check dam with an hyperbolic cross-section.

For solving the *design problem* by eq.(35) the mean flow velocity V_1 can be set equal to the value V_c calculated by eq. (21). Eq. (35) allows to solve the control problem, i.e. to determine the relationship between the mean flow velocity V_1 and the critical depth in the slot for a slit hyperbolic check dam having a known profile $b = b(k)$.

Squaring each member of eq. (35) and solving respect to V_1 the following relationship is obtained:

$$V_1 = \frac{b(k)}{\left(B^2 - b^2(k)\right)^{1/2}} \sqrt{2\,g\,k} \tag{36}$$

The function (35) assumes real values of $b(k)$ for critical depth values k greater than or equal to $V_1^2/2g$. Using eq.(35) the value of the critical depth, k^*, for which the slot width is equal to the channel width can be determined:

$$B = \frac{B\,V_1}{\left(2g\,k^* - V_1^2\right)^{1/2}} \tag{37}$$

From eq.(37) we obtain:

$$k^* = \frac{V_1^2}{g} \tag{38}$$

The slot will have an hyperbolic profile for water depths greater than k^* and a rectangular cross-section, with a width equal to the channel width B for water depth values less than k^*.

From the hydraulic behaviour point of view, supercritical uniform flow will cross the slot inside the check dam for water depth less than or equal to k^* because the slot cross-section is coincident with the channel cross-section (Figure 9). For water depth greater than k^* the narrowing of the hyperbolic slot determines a flow slowing down and the mean velocity V_1 will be constant and independent of the water depth, i.e. the velocity V_1 will be independent of the discharge.

Usinq eq. (34) and the relationship (33), for a hyperbolic slot the critical stage-discharge relationship assumes the following expression [8]:

$$Q = \frac{B V_1}{g} \left(2gk - V_1^2\right)^{1/2} \sqrt{g\,k \left(2 - \frac{V_1^2}{g\,k}\right)} \tag{39}$$

Simplifying eq.(39) we obtain the following equation:

$$Q = \frac{B V_1}{g} \left(2gk - V_1^2\right) \tag{40}$$

Taking into account that in the cross-section 1 just upstream the check dam the discharge is equal to the product $B\,h_1\,V_1$, using eq.(40) the following relationship is obtained for calculating the water depth h_1:

$$h_1 = \frac{2\,g\,k - V_1^2}{g} \tag{41}$$

Eq.(41) allows to test if for the design discharge, corresponding to a given critical depth k inside the slot and for the fixed value of the mean flow velocity V_1, the water depth h_1 is less than or equal to the geometrical depth of the cross-section into which the check dam will be constructed.

HYDRAULIC BEHAVIOUR OF A W-WEIR CHECK DAM

The need to provide a suitable habitat for aquatic flora, fauna, vertebrate organisms and invertebrates suggest to maintain stable a fluvial reach in order to maximize the habitat diversity in terms of wetted perimeter, flow velocity and turbulence, and substrate [5].

In order to maintain or restore the biodiversity of a fluvial habitat, non-structural and structural approaches have been proposed [43], [44], [45].

Structural measures have to include weirs able to vary both the turbulence and aeration performance of a flow. The design and implementation of check dams and sills have been used to promote or improve the flow aeration [38], [63].

Generally, many river miles are required to obtain a significant air-water transfer of oxygen [42], but this same oxygen transfer can be obtained by the short crossing of an hydraulic structure as a spillway, a free overfall, an unsubmerged hydraulic jump or at the base of a spillway. The primary reason for this accelerated oxygen transfer is that air is entrained into the flow, which produces a large number of bubbles. These air bubbles increase the surface area available for mass transfer.

The American Government Agencies which deal with the improvement of fluvial water quality have established a 5 g m^{-3} daily average dissolved-oxygen standard for a fluvial reach. From this point of view the construction of a check dam has also the objective to improve flow aeration [42]: the outflow process from the weir determines an increase of the flow surface whose perimeter has a contact with the atmospheric environment, the entrainment of air due to the impact of the flow with the pool downstream the free overfall.

A W-weir has a width L_s which is greater than the width L of the channel into which is located. The W-weir is constructed using a plan scheme of a W crest having four equal side. The ability of a W-weir to pass large discharges at a comparatively low head makes it an ideal structure in situations where the maximum allowable head is limited.

Figure 10 shows the scheme of a W-weir in which the main geometrical elements are reported and allows to establish the following relationship:

$$L_S = \frac{L}{sen\alpha} \tag{42}$$

in which α is the angle, named obliquity, of the weir side with the channel bank.

The shape of the structure provides better hydraulic performance and enhances oxygen transfer by increasing flow turbulence levels.

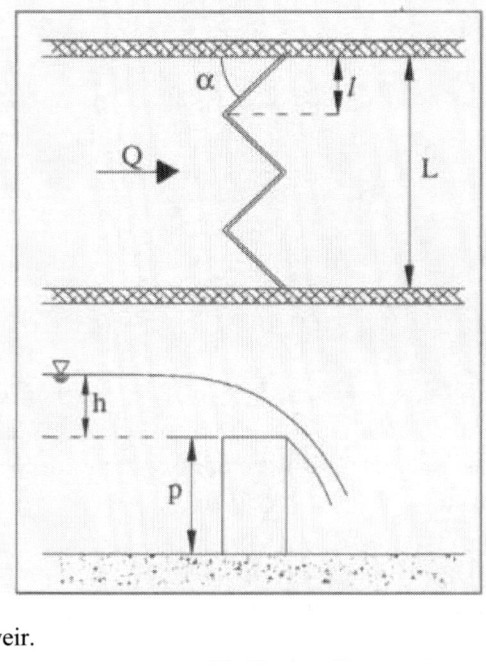

Figure 10. Scheme of the W-weir.

The first experimental study on thin weir having a linear crest, realized with a number of sides n ranging from 2 to 8, was carried out by Gentilini [39]. This Author used thin weir having different obliquity values (Table 1) and an height equal to 0.34 m. Gentilini [39] measured the pairs head h – discharge Q for weirs having a different number n of sides inserted in flumes having different width values.

Table 1. Geometrical characteristics of the weirs used in the experimental investigation of Gentilini [39] and Carollo et al. [12]

Authors	Weir	α [°]	n	L [cm]	P [cm]
[39]	thin	30	2	19.9	34.00
[39]	thin	30	2	20.1	34.00
[12]	thin	30	4	30.4	10.04
[12]	broad	30	4	30.4	10.18
[39]	thin	45	2	10.0	34.00
[39]	thin	45	2	20.0	34.00
[39]	thin	45	2	40.0	34.00
[39]	thin	45	4	20.0	34.00
[12]	thin	45	4	30.4	9.99
[39]	thin	45	4	40.0	34.00
[39]	thin	45	8	40.0	34.00
[12]	broad	45	4	30.4	10.18
[12]	thin	60	4	30.4	10.11
[39]	thin	60	2	26.0	34.00
[12]	broad	60	4	30.4	10.18
[12]	thin	90	4	30.4	10.10
[12]	broad	90	4	30.4	10.18
[12]	broad	120	4	30.4	10.18
[39]	thin	135	2	10.0	34.00
[39]	thin	135	2	20.0	34.00
[12]	thin	135	4	30.4	9.99
[12]	broad	135	4	30.4	10.18
[12]	thin	150	4	30.4	10.04
[12]	broad	150	4	30.4	10.18

Recently Ferro [33] theoretically deduced the stage-discharge relationship for a W-weir.

For a turbulent flow condition, the outflow process from a W-weir can be expressed by the following functional relationship [12], [33]:

$$F(h, Q, p, L, sen\,\alpha, l,\ g, \mu) = 0 \tag{43}$$

in cui μ is the water viscosity, g is acceleration due to gravity and l is the ratio between the channel width and n the number of sides of the W-weir (Figure 10).

Since the functional relationship (43) represents a physical phenomenon which does not depend on the measurements units used for the eight variables, according to the Π theorem of dimensional analysis [4], choosing l, g and μ as dimensional independent variables, the functional relationship (43) can be expressed using only five dimensionless groups Π_i (i = 1,..,5):

$$F_1(\Pi_1, \Pi_2, \Pi_3, \Pi_4, \Pi_5) = 0 \tag{44}$$

in which F_1 is a functional symbol.

The dimensionless group Π_1 has the following expression:

$$\Pi_1 = l^{\delta} g^{\beta} \mu^{\gamma} h \tag{45}$$

in which δ, β and γ are constants. Substituting into eq. (45) the measurement units of each variable the following relationship is obtained:

$$\Pi_1 = m^{\delta} m^{\beta} s^{-2\beta} kg^{\gamma} s^{2\gamma} m^{-4\gamma} m \tag{46}$$

Since the group Π_1 is dimensionless, the numerical values of the constants δ, β and γ can be deduced solving the following eqution system derived by eq. (46):

$$\begin{cases} 0 = \delta + \beta - 4\gamma + 1 \\ 0 = -2\beta + 2\gamma \\ 0 = \gamma \end{cases} \tag{47}$$

The solution $\delta = -1$, $\beta = 0$ and $\gamma = 0$ allows to establish that the group Π_1 is equal to the ratio h/l.

This procedure can be repeated for obtaining the expression of the other groups:

$$\Pi_2 = \frac{Q^2}{g l^5} \tag{48a}$$

$$\Pi_3 = \frac{p}{l} \tag{48b}$$

$$\Pi_4 = \frac{L}{l} = n \tag{48c}$$

$$\Pi_5 = sen\alpha \tag{48d}$$

Rearranging the dimensionless groups Π_2, Π_4 and Π_5 we obtain

$$\psi_2 = \frac{\Pi_2^{\frac{1}{3}}}{\Pi_4^{\frac{2}{3}}} \Pi_5^{\frac{2}{3}} = \frac{k_s}{l} \tag{49}$$

Rearranging ψ_2 and Π_3 we obtain

$$\frac{\Pi_3}{\psi_2} = \frac{p}{l}\frac{l}{k_s} = \frac{p}{k_s} \tag{50}$$

Finally eq. (44) can be expressed by the following functional relationship:

$$\phi\left(\frac{h}{l}, \frac{k_s}{l}, \frac{p}{k_s}, n, sen\alpha\right) = 0 \tag{51}$$

EXPERIMENTAL INVESTIGATION ON W-WEIR

Recently Carollo et al. [12] carried out some experimental runs in a flume located at the Dipartimento di Ingegneria e Tecnologie Agro-Forestali of the University of Palermo. The flume, having the walls and the bed made of glass, was 4.90 m long, 0.304 m wide and 0.24 m deep; it was connected to an hydraulic circuit allowing for a recirculation of discharge. The discharge was measured by an electro-magnetic gauge which is characterized by a maximum error equal to 0.2% of the measured value.

The experimental runs were carried out using six thin W-weirs characterized by a number of sides equal to 4, a thickness equal to 2 mm and an height p quasi equal to 10 cm (Table 1). The six W-weirs are also characterized by a α value equal to 30°, 45°, 60°, 90° (Bazin weir), 135° and 150°. In the experimental runs these last obliquity values were obtained overturning by 180°, respect to the direction normal to the flow motion, the W-weirs having $\alpha = 45°$ and $\alpha = 30°$.

Carollo et al. (2007) also tested 7 wood broad weirs having four sides with a thickness equal to 4 cm. This thickness was selected imposing that the weir thickness for all experimental runs had to be greater than 0.65 times the maximum head.

The W-weir height was always 10.18 cm and the obliquity was set equal to 30°, 45°, 60° 90°, 120°, 135° and 150°.

For each experimental run the discharge Q and the head h were measured. The head h was measured at the channel axis by a point gauge (accuracy 0.1 mm).

For the thin W-weirs the measurements were characterized by h values ranging from 1.22 to 5.88 cm and discharge values varying in the range 1.4 – 8.8 l s^{-1}.

For the broad W-weirs the measurements were characterized by h values ranging from 1.73 to 5.78 cm and discharge values varying in the range 1.2 – 8.5 l s^{-1}.

For the *thin weirs*, the analysis was carried out by the pairs (h, Q) measured by Carollo et al. [12] and the ones measured by Gentilini [39]. Table 1 lists the geometrical characteristics (obliquity, number of sides, channel width, weir height) of the weirs used in these investigations. Figure 11 shows, as an example for $\alpha = 30°$ and $\alpha = 150°$, that the stage-discharge relationship is the same; in other words the relationship $h(Q)$ is independent of the way of weir position.

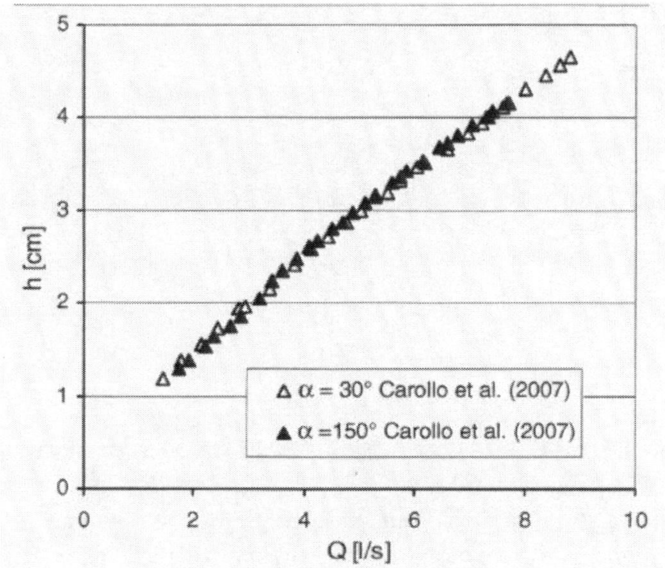

Figure 11. Stage-discharge relationships corresponding to two different weir positions.

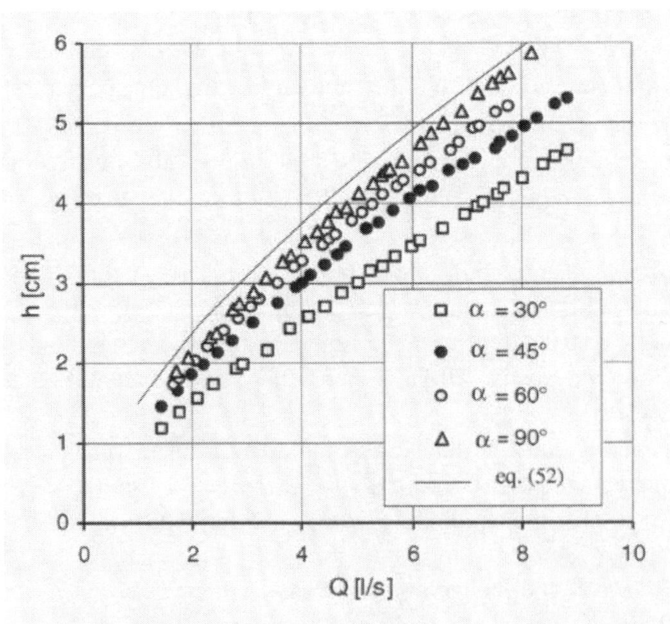

Figure 12. Stage discharge relationships for different obliquity values.

In Figure 12 the pairs (h, Q) corresponding to obliquity less than or equal to 90° and the stage-discharge relationship for a Bazin weir are plotted

$$Q = \mu L h \sqrt{2gh} \tag{52}$$

in which μ, which is the discharge coefficient, was set equal to 0.41.

Figure 12 shows that for a given discharge Q the head h decreases with the obliquity α. This reduction of the head is noticeable in the comparison between the condition $\alpha = 45°$ and $\alpha = 30°$.

The comparison between the experimental pairs (h, Q) corresponding to the run $\alpha = 90°$ and the stage-discharge (52) shows that, for a each head h, the measured discharge is greater than the value calculated by eq. (52). This circumstance can be explained taking into account that eq. (52) neglects the effect of the flow velocity upstream the weir; in other words eq.(52) is deduced for an hydrostatic condition upstream the weir.

When the weir is located inside a channel the effect of the upstream flow velocity is not negligible [17] and this effect is more and more remarkable for decreasing values of the weir height; in other words, the stage-discharge relationship of the weir will be affected by the weir height.

Figure 13, as an example for $\alpha = 45°$, plots the experimental pairs (k_s/l, h/l) of Gentilini's investigation for different n values. These measurements were carried out using weirs characterized by a single weir height equal to 34 cm. All experimental pairs collapse in a single curve demonstrating that the number of sides n does not affect the stage-discharge relationship.

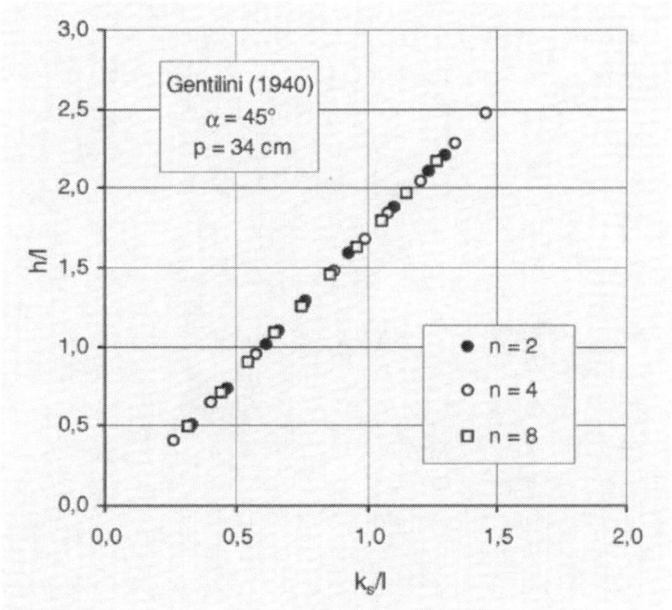

Figure 13. Experimental data of Gentilini [39] for different n values.

The experimental pairs (k_s/l, h/l) of Carollo et al. [12] corresponding to each investigated obliquity value, demonstrated that, for each k_s/l value, the ratio h/l decreases when α increases and this trend is particularly appreciable for the highest k_s/l values. The pairs (k_s/l, h/l) tend to overlap for the lowest k_s/l values. The same trend is recognized for the experimental data of Gentilini [39].

For given values of obliquity, the comparison between the pairs (k_s/l, h/l) measured by Carollo et al. [12], characterized by a weir height p = 10 cm, and the ones measured by Gentilini [39] for the condition p = 34 cm showed that the pairs of Carollo et al. [12] are located just below the measurements of Gentilini; in other words, the weir height can affect the stage-discharge relationship of the W-weir. Using these experimental results, Carollo et al. [12] proposed for the functional relationship (51) the following expression:

$$\frac{h}{l} = a\left(\frac{k_s}{l}\right)^b \left(\frac{p}{k_s}\right)^c \tag{53}$$

where a, b and c are constants to be experimentally determined.

Carollo et al. [12] using the available measurements established that c coefficients can be assumed independent of obliquity while a and b coefficients are power function of the obliquity.

$$\frac{h}{l} = a_o(sen\alpha)^{a_1} \left(\frac{k_s}{l}\right)^{bo(sen\alpha)^{b_1}} \left(\frac{p}{k_s}\right)^c \tag{44}$$

where a_o, a_1, b_o, b_1 and c are constants.

Eq. (54) was calibrated by Carollo et al. [12] using the measurements obtained for α values less than or equal to 90°, and the following equation was obtained:

$$\frac{h}{l} = 1{,}327(sen\alpha)^{-0{,}541} \left(\frac{k_s}{l}\right)^{1{,}017(sen\alpha)^{-0{,}197}} \left(\frac{p}{k_s}\right)^{0{,}026} \tag{55}$$

Figure 14 shows that eq. (55) also allows an accurate estimate of the ratio h/l for the experimental data corresponding to α values greater than 90°. Eq.(55) is characterized by a mean error equal to 2.1% and only 2 pairs of the 291 experimental ones are out the ±10% error bands. For the *broad weirs* the experimental runs confirmed that the relationship $h(Q)$ is independent of the way of weir position. For the broad weirs the experimental data pairs (Q, h) corresponding to the investigated obliquity values (30°, 45°, 60°, 90°, 120°, 135°, 150°) demonstrated that, excluding the obliquity equal to 30° and 150°, all data pairs collapse on a single curve. This result underline that the hydraulic behaviour of the W-weir is useful, i.e. for each discharge the W-weir determines a reduction of the head respect to a conventional weir, for very low obliquity values or very high □ values. In other words the length L_s of the W-weir has to be much longer than the channel width into which the weir is located.

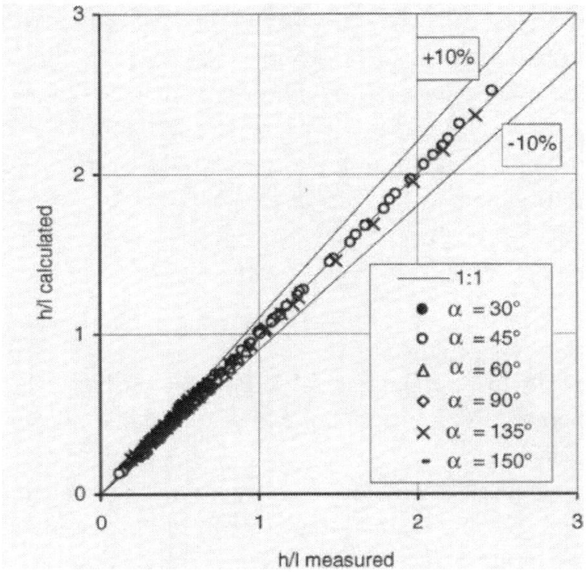

Figure 14. Comparison between the experimental data of Carollo et al. [12] and eq. (55).

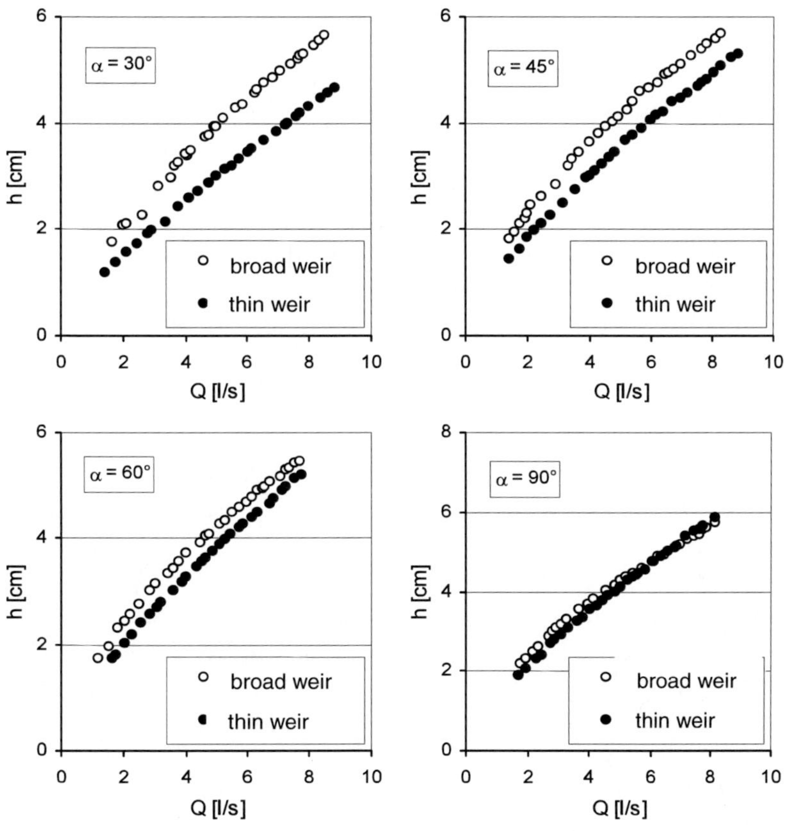

Figure 15. Comparison between the experimental data measured by Carollo et al. [12] for thin and broad W-weirs.

For a given obliquity (α = 30°, 45°, 60°, 90°), Figure 15 shows the comparison between the experimental pairs (Q, h) measured by Carollo et al. [12] for a thin and a broad W-weir.

For a given discharge, the reduction of the head h which can be obtained using a thin W-weir respect to the broad one increases for decreasing obliquity values. Therefore, for the aims of the river restoration the use of a thin W-weir can be useful, especially for low values of the obliquity, in order to obtain low values of the head.

CONCLUSION

The construction of open check dams allows to establish a selective retention of the sediment particles transported by flow and to preserve the bed equilibrium of the river reaches downstream the check dams.

In this Chapter the equations available to solve technical problems for a slit check-dam with a linear cross-section (rectangular and trapezium) were presented. For the particular case of an inverse trapezium slit cross-section (the wider width of the trapezium is placed on the bed), the analysis showed that the trapping efficiency is independent of the flow discharge.

The developed analysis also demonstrated that the slot which allows to establish a constant value of the mean flow velocity upstream the check dam, i.e. independent of the water depth in the slot, has an hyperbolic shape. Then the equations available for solving technical problems of a slit check-dam with a hyperbolic cross-section were also presented.

Check dams are structural measures able to vary both the turbulence and aeration performance of a flow. The W-weir is characterized by a length of the sill greater than a linear weir and this circumstance produces the decrease of the head corresponding to a given discharge. Furthermore, the length of the weir and the increased turbulence of the stream downstream the check dam will improve flow aeration. The wide experimental investigation, carried out using W-weir of different height and obliquity, allowed to determine a power stage - discharge relationship useful for solving the design problem.

REFERENCES

[1] Armanini, A., and Benedetti, G. (1996). Sulla larghezza di apertura delle briglie a fessura. *Proc. of XXV Convegno di Idraulica e Costruzioni Idrauliche*, Torino, 16-18 settembre, Vol. III, 13-24.

[2] Armanini, A., and Larcher, M. (2001). Rational criterion for designing opening of slit-check dam. *Journal of Hydraulic Engineering*, Proc. ASCE, Vol. 127, 94-104.

[3] Avery, S.T., and Novak, P.(1978). Oxygen transfer at hydraulic structures. *Journal of Hydraulic Engineering*, Proc. ASCE, Vol. 104, 1521-1539.

[4] Barenblatt, G.I. (1987). *Dimensional Analysis*. Gordon and Breach, Science Publishers Inc., Amsterdam.

[5] Bhuiyan, F., Hey, R.D., and Wormleaton, P.R. (2007). Hydraulic Evaluation of W-weir for River Restoration. *Journal of Hydraulic Engineering*, Proc. ASCE, Vol. 133, 596-609.

[6] Benedetti, G. (1997). Analisi teorica e sperimentale del funzionamento delle briglie aperte. *Dendronatura*, 2, 53-65.

[7] Benini, G. (1984). Antichi contributi italiani alla sistemazione dei torrenti. *Proc. of Workshop "Idraulica del Territorio Montano"*, Bressanone, 481-488.

[8] Bonsangue S., Di Stefano C., Ferro V. (2008): *Contributo al dimensionamento delle briglie a fessura iperbolica*, Quaderni di Idronomia Montana, 28/2, 239-250.

[9] Brath, A., Milano, V., Pagliara, S., and Viti, C. (1992). Ricerca sperimentale sulle briglie selettive del tipo a fessura. *Idrotecnica*, 4, 229-244.

[10] Busnelli, M., Stelling, G., and Larcher, M. (2001). Numerical morphological modeling of open-check. *Journal of Hydraulic Engineering*, Proc. ASCE, Vol. 127, 105-114.

[11] Catella, M., Paris, E., and Solari, L. (2002). Risultati preliminari sul comportamento delle briglie a fessura nel bacino montano dell'Alta Versilia, *Proc. of XXVIII Convegno di Idraulica e Costruzioni Idrauliche*, Vol.V, 205-214.

[12] Carollo, F.G., Ferro, V., and Pampalone, V. (2007). Processo di efflusso da uno stramazzo a profilo W. Quaderni di Idronomia Montana, 27, 195-207.

[13] Carollo, F.G., Ferro, V., and Pampalone, V. (2008). Briglie con savanella a profilo W. In V. Ferro Ed., *Sistemazione dei bacini montani e difesa del suolo*, Cosenza: Ed. Bios, 427-446.

[14] Cavinato, A., Fistarol, A., Fornari, E., and Piccoli, G. (1996). Laminazione dell'idrogramma solido mediante briglie selettive a fessura aperta. *Interpraevent 1996*, Band 5, 99-108.

[15] Cerato, M. (1995). Sulla progettazione ed il funzionamento delle briglie filtranti: alcune esperienze in provincia di Trento. *L'Italia Forestale e Montana*, 50, 147-169.

[16] Cerato, M. (1996). Sulla progettazione ed il funzionamento delle briglie filtranti. In S. Fattorelli S., Fedeli E.C. e Nardin D. (Eds). *Proc. of Workshop "Sviluppo e Gestione dei Bacini Idrografici"*, Istituto Italo Latino Americano-AIDI-PAT, 73-87.

[17] Citrini, D., and Noseda, G. (1987). *Idraulica*, seconda edizione, Ed. Ambrosiana, Milano.

[18] Clauzel, L., and Poncet, A. (1963) . Barrages filtrantes et correction torrentielle par ségrégation des matériaux charriés. *Revue Forestière Française*, Vol. XV, 512-522.

[19] Cola, R. (1970). Sul funzionamento delle briglie selettive. *Rassegna Tecnica del Friuli-Venezia Giulia*, 5, 3-10.

[20] Cola, R. (1972). Criteri di progettazione esecutiva delle opere idrauliche del torrente Pesarina. *Proc. of Workshop "Nuovi criteri di progettazione e nuove opere di sistemazione dei torrenti"*, Udine, 48-55.

[21] Della Giacoma, F., Ferrari, L., and Armanini, A. (1989). From the check dam to the development of functional check dams for a most effective torrent control. *Proc. of International Workshop on "Fluvial Hydraulics in Mountain Regions"*, Trento, 3-6 October, B169-B182.

[22] Di Stefano, C., Ferro, V., and Bonsangue, S. (2007). Contributo al dimensionamento idraulico delle briglie a fessura lineare. *Quaderni di Idronomia Montana*, 27, 403-420.

[23] Di Stefano, C., and Ferro, V. (2008). Briglie a fessura. In V. Ferro Ed., *Sistemazione dei bacini montani e difesa del suolo*, Cosenza: Ed. Bios, 403-424.

[24] Dragogna, G. (1967). Sistemazione dei torrenti: le briglie selettive. *Monti e Boschi*, XVIII, 47-55.

[25] Dragogna, G. (1970). Briglie selettive e briglie di salvaguardia. *Annali Accademia Italiana di Scienze Forestali*, XIX, 441-479.

[26] Dragogna, G. (1970). La briglia selettiva di salvaguardia. *Il Frantoio*, VII (9), 5pp.

[27] Dragogna, G. (1975). Trasporto di massa e briglie selettive. *Proc. of "Interpraevent"*, Innsbruck, 75-89.

[28] Ferro, V. (1988). Contributo al dimensionamento di briglie selettive a reticolo con elementi longitudinali. Ricerca sperimentale. *Proc. of XXI Convegno di Idraulica e Costruzioni Idrauliche*, L'Aquila, Parte I, 539-557.

[29] Ferro, V. (1990). Un criterio di dimensionamento per le briglie selettive a reticolo con elementi trasversali. Ricerca sperimentale, *Rivista di Ingegneria Agraria*, 21, 225-240.

[30] Ferro, V. (2000). Contributo al dimensionamento delle briglie a finestra. *Rivista di Ingegneria Agraria*, 31, 207-215.

[31] Ferro, V. (2000). Simultaneous flow over and under a gate. *Journal of Irrigation and Drainage Engineering*, Proc. ASCE, 126, 190-193.

[32] Ferro, V. (2006). *La sistemazione dei bacini idrografici – seconda edizione*, McGraw-Hill, Milano, 848 pp.

[33] Ferro, V. (2006). Contributo al dimensionamento di una briglia con savanella a profilo W. *Quaderni di Idronomia Montana*, 25, 309-323.

[34] Ferro, V., and Ferreri, G. (1988). Sul funzionamento idraulico di alcuni tipi di briglie non convenzionali. *Rivista di Ingegneria Agraria*, 19, 44-55.

[35] Ferro, V., and Bonsangue, S. (2007). Sul dimensionamento idraulico della briglia a fessura trapezia rovescia. *L'Italia Forestale e Montana*, 2, 99-115.

[36] Fiebiger, G. (1984). Functional Bautypen der Wildbach- und Lawinenverbauung als Grundlage der Geschiebebewirtschaftung. Proc. of Interpreavent, Band 1, 123-133.

[37] Fiebiger, G. (1986): Bedload management on torrent control by functional check dams. *Beitrüge ur Wildbacherosions und Lawinenforschung*, Mitt. der FBVA, Vol. 159, 325-332.

[38] Gameson, A.L.H. (1957). Weirs and the aeration of rivers. *Journal of the Institute of Water Engineers*, Vol.11, 477-490.

[39] Gentilini, B. (1940). Stramazzi con cresta a pianta obliqua e zigzag, *Memorie e Studi del Regio Politecnico di Milano*, Vol. 48.

[40] Genet, E. (1953). Un noveau type de barrage de correction : le barrage-peigne. *Compte rendu général du voyage d'etudes dans les Alpes Fracaise du Groupe de Travail FAO/EFC/TORR"*, 1° Session, 1952, Nancy.

[41] Gorfer, G. (1991). *Per una difesa del territorio. La sistemazione dei bacini montani in provincia di Trento attraverso i secoli.* Azienda Speciale di Sistemazione Montana, Provincia di Trento, Edizioni Arca s.r.l.

[42] Gulliver, J.S., Rindels, A. (1993). Measurement of air-water oxygen transfer at hydraulic structures. *Journal of Hydraulic Engineering*, Proc. ASCE, Vol. 119, 327-349.

[43] Hey, R.D. (1992). *River mechanics and habitat creation*, In Fisheries in the year 2000, O'Gardy K.T., Butterworth A.J.B., Spillett R.P., Domaniewski J. (Eds.), Institute of Fishery Management, Nottingam, U.K., 271-285, (1992).

[44] Hey, R.D. (1994). *Restoration of gravel bed rivers: principles and practice*, In Natural channel design: perspective and practice, Shrubsole D. (Ed.), Canadian Water Resources Association, Cambridge, Ontario, 157-173.

[45] Hey, R.D. (1996). *Environmentally sensitive river engineering*, In River restoration, Pets G. e Calow P. (Eds.), Blackwell Science, Oxford, 80-105.

[46] Jansen, P.Ph., van Bendegom, L., van de Berg, J., de Vries, M. and Zanen, A. (1979). *Principles of River Engineering*, Pitman, London.

[47] Johnson, P.A., and McCuen, R.H. (1989). Slit dam design for debris flow mitigation. *Journal of Hydraulic Engineering*, Proc. ASCE, 115, 1293-1296.

[48] Kasai, S., Ohgi, Y., Mizoguchi, I., Matsuda, A., Aramaki, H., and Tanami, M. (1996). Structural characteristics of wood-debris entrapment facilities. *Proc. of Interpraevent*, Band 5, 271-277.

[49] Kettl, W. (1989). Evoluzione delle sistemazioni idraulico forestali: dallo scopo della sistemazione allo sviluppo delle tipologie costruttive. *Quaderni di Idronomia Montana*, 8, 35-65.

[50] Kim, J., and Walters, R.W. (2001). Oxygen transfer at low drop weirs. *Journal of Environmental Engineering,* Proc. ASCE, Vol. 127, 604-610.

[51] Larcher, M., and Armanini, A. (2000). Dimensionamento della larghezza dell'apertura delle briglie a fessura. *Proc. of XXVIII Convegno di Idraulica e Costruzioni Idrauliche*, Genova, Vol. I, 289-297.

[52] Maione, U., Cavalazzi, F., and Simonetti, M. (2000). Le briglie a fessura. *L'Acqua*, 3, 77-91.

[53] Matsamura, K., Nakamur, A. T., and Hirose, T. (1996). New development of sabo dams filled with sediment from construction field. *Interpraevent 1996*, Band 5, 47-58.

[54] Milano, V. (2003). Criteri di progettazione delle briglie selettive a fessura. *Proc. of XXIV Workshop on "Tecniche di difesa dall'inquinamento"*, Ed. BIOS, Cosenza.

[55] Milano, V., Pagliara, S., and Venutelli, M. (1998). Criteri progettuali sulle sistemazioni torrentizie con briglie selettive a fessura. *Idrotecnica*, 3, 45-52.

[56] Mizuyama, T., Kobashi, S., and Mizuno, H. (1996). Development and improvement of open dams function of structures against debris flow. *Interpraevent 1996*, Band 5, 59-65.

[57] Nakasone, H. (1987). Study of aeration at weirs and cascades. *Journal of Environmental Engineering*. Proc. *ASCE,* Vol. 113, 64-81.

[58] Passerini, G. (1957). Per la difesa degli abitati dai ' trasporti in massa' . *L'Acqua e il Suolo*, IX, 40-60.

[59] Poggiolini, G. (1961). Briglie filtranti nella sistemazione dei torrenti. *L'Acqua e il Suolo*, XIII, 28-31.

[60] Puglisi, S. (1967). L'impiego di dispositivi filtranti nella correzione dei torrenti. Nota introduttiva. *L'Italia Forestale e Montana*, 22, 12-24.

[61] Puglisi, S. (1968). Resoconto delle esperienze in corso con dispositivi filtranti in alcuni torrenti dell'Appennino Lucano. *L'Italia Forestale e Montana*, 23, 26-73.

[62] Tebbutt, T.H.Y., Essery, I.T.S., and Rasaratnam, S.K. (1977). Reaeration performance of stepped cascades. *Journal of the Institute of Water Engineers and Scientists*, Vol.31, 4.

[63] Tullis, J.P., Amanian, N., and Waldron, D. (1995). Design of labyrinth spillways. *Journal of Hydraulic Engineering*, Proc. ASCE, Vol. 121, 247-255.

[64] Watson, C.C., Walters, R.W., and Hogan, S.A. (1998). Aeration performance of low drop weirs. *Journal of Hydraulic Engineering*, Proc. ASCE, Vol. 124, 1, 65-71, (1998).

[65] Wu, S., and Rajaratnam, N. (1996). Submerged flow regimes of rectangular sharp-crested weirs. *Journal of Hydraulic Engineering,* Proc. ASCE, Vol. 122, 412-414.

[66] Zolin, G. (1995). Impatto ambientale delle briglie filtranti. *L'Italia Forestale e Montana*, 50, 329-340.

In: Check Dams, Morphological Adjustments…
Editors: C. Conesa and M. A. Lenzi
ISBN: 978-1-62618-856-3

Chapter 4

CHECK DAMS IN THE TORRENT CONTROL PRACTICE IN SMALL MOUNTAINOUS CATCHMENTS

Stanimir Kostadinov* and Nada Dragović
University of Belgrade, Faculty of Forestry
Kneza Višeslava 1, 11030 Belgrade, Serbia

ABSTRACT

Torrent control includes the control of deep and lateral channel erosion and the control of floods. Longitudinal and transversal structures are constructed to this purpose. Check dams are the basic type of structures in torrent control practice. They are the main elements of the classical torrent control systems in small catchments. There are several types of check dams and they can be classified according to different aspects.

By construction of check dams in the torrent bed, the bedload is retained in their accumulation space, forming the siltation. The slope of the upper surface of retained bedload is called the slope of siltation. It practically represents the specific case of the equilibrium bed slope. The positions of the check dams in the stream bed are determined by the calculations of the slope of siltation The effects of check dams depends of many factors, but the most important is slope of siltation.

The tendency to make the dam construction cheaper and more effortless and to prolong the effect, i.e. to increase the functionality, led to various grid dam constructions in many countries. They have a series of advantages compared to massive check dams. The researchers who delt with these problems have tended to invent a check dam which would affect the longest possible stretch of a streambed, i.e. by the least possible financial investments to control erosion processes in the bed and to check the moved sediment, etc.

In Serbia, during the fifties of the 20th century, Prof. S. Rosić introduced new types of the special filtration check dams with a filter system "for the separation of water from the sediment". Their use in the torrent control practice of Serbia showed a series of advantages regarding the efficiency and economicity, compared to classical torrent-control check dams.

* E-mail: kost@EUnet.rs

In the sixties of the 20th century in Tbilisi (Georgia), I. I. Herheulidze launched the prefabricated grid (filtration) check dams made of reinforced-concrete components and he named them "sediment traps".

In direct torrent control works during the last 25 years in the developed countries of Europe (first of all in Austria) a method (system) is being increasingly applied known as "Bedload management", based upon construction of so-called "Functional check dams", differing from the classical check dams by the active role the former play in the torrent control.

This paper presents a different types of check dams were used in the torrent control practice in small catchments of the different countries, method of design and function as well as the effects of the check dams, based on the experience of their implementation and research in the torrents of Serbia, Georgia and FYR Macedonia. On the basis of these experiences and achievements, the proposals for improvement of the activities of torrent control in small catchments are presented.

Keywords: check dams, small catchments, filtration check dams, sediment traps, functional check dams, effects of check dams.

1. Introduction

1.1. Basic Characteristics of Torrents

The fundamental property of torrent streams is an direct link among phenomena in the catchment and in the stream. The processes taking place within the catchment are directly and nearly simultaneously manifested in the hydrograph network. The most important ones are the surface runoff and the soil erosion processes, the product of which (water and sediment) enter the catchment hydrographic network and continue their movement as a two-phase fluid. Since the intensive precipitation in a catchment represents the main cause of both processes, the geneses of flood waters and the large masses of eroded sediment are simultaneous. This is why the largest part of the yearly inflow of water and sediment takes place in torrential floods. Therefore, the character of torrent streams sediment and hydrological regime is torrential. The torrential character of hydrological regime is manifested in two ways: by a large range of yearly discharge and by a short duration of flood waves. The discharge ratio of flow and flood waves is of the order of magnitude 1:1000. As to the duration of flood waves, it is only few days a year, i.e. 1–5% of the total yearly duration.

The main factor of hydraulic regime of torrential streams is the great channel slope, ranging from 1 to 15 per cent or more. The torrential streams are generally characterised by high flow velocities and tractive forces.

1.2. Torrent Control Systems

The torrent control essentially consists of a complex of erosion control measures and works on the catchment slopes and in the streambeds, as well as the defense from the torrential floods. Practically, it is dealt with the management of the torrential catchments, the catchments attacked by erosion and with the incidence of torrential flows being defined in such a manner. A transverse construction (check dams, sills, and submerged sills) as a method of torrent control has been known since ancient times. Archaeological excavations prove that there were such constructions in ancient civilizations. The organized and scientifically based erosion control started in mid nineteenth century in France, and later on it was widespread throughout Europe and the world. Depending on the type of transverse constructions (check dams), there are various systems of torrent control: European, French, German and later on more modern systems: Italian, Soviet (ex USSR) integral meliorations (USA), Prof. Rosić's system (former Yugoslavia), etc. With the development of science in the fields of torrent and erosion control, new types of check dams were invented distinguished from classical ones by their construction, intention, and of course effects.

2. Check Dams for Torrent Control in Mountainous Catchments

2.1. The Role and Types of Check Dams

The main idea of torrent control in the classical systems of torrent control is to prevent deep erosion by constructing various types of barriers (transverse construction) along the stream, so that the usually steep torrent channel is transformed into a stepwise one with a gentler slope angle. In such an artificial torrent bed, the velocity of the torrent is decreased, as well as the destructive force of the torrent; i.e. to scour the bed and to transport sediment. Regarding their effective height, all the transverse construction is classified into three groups (Kostadinov, 1996 b):

- submerged sills, without effective check dam height;
- sills, effective height up to 2.0 m;
- check dams, effective height more than 2.0 m.

In USA all transverse construction even with effective height les than 1.0 (0.3 to 0.8 m) m they call them check dams.

The beneficial effect of sills and check dams is a multiple one (Kostadinov, 1996 b):

1. They protect transverse profiles of torrential beds from further action of deep erosion processes.
2. They retain the bed load in their storage area. This effect is most expressed and continues until the check dams are completely silted up (silting up of the storage behind the check dam by deposition of the sediment). In this way, they also contribute to conservation of unstable hill slopes upstream.
3. Check dams obstruct further deepening of the channel, owing to a series of stable points which form a new (artificial) erosion basis in the torrential stream.
4. Due to reduction of the longitudinal bed slope of a torrent, the velocity of the torrential flood water is reduced. This phenomenon affects the reduction of the tractive force of water, and therefore the coarser rock parts which can no longer be

transported get stuck and thus buried promote soil conservation and stream bed consolidation.

5. Check dams are constructed to correct the direction of the water flow, directing it always towards the centre of the stream bed. This effect is achieved by erecting overflow sections, i.e. weir overflows of different shapes.
6. Check dams function as regulators of sediment transport. During torrential floods they retain great quantities of coarse sediments, whereas the finer material is carried away by medium and low waters.

Figure 1. System of classical check dams, Austria.

There are several types of check dams and they can be classified according to different aspects:

(a) Regarding the shape, check dams can be:
- rectilinear check dams;
- partially arched check dams;
- arched check dams.

(b) Regarding their purpose, check dams can be:
- deposition check dams (to retain sediment)
- consolidation check dams (for channel bed and slope stabilization)

(c) Regarding their resistance to external forces, check dams can be:
- gravity check dams

- gravity arch check dams
- arch check dams
- fixed check dams.

(d) Regarding their building material, they can be:
- concrete check dams;
- stone masonry check dams;
- mortarless stone check dams (dry laid masonry check dams);
- gabion check dams;
- reinforced concrete check dams.

Figure 2. Arched check dam, Montenegro.

Figure 3. Gabion check dam in the Župska reka torrent, Serbia (Photo: B. Radić).

For complete affirmation of the works in deep erosion processes control and retaining of the bed load in the torrent bed, the theory has been developed of the equilibrium bed slope and its ideal value known as the normal slope (Gavrilović, 1972). The first formulae for calculation of these slopes were given by the French torrent control expert E. Thiéry by the end of the XIX century (Thiéry, 1914).

By construction of check dams in the torrent bed the bed load is retained in their accumulation space, forming the siltation. The slope of the upper surface of siltation is called the slope of siltation. It practically represents the specific case of the eqilibrium bed slope.

Check dams and sills are being constructed in the system in such a manner that at the end of the siltation of a downstream check dam the next upstream check dam is constructed (Figure 4). In this way, the siltation of the downstream check dam prevents undermining of the foundations of the upstream check dam and provides its stability, so that a destruction of a check dam endangers the whole system upstream from it. The positions of the check dams in the stream bed are determined by the calculations of the slope of siltation.

All types of works performed in the stream bed and in the catchment require regular maintenance in order that destructions or some other damages can be prevented, and the necessary care should be foreseen for the newly grown cultures in the catchment.

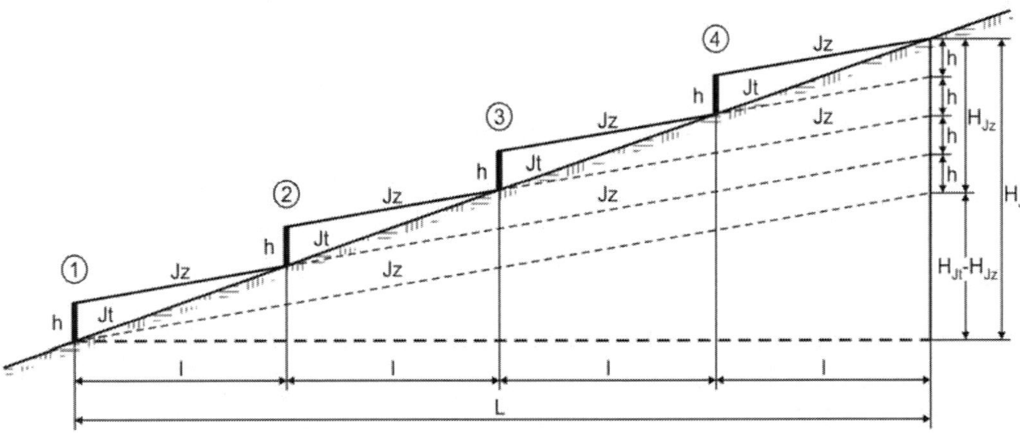

Figure 4. Scheme of torrent control by system of check dams.

2.2. Check Dams in Classical European System for Torrent Control

The erosion processes take place most intensively in the hydrographic network of torrents, where a watercourse is formed which following intensive rainfall (showers) can, in the form of a floodwave, possess immense destructive power, proportional to the water velocity and discharge.

In the upper parts of the hydrographic network (upper, source parts of the catchment), where gullies with deep erosion processes prevail, the regulation purpose is to create the resistant sections, especially around the headwall developing, which would help ground stabilization and establishment of vegetation. The regulation of gullies is performed by the low objects, whose height is most frequently around 0.5 m and seldom exceeds 1 m. Depending on the conditions, wattlings, dry laid masonry sills or gabion sills are applied.

Figure 5. Gully control with stone masonry and wattling sills, Croatia.

The wattling are convenient for regulation of gullies in which there are conditions for rooting of willow scions. Such are the gullies having permanent moisture and deep enough layer of soil. Dry laid stone masonry sills are applied in the gullies where coarser bed load is carried. In some gullies it is necessary to construct several hundreds of such sills. In the cases when transversal objects of longer duration and with the effective height above 1.0 *m* are necessary, good results are achieved with the gabion sills.

A significant characteristic of the gully regulation works in Bulgaria, along with small gabion check dams (sils) is the wide application of afforestation in the catchment and also in the gully itself by the appropriate kinds of trees. Considerable successes have been achieved in the region of the city of Kardjaly in the southern parts of Bulgaria by afforestation of gullies by alder (*Alnus glutinosa*). This species has the power of survival under unfavourable conditions, and by its dense root system it assists soil stabilization (Zakov, D., 1995).

For gully control in USA were used small check dams by rock and brush, log and rock, and log, wire and rock with effective height range 0,5 to 0,8 m (Sonett, and Stovall, 1996).

The feature of the middle part of the hydrographic network is large height of the banks, going up to several tens of meters. Land and rock slides are frequent. Regulation of this part of the hydrographic net is usually performed by construction of the check dams. Earlier, stone masonry check dams and concrete check dams were built most frequently. Recently, more and more the mountable check dams from reinforced concrete elements or reinforced concrete check dams have been applied. The effective height of the check dams ranges mainly from 2.0 to 5.0 m. but can be sometimes more. It is with a series of such check dams that large quantities of bed load can be retained, the bed slope and the destructivity of the torrent flow reduced, and the sliding processes on the slopes stopped. Such heights of the check dams are adopted for reasons of their stability and economy. Lower height can be adopted also because of the fact that together with the construction of the check dams intensive afforestation is conducted in the catchment, considerably reducing the bed load transport (due to reduced erosion in the catchment).

3. Some Problems of Check Dams Design

In the course of the torrent control activities and check dam application, the designers had to solve a sees of problems arising from the facts that the check dams are built under very unfavorable conditions for the further function, and most of the input data for their design were, or still are, undetermined i.e., undefined, The most important problems which arise

1. The determination of disposition of the check dams in the torrent bed, i.e., determination of the distance between the check dams built as a part of a system.
2. The structural analysis design (static dimensioning) of the check dams.
3. Effects of check dams.

3.1. The Determination of the Distance between the Check Dams in the Torrent Bed

One of the most important factors affecting the determination of distances between the check dams within a system is the forecast of the newly-formed slope of siltation -J_z deposited behind the check dam. In the theoretical studies, as well as in the practice of torrent control, the slope of siltation is distinguished the equilibrium bed slope which is being created by the deposition of sediments along the hydrographic network of torrents which are not regulated by the check dams.

Slope of siltation depends on a lot of factors: hydraulic elements of the channel and flow, geological and petrographic composition of bedload in the channel, type of check dam, hydrological conditions, etc. The most important ones being the natural torrent bed slope- J_t and the grain size of the sediment. Also, important role in the process of slope of siltation formation has the type of check dam. It should be mentioned that due to the periodicity of torrential incidents, it is not possible to determine precisely the size of sediment which will create the silting-up area. This is why here the question is not of precise determination of the slope of siltation but rather of indicating the forecast of its most probable value.

Some Formulas for the Calculation of the Slope of Siltation

The best results are achieved by direct study of the conditions in the field and by the equations for the calculation of the slope of siltation. Namely, the calculation (investigation) is done from case to case, i.e. separately for each section, primarily because of the different sediment grain sizes. If field investigations are impossible, the slope of siltation is calculated (estimated) by the formulas based on the determination of critical velocity and critical tangential stress.

Most of the formulas for the calculation of the slope of siltation are based on Thiéry and Valentini's formulas.. These are some of the well-known formulas (Kostadinov, 1996 b):

a) Methods based on Thiêry's equation

1. Thiéry's formula:

$$J_z = \frac{fb(\rho_s - \rho_0)}{0.03\rho_0 K^2 C^2 R} \tag{1}$$

2. Pimpirev's formula:

$$J_z = 14\frac{\rho_s - \rho_0}{\rho_0}\frac{b}{C^2 R} \tag{2}$$

b) Methods based on Valentini's formula

3. Valentini's formula:

$$J_z = 0.093\frac{d_{sr}}{R} \tag{3}$$

4. Novak's formula:

$$J_z = 0.093\frac{a}{R}; \quad a = k_m d_{sr} \tag{4}$$

c) Methods based on threshold velocity of water for bed-load movement or deposition

5. Formula based on Chezy's mean velocity:

$$J_z = \frac{V_{gr}}{C^2 R}; \quad V_{gr} = 0.7V_{sr} ... ms^{-1} \tag{5}$$

6. Formula based on Thiéry's equation for maximal velocity of water:

$$J_z = \frac{V_{max}^{\;2}}{C^2 R}; \quad V_{max} = 4.5\sqrt{b} ... ms^{-1} \tag{6}$$

d) Methods based on critical potential stress for the movement of bed-load

7. Lane's formula:

$$J_z = \frac{\tau_{kr}}{\rho_0 R}; \quad \tau_{kr} = 0.071d_e ... kNm^{-2} \tag{7}$$

e) Regional analytical dependencies

The calculation of the slope of siltation is best done in situ, i.e. the investigations should be done in the field on the formed silting-up area. This will enable the more precise calculation of the slope of siltation.

8. Biolčev et al.'s dependence for the torrents in Bulgaria (1975):

$$J_z = 0.51 \cdot J_t - 20 \tag{8}$$

Symbols used in above equations denote:

f – coefficient of friction
b –the largest size of sediment grain in the torrent bottom (m)
ρs – volume weight of sediment (kg·m^{-3})
ρ_0 – volume weight of torrential water (kg·m^{-3})
K – coefficient of torrentiality (Thiéry, 1914)

This is the relation between the mean density of clean water and mean density of torrent water.

$$K = \frac{\rho}{\rho_0} \tag{9}$$

ρ – volume weight of pure water (without sediment) [kg·m^{-3}]
ρ_0 – volume weight of torrential water (water plus sediment in it) [kg·m^{-3}]

The mean density of torrential water is a very much variable quantity depending on the quantity of sediment carried by water at that moment, i. e. depending on the concentration of sediment in water. According to the investigations carried out by some authors in the former Soviet Union (Herheulidze, 1970 a), in the extreme situations of the inflow of torrential flood waves, the amounts of volume weight of torrential water ranges between 1,620 kg·m^{-3} (Sokolovski) up to 2,020 kg·m^{-3} (Fleishman) (Kostadinov, 1996b). Consequently, if the torrent water carries more sediment, the value of the coefficient of torrentiality –*K* will be lower and vice versa. Pure water without sediment will have the coefficient of torrentiality *K* – 1.0 (for the streams without sediment).

Owing to the difficulties in the direct measurement of volume weight of torrential water, it is calculated by the formula:

$$\rho_0 = \rho + S\,(\rho_s - \rho)\ [\text{kg·m}^{-3}] \tag{10}$$

S – volume concentration of sediment in water
C – Chezy's coefficient of velocity
R – hydraulic radius (m)
d_{sr} – mean diameter of grain (m)

a – constant
k_m – constant depending, on geological composition of sediment
V_{sr} – limiting velocity of water necessary for bed-load movement from torrent bed [ms^{-1}]

That is the velocity of water required to move a grain of sediment of definite size from the stream bed. There are many empiric formulas by different scientists, for the determination of limiting velocity of water (Shamov, 1959).

When starting these works in Serbia, the designers dealing with torrent control determine the slope of siltation on the basis of their experience (only as an orientation), by using the formulae and analytical correlations. The largest number of the formulae proposed for determination of the slope of siltation is based upon Thiéry 's formula of boundary water velocity necessary for bedload movement. The great difficulties in calculation of the acceptable values of the constituent parameters make these formulae rather unsuitable for practical estimations (Kostadinov , 1993).

The obtained values often were not the realistic ones. The distances obtained were either too short or too long (depending on the adopted bedload size entering the formulae), which was not good for the check dam function. The too-short distances were irrational, since the effect of every check dam was not fully utilized, and the too-long distances between the check dams can jeopardize the stability of the check dam as well as of the who system of the check dams within the particular torrent.

The research performed in the torrents in southeastern Serbia (in the torrents of Grdelička Klisura gorge and Vranjska Kotlina value) had as a basis the fact, that for determination of the slope of siltation it is most appropriate to utilize the analytical relationships (the dependence of the slope of siltation – J_z on the natural bed slope – J_t), obtained on the basis of direct field research on already formed slopes of siltation behind of the erected check dams . Such an analytical dependence, obtained based upon the research into the torrents of southeastern Serbia (Kostadinov , 1993) is:

$$J_z = 0{,}469\, J_t + 0{,}212 \qquad (11)$$

J_z – slope of siltation in %
J_t – natural bed slope in %

This is a relatively convenient way of determination of the slope of siltation, since for any region with a certain type of the geological bedrock in the catchment and the size and uniformity coefficient of the bedload sediment, the analytical dependence can be found of the slope of siltation on the natural bed slope, and thereafter the necessary slope of siltation and following it the distance between the check dams can be determined relatively reliably, which is necessary with check dam design. This regional analytical dependence for south-east Serbia (Eq. 11) was used for calculation of slope of siltation for the check dams in other region of Serbia, in the torrents with similar grain size of the bedload sediment like in south-east Serbia, and the results were very good (Kostadinov, 1996 b).

Based on the research performed in the torrents in River Nišava catchment (East Serbia) an analytical dependence, was obtained for slope of siltation calculation (Velojić,M., 2000)

$$J_z = 0.52564\, J_t - 0.7053 \qquad (12)$$

Based on the research performed in the torrents in River Drina catchment (West Serbia) similar analytical dependence, was obtained for slope of siltation calculation (Todosijević, M., 2005).

$$J_z = 0{,}337\ J_t + 2{,}037 \qquad (13)$$

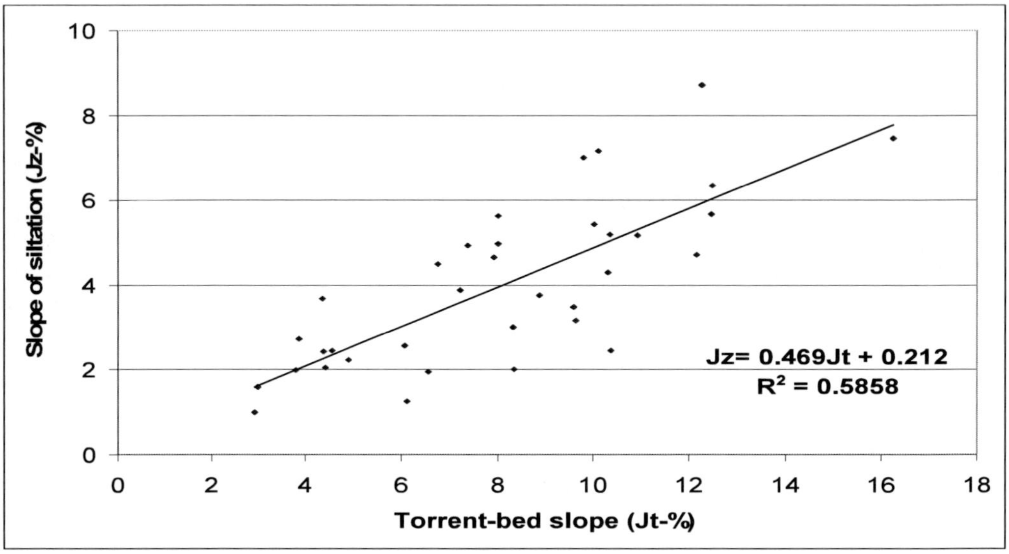

Figure 6. Correlation of the slope of siltation and torrent-bed slope for the torrents in South-East Serbia.

3.2. The Structural Analysis Design of the Check Dams

The aim of static evaluation of check dams is to determine dimensions which will enable the construction to resist all the external forces acting upon it, i.e., to be stable and to perform its function, but at the same time, the calculated dimensions should be the optimum economical solution in given circumstances.

In torrent control practice, much debate takes place about the choice of design load scheme for static evaluation of gravity check dams, and consequently there are differences in the design solutions of this problem. The dilemmas occur because in the course of time the dominant forces which act on check dams change and endanger their stability.

At the first stage, immediately after construction, the storage area of the check dam is not filled, and thus at the time of the first flood peak both the dynamic and the hydrostatic pressures of torrential water act upon the check dam. After each flood wave (and sometimes even after the first one), the storage is filled with the bedload, so that the loads which act on the upstream area of a check dam reflect themselves in the decreased hydrostatic pressure and the increased active pressure of the deposited bedload (earth pressure).

When the storage is completely filled up with sediment, i.e., after silting up, the check dam is practically loaded with the deposited sediment, i.e., the load it is subjected to is much lower than the hydrostatic pressure of water. Consequently, at the first stage after construction, the check dam resists both hydrodynamic and hydrostatic pressure of water, and the check dam dimensions should be greater, whereas after silting up, a check dam acts as a retaining wall, and much smaller dimensions are necessary. This is the reason for the designer's dilemma of load evaluation in the check dam design. In addition, there is another dilemma: whether the initial standpoint should be the scheme of full loading of the construction, taking into consideration all the forces that affect its stability, or to adopt the so-called reduced load scheme (bearing in mind the significance of the construction, its position, etc.). If all the forces are considered, the size of the check dam is too large, and its construction is expensive and uneconomic. On the other hand, in torrent control practice, some of the forces acting on the check dam are neglected, and the reduced load scheme is taken into consideration in order to get significantly more economic dimensions of the structure.

This is the reason why in the torrent control practice there is a series of methods of static evaluation of check dams, in which different standpoints (different load schemes) result in different dimensions of check dams for the same effective height. (Kostadinov, 1990).

Under the influence of Thiéry, who in the second half of the Nineteenth Century established the basic theories of torrent control, in Serbia *ever* since the beginning of works with static dimensioning of the check dams so-called reduced load scheme has been applied. The check dams were evaluated according to three formula of Thiéry, taking into account as the check dam load merely the hydrostatic pressure of water at the useful check dam height (Kostadinov, 1996 a). Later on, on basis of this formula of Thiéry, many other formulae were developed, their basis however being this equation.

Due to application of different formulae for static dimensioning of the check dams (with different adopted loading schemes) by different designers, in practice very different static dimensions of the check dams are present for the same effective height. In Table 1 thicknesses of check dams are presented at the spillway-overflow sections for certain effective check dam heights, for the check dams constructed in the torrents of southeastern Serbia (Kostadinov, 1996 a).

As can be seen, for the same effective heights of check dams, very different thicknesses on spillways were measured. In the table, each thickness value is quoted only once, although it often occurred several times. Most frequently check dams are 2.5–5.0 m in height (most often 4.0 m), but there are also great differences in check dam thickness.

The differences occur because the check dams were erected over a long period of time, in which the designers attitude changed regarding the implementation of the design load scheme in the static evaluation of check dams. Some check dams were evaluated as per earth pressure, as abutment walls, resulting in their very low thickness, whereas the other check dams of the same height were over-evaluated.

Table 1 also gives the thickness of check dams at the spillway (for each effective height) calculated according to method of Lj. Jevtić (Jevtić , Kostadinov , 1992). According to this method, static evaluation of a check dam is determined on the basis of reduced load scheme. By application of this method for static dimensioning, considerably more economic dimensions of the check dams are obtained with satisfied stability conditions (when compared with the formula of Thiéry 's) and the method is used in the torrent control practice in Serbia.

The static dimensioning of the arch check dams in practice up to now was performed mostly according to the formula of Thiéry 's (for arch check dams) and considerably less frequently, or rather only exceptionally, according to the cylindrical (cauldron) formula.

Table 1. Review of Actual Dimensions of Constructed Check Dams

Effective check-dam height	Thickness of check dam at the spillway (overflow sections)	Required thickness *m*	
m	*m*	calc.	adopted
1.50	0.60; 1.00	0.33	0.35
2.00	0.60; 0.85; 1.00	0.41	0.45
2.50	0.60; 0.80; 0.90; 1.00; 1.20	0.56	0.60
3.00	0.70; 0.80; 0.90; 1.00; 1.30	0.67	0.70
3.50	1.10; 1.20; 1.30; 1.40; 1.80	0.78	0.80
4.00	0.60; 1.00; 1.10; 1.20; 1.30; 1.40; 1.50; 1.60; 1.80	0.88	0.90
4.50	1.00; 1.30; 1.40; 2.00	1.00	1.00
5.00	0.70; 1.00; 1.10; 1,40; 1.60; 1.70	1.11	1.15
6.50	1.80; 2.30	1.45	1.45
7.00	1.00; 1.20	1.53	1.55
7.50	1.50; 2.00	1.67	1.70

3.3. Effects of Check Dams

The evaluation of criteria and methods in the assessment of the effects of check dams for torrent control should start from their main role: bedload retention, reduction of bed slope and stabilisation effect.

a) Bedload Retention

The most impressive role of check dams is bedload retention in the silting-up area (storage area) upstream of the dam. The effect in this respect is based on the amount of the retained sediment and the cost needed to construct the dam. The amount of the trapped sediment behind the dam (volume of the silting-up area) can be calculated by the formula (Kitin , 1975):

$$W = \frac{m \cdot h_k^2}{2(J_t - J_z)} \quad [\mathrm{m}^3] \qquad (14)$$

W – volume of the retained sediment in the silting-up area in m^3
m – mean width of the silting-up area in m
h_k – dam effective height in m
J_t – natural torrent bed slope in decimal form
J_z – slope of siltation.

The dam price is determined by the precise bill of quantities. Clearly, the more economical structure is the one that retains a greater quantity of sediment under the lower financial investment. Mean width of the silting-up area depends on the width of the valley upstream of the dam. Namely, the silting-up area upstream of the check dam is formed over the entire width of the stream channel, i.e. the valley. As the width of the valley cannot be changed, (in the design) narrow profiles should be chosen (low consumption of material for the dam) for the dam position from which the valley widens upstream. Naturally, bed slope affects the economy of the check dam. Silting-up areas are shorter on steeper slopes and the amount of trapped sediment is smaller. For this reason, check dam construction is avoided on too steep slopes. The check dam effect on sediment retention can be expressed in natural form (the amount of sediment in m^3 that is retained by 1 m^3 of check dam) or in financial form (the cost of 1 m^3 of retained sediment). It should be taken into account that 1 m^3 of sediment in a storage can mean also 1 m^3 of water that would be lost because of it.

The cost of check dam construction depends on:

- type of check dam;
- quantity of applied material;
- torrent channel geometrical characteristics.

The greatest economic (financial) effect can be realised by the proper selection of the position for check dam construction, type of check dam and type of material. A significant saving in the construction can be reached by the correct choice of the profile for check dam construction. Also, significant effects can be achieved by the correct choice of the check dam type and material. The choice of check dam type can affect the formation of the slope of siltation which was said to have a significant role in the assessment of check dam economy.

b) Reduction of Bed Slope

Check dam effect on the bed slope reduction and, in this way, the changes in hydraulic elements of flow can be expressed by the change of tractive force. Based on Du Boys equation, the mean profile tractive force is (in Kostadinov,1996b):

$$\sigma_o = \rho \cdot R \cdot i \cdot g \text{ [Pa]} \tag{15}$$

ρ – density of torrent water in kg·m^{-3}

R – hydraulic radius of flow in m

i – torrent bed slope

g – acceleration of gravity, g = 9.81 m·s^{-2}

ρ_0 – tractive force in Pa.

As by check dam erection bed slope and hydraulic radius decrease, in this way also tractive force i.e. bedload transport competency also decrease. This means that the decrease in transport competency will cause the bedload deposition upstream of the check dam, which prevents the siltation of the storage or other structures. The effect of bed slope reduction can also be expressed by the reduction in channel capacity for sediment transport (Kostadinov, 1996 b).

c) Stabilisation Effect

It is reflected in the formation of local erosion basis (submerged sils, sills and check dams), stabilisation of sliding slopes (sills and check dam) and the correction of flow direction (submerged sils, sills and check dams).

The effect can be expressed by the estimation of the quantity of material fixed or stabilised by the given structure, to prevent its detachment and reaching the stream in the form of erosion sediment. By check dam erection downstream of the sliding slope, the check dam silting-up area could have the function of a toe which supports the slope and prevents its further sliding and the movement of a great quantity of material (sediment) to the stream. In addition to the natural effect, stabilisation effect of check dams can also be expressed by its financial effect.

d) Summary Effect of Check Dams in Torrent Channels

Check dams in torrent channels most often demonstrate several effects at the same time. The evaluation of effects of individual structures should consider, on the one hand, all the positive effects (sediment retention, bed slope reduction, and stabilisation effect), and on the other hand, the cost of the structure and the negative effects, if any. In this way it can be seen whether the construction of the selected type is feasible or it is necessary to search for another type of structure, or another type of material, etc.

The above analyses have to be done during the project design, so as to find the optimal solution for the problems in the treated torrent. Naturally, there are cases when, because of the great importance of the infrastructure protected against torrential floods (roads, settlements, industry, etc.), the above criteria are irrelevant, and the price of the check dam is not a decisive element. This means that the criterion of the significance of the object protected by check dams or other transversal structures should also be considered.

4. Contemporary Trends and Types of Check Dams for Torrent Control in Europe

The tendency to make the dam construction cheaper and more effortless and to prolong the effect, i.e. to increase the functionality, led to various grid filtration dam constructions in many countries. They have a series of advantages compared to massive check dams. The researchers who delt with these problems have tended to invent a check dam which would affect the longest possible stretch of a stream, i. e. by the least possible financial investments to control erosion processes and to check the moved sediment, etc.

The length of the zone of check dam influence depends on the slope of siltation to be formed behind the dam after its construction. On the other hand, the slope of siltation affects the quantity of checked bed-load in the impoundment, which is an essential indicator of the efficiency and the economy of the check dam.

4.1. Deposit (Settling Basin) Check Dams by Prof. Rosić

Along with the other factors the slope of siltation can be affected by the choice of check dam types. In these circumstances, in the fifties of the 20th century, Professor Rosić introduced new types of check dams: deposit (settling basin) check dams, filter check dams, and auxiliary check dams. All these belong to a group of filtration check dams. In the period

1954–1970 in Serbia, several deposit (filtration) check dams were built in various torrents with different characteristics.

Filtration (grid) construction annuls the main part of the kinetic energy accumulated by the torrent and retains a considerable part of its solid phase – bed load sediment. For this reason, the torrent which passes through such a dam is no more a danger to the target protected structure. In addition, the period of aggradation is prolonged, compared to massive check dams.

Deposit check dams are specific check dams being an element in the system of torrent control proposed by S. Rosić (Rosić, 1952). While classical check dam effect is to lower the gradient of natural channel slope, deposit check dams function is the physical separation of water and sediment through the process of filtration and further direction of water through the newly- formed siltation (made of coarse particles of deposited bed-load) by filtration. In this way, the surplus of kinetic energy is eliminated thus preventing the destructive effect of water on the bottom and torrent banks, or on the structures in the watercourse.

Although the siltation zone does not reach the catchment headwaters (according to Rosić's theory), it is much longer than the siltation behind a classical check dam of the same effective height constructed under equal conditions. This is the consequence of deposit check dam construction whish consists of (Figure 7):

- check dam with apron and submerged sill;
- filter system.

The check dam is made of concrete or stone masonry. It is a gravity structure with one, two or three large openings by which it differs from the special check dams. The sizes of openings are height 1.40 m, width 0.70 m. The number of openings depends on profile width where the check dam is erected. The effective height of the check dam is 2.0 to 2.5 m.

Figure 7. Deposit check dam in the torrent Želebić Potok, Serbia, (Foto:A.Obradović).

The filter system is composed of:

- central channel;
- lateral channels (omitted if the profile is narrow)
- boulders of durasol balls, diameter 20 cm, stacked to as to enable the best possible porosity.

Deposit check dams as check dams with a specific construction are more expensive than classical check- dams of the same height. Thanks to the filtration system, the deposited bed-load forms a much longer zone of siltation (although the siltation zone does not reach the spring zone of the catchment, according to S. Rosić), and in this way the beneficial effect is double:

- they deposit a significantly greater amount of bed-load in their storage area,
- they protect greater length of torrent channel.

The effect of the deposit check dams in comparison with the classical check dams has been a subject to general laboratory and field research and consideration, leading to a conclusion that deposit check dams are more efficient and economic over all parent rock producing sediment of uniform particle sizes (uniformity coefficient after Hazen less than 10.0) and where clay is not present in the sediment. Such geological substrates are limestones, dolomites, serpentines, and somewhat less, Paleozoic schists (Kostadinov, 1979).

Despite of all its positive characteristics, this type of check dams did not find wide application in the torrent control practice in Serbia. The reason for this is that the designing experts mostly utilized the lower-resistance line, due to the fact that design and building of the classical check dams are much easier and simpler. It was also in accordance with the inadequate financing system effective until now in the conditions of the undefined social ownership of the capital and production facilities (neither private nor state ownership). Under such conditions, the consciousness of an efficiency, rationality and economy found it hard to break through and take part in decision making. The development of the scientific thought in Europe in the region of erosion and torrent control over the last 25 years has completely confirmed the ideas of Prof. Rosić and his type of check dams.

Based on the analysis of the effects classical and deposit (settling basin) check dams in the torrents of Serbia we can conclude that: deposit check dams after S. Rosić, were the initial stage, the origin of the idea which was 25 years later, transformed into a strategy of torrent control called "Bed load management". It was already the beginning of the fifties when Prof. Rosić realized the significance and the advantages of "active" struggle (separation of water and sediment) against torrent floods and torrent erosion, compared to previous "passive" control by classical check dams. (Kostadinov,1995).

For this reason, these types of filtration check dams should be used in all cases of the regulation of torrents with strong torrential-erosion processes and unfavourable (unstable) bedrock.

4.2. "Sediment Traps" as a New Types of Structures for Torrent Control

Pursuant to the actual construction of torrent-control hydrotechnical structures, between 1964 and 1968, Caucasian Scientific Research Hydro-meteorological Institute (CSRHMI) in Tbilisi, Georgia (former USSR) developed the principal schemes, explained the general dimensions and drew up the constructive and detail drawings for a series of different types of grid torrent control structures made of precast reinforced concrete, of different heights and sectional and base configuration. These types of grid dams were named "sediment traps" (Herheulidze, 1970).

The minimal number of prefabricated components in these constructions is four, i.e.: 1) main long component No.1 (beam); 2) auxiliary short component No. 2 (disk, ring) for levelling the plane for mounting the components No.1 in individual assemblies; 3) component No. 3 (insertion, reinforced skeleton) which fastens together the construction into a whole; 4) component No. 4 – prefabricated shoe, serving as a support for the component No. 3.

The constructions can be mounted both on plate-shaped or on strip, or on isolated (dot) foundations (shoes). The latter are the most suitable ones and they enable very fast mounting.

The same components can form different constructions of simple, double, two-stepped, three-stepped and several-stepped dams of different height and length, with different arrangement of sides for torrent control. The ends of the dams which are built into the gorge (canyon) sides can have a developed section and a configuration which prevents the unfavourable torrential impact on the channel slopes in the dam.

Prefabricated components are 3-6 m long and calculated both for high load of 100–120 KN/m^2 torrents and for the impacts of individual rocks, diameter 1m, and also for lower loads in the conditions of grid dams applied in small torrents or ordinary gullies. In practice, the most frequently applied components are No.1, length 3.0m.

4.2.1. Experience in the Implementation of "Sediment Trap" Dams in Georgia

In 1969, an experimental grid dam was built on a strong torrential river Durudzi in east Georgia. Maximal dam height was 9.9 m, average – 7.1 m. The dam was built without the channel protection against undermining from the downstream side, because it was expected that water energy after passing the grid construction would be completely stilled. However, the sediment coming from the drainage basin contains clay material and relatively soon the openings on the grid construction in the central part were completely closed. (It would not have happened, or it would have been much lower if it was limestone sediment or similar material). Water started to overflow above the dam in that part of the channel and to fall from the height of more than 7 m, so it caused the undermining of the bottom. In this way, the grid construction in that part of the channel remained without a support and the concrete beams were broken (component No. 1).

The experimental grid dam, height 5 m, in a severely eroded gully near the railway station Gachiani on the Caucasian railway was built in 1967. It should be mentioned that the mounting of this structure, including three special grid frames in the lower, downstream part of the dam, consumed only 16 working hours of a motor crane.

4.2.2. Experience with "Sediment Trap" Dams in FYR of Macedonia

The first structures of this type were built in the channel of Džepčiška River and Neproštenska River, with drainage basins on the slopes of Mt. Šarplanina. The structures were built in 1967.The dams were made of local material (ungranulated gravel, unsieved). Reinforced concrete components were cast at the site of installation in relatively poor conditions, so the desired concrete quality could not be achieved. As the reinforced concrete components were not adequate, they were damaged soon. The damage was not repaired so later on the dams collapsed.

During the period 1980-1982, 6 "sediment trap" dams after Herheulidze were built in the torrents – Poroj 1 and river Pena.

During 1988, another 4 structures of that type were built, i.e. two in torrents Dolgaš and Avmagica – Kosovrasti (also near Tetovo).

These structures had less damage than dam no.1 in the river Pena. Based on the experience of Džepčiška River and dam no.1 in the river Pena, the damage to the new dam was reduced to the minimum. After 2000, in FYR Macedonia, several dams were designed for torrent management in the region of Šar Planina.

In table 2 are presented the main characteristics and damages of the "sediment trap' dams built in Macedonia.

Figure 8. "Sediment trap" dam in the torrent Pena River, FYR of Macedonia.

Based on the constructed reinforced-concrete grid dams in the rivers in Georgia and FYR Macedonia, we can conclude as follows: (Đorđević, et al. 1990)

- The dams affect the longitudinal profile of the channel upstream of the structure depending on the height of the siltation, as well as the gravity dam. Their advantage is that during the silting up, the finer fractions are carried downstream, and only the coarser fractions accumulate in the siltation. In this way, the siltation has a considerably grater slope and length than it is the case in gravity dams. Therefore, they protect against undermining a longer stretch of the torrent channel.

- The longitudinal channel slope downstream of the dam is mainly formed depending on the natural channel slope and the decreased flow energy after passing through the dam.
- When the flood wave passes the grid dam, its flow energy is reduced which also reduces the flood hazard downstream.
- Unstable profiles in which the dams are built stabilise in time, both in the dam zone and upstream. This proves that they can be built also on unstable profiles and stretches;
- The weaknesses and the damage observed during the construction and utilisation can be eliminated in future;
- The results of research point out that these dams also should be protected from the downstream side by a type of apron: large stones, gabion concrete blocks, sill at the end of the apron;
- It is necessary to continue the research of the hydraulic characteristics of grid dams, as the barriers in torrent channels and the dependence of dam efficiency on the size of grid openings, hydraulic characteristics of flow and bedload characteristics.

Taking into account the previous experience on the implementation of "sediment trap" grid dams, it can be concluded that they are very efficient structures for the control of debris flow and for the stabilisation of flow energy, as well as for the rehabilitation and prevention of landslide and rockfall upstream of the dams. By its effect, this type of structures meets all the requirements and can be classified in the group of the so-called functional dams for torrent management by the system of "bedload management''.

Table 2. Main characteristics of the "sediment trap" dams built in Macedonia

Torrent	No of the dam	hm	h_pm	Type	Damages	Location of building
Poroy 1	1	11,50	1,5	stepped in 4 rows	Abrasion of concrete	Downstream of a landslide
Pena river	1	6,75	without spillwey	stepped in 4 rows	Abrasion of concrete	
Pena river	2	7,50	3,5		Damaged spillway Damaged apron Restored	Downstream of a landslide
Pena river	3	6,00	2,5	stepped in 4 rows	No damages	Downstream of a landslide
Pena river	4	7,00	4,5	stepped in 4 rows	Damaged spillway Restored in 1986	Profile with unstable banks
Pena river	6	8,00	without spillwey	stepped in 4 rows	Damaged the right side Restored	Profile with unstable banks

h - effective height, m;
h_p – depth of the spillway, m.

4.3. Functional Check Dams for Bedload Management

The analyses performed by the beginning of seventies in Western Europe on effect of the check dams upon the torrents result in the change of attitude and new orientation in the design and building of the check dams. The development of the new types of check dams and the new torrent control strategy was possible due to appearance of new materials and building technologies, first of all concrete and reinforced concrete, and thereafter metallic and network constructions. The conclusion has been reached that by classical check dams (retaining bed load sediment by siltation forming) it is not possible to perform bed load management, because most often the accumulation space of a heck dam is not large enough. The new strategy in preventing of torrential floods, "*The bed load management*", is based upon the assumption not to provoke the perturbation of the transport of the bed load.

Stabilization of the transport of the bed load, supported by the new types of the check dams called functional check dams, has a purpose of stabilizing the lower course and the receiving water course (into which the torrent is flowing) and preventing the incidence of the torrential floods. By the new strategy of torrent control, modern, up-to-date bed load management is achieved. According to Fiebiger, G., the basic principles of torrent control strategy of bed load management are (Fiebiger, 1986):

1. Stabilizing (fixation of torrent bed in a wanted level to stop and/or prevent depth erosion);
2. Consolidating (elevating of torrent bed to support and/or prevent slides and slopes);
3. Sorting (filtrations and/or storage of undesirable bed load components during a flood):
   - bed load sizing (filtration and storage the large pieces of bed load during a flood or debris flow);
   - wood grading (filtration of undesirable wood during a flood);
4. Retaining (storage and deposition of bed load until aggravation the retention capacity);
5. Dosing (parting of a large mass in small amounts):
   - bed load dosing (quantitatively dosing transport of intermediate stored bed load by declining floods and medium waters);
   - water dosing (quantitative change of the water discharge by qualitative change the hydrograph at a specified point);
6. Breaking of debris flow (declining the high energy level of a debris flow to a lower level under particular energy change).

The more the development goes on, the more check dams were constructed with multiple functions as consolidating and sorting or retaining and dosing, and so on.

In Table 3, bed load management principles and functional check dams related to bed load management are shown, according to Fiebiger (1986).

The practical task of the functional check dams is to prevent torrential floods not retaining permanently the transported bed load, i.e. they perform dosation and sorting of the bed load as well as the dosation of the water flow. In order to perform this function, they have to have a free accumulation space, and also over a time interval as long as possible. During the flood waves retaining of the bed load of certain sizes takes place, later on to be carried away by low flow waters gradually downstream through the check dam openings, thereby releasing space necessary in case of new flood wave incidence.

It should be mentioned that, when the task is the protection of artificial water reservoirs (artificial lakes) or waterways from the bed load sediment, the functional check dams cannot be applied as the only objects, but they have to be combined with the classical check dams, or the classical ones only should be applied.

Table 3. Bed load management principles and functional check dams

Bedload management principles	Functional check dams to bed load management
1. Stabilizing	1. Bed sills
2. Consolidating	2. Consolidating dam
3. Sorting: - bed load sizing - wood grading	3. Sorting dam: - bed load sorting dam - wood sceens and gratings
4. Retaining	4. Deposit dam (retention dam)
5. Dosing: - bed load dosing - water dosing	5. Dosing dam: - bed load dosing dam - flow (flood) dosing dam
6. Breaking of debris flow	6. Debris flow debreaker

Source: Fiebiger, G. (1986).

Figure 9. Functional check dam for bedload sorting, Dientenbach torrent, Austria.

4.4. Torrent Control and Environment

In the middle and upper courses of torrents, transverse structures were usually built, such as sills and check dams for streambed and bank consolidation and bed load deposition. Check dams were built mainly of concrete or stone masonry. The main disadvantage from the aspect of ecology is that they cut all the migration directions of aqueous animals, especially fish. This refers to torrents which have water throughout the year, whereas this is not a problem for those temporary watercourses which have water only after heavy rains. That is the reason why so far we have not a case of torrent chec dam with a provision of fish paths. Attention has never been paid to this problem, although in many cases, at places where such check dams would be necessary, it is difficult to think of fish paths construction because of the high energy of flow, large quantities of sediment and also due to the fact that these watercourses often go dry, so that fish paths are irrelevant. Nevertheless, in building check dams across the watercourses with permanent supply of water, fish paths should also be provided for. This also refers to ponds and rock-fill dams (Kostadinov,1996c).

CONCLUSION

Check dams are very important structures in the torrent control practice. They directly protect the river bed, i.e. fix the cross profile and retain bedload sediment. The retained bedload forms the silting-up area upstream of the dam, which reduces the longitudinal slope of the stream, by which stream velocity and its transport competency are reduced, i.e. the destructive power of water. In this way, the development of deep and lateral erosion is prevented and the torrent bed is stabilised. This is the main objective of check dams.

In the torrent control practice in small mountainous catchments up to now the classical stone masonry and concrete check dams have been mostly applied. Regarding their resistance to external forces, the gravity check dams have been mostly applied, arch check dams being considerably less frequent. In the course of sixties, rather frequent application of the gabion check dams started.

Due to the undefined conditions of the load to which check dams are exposed, as well as due to difficulties with determination of the slope of siltation, many doubts accompanied design and construction of the check dams. On the basis of long-lasting practical and scientific and research work it can be concluded that in solving these problems regional analytical dependencies should be applied for determination of the slope of siltation, and the distances between the check dams on its basis. Concerning the structural analysis design of the check dams, the reduced load scheme should be applied, because it results in more economic dimensions, satisfying the stability conditions.

In the course of fifties the authors in Serbia were proposed new types of check dams, with essentially different functions and torrent control roles. Deposit check dams by Professor Rosić have proved as considerably more efficient and economic than the classical ones over all parent rock producing sediment of uniform particle sizes (uniformity coefficient after Hazen less than 10.0) and where clay is not present in the sediment. Such geological substrates are limestones, dolomites, serpentines, and somewhat less, Paleozoic schists. On the basis of results of the research, a much more frequent application of this type of check dams is recommended.

Contemporary approach to torrent control is characterized by implementation of new types of check dams. Bearing in mind the new strategies in the torrent control practice in the developed countries of Western Europe, known as "the bedload management", these ideas should be accepted and applied in a corresponding manner, utilizing the new types of so-called functional check dams. Into this type of check dams also a new type of check dam should be listed, proposed by I. I. Herheulidze from Georgia (former USSR), and named "Sediment traps", deposit check dams by Prof. Rosić, and all types of filtration check dams. It should emphasize that in the case of water reservoirs protection against sedimentation with bedload it should use the classical check dams.

Torrent control practice should open more widely to the new ideas, also applying new types of check dams, utilizing new materials and technologies for purpose of more efficient and economic work. In doing so, each and every torrent should be approached studiously, not being a slave of the accepted schemes, and looking for the best solution for particular conditions by combining the present knowledge and searching for the new solutions.

REFERENCES

Biolčev,, Kitin, and Angelov, (1975). Opredeljane naklona na zadbaražnia nasip. *Gorskostopanska nauka, no. 1*, Sofia, 64-71.

Đorđević, Trandafilov, Stoilov, Đorđević, and Minovski, (1990): *Dejstvo i funkcionisanje na armirano-betonski rešetkasti pregradi.* Univ. "Kiril i Metodi", Zavod za vodostopanstvo SRM. Skopje.

Feibiger, (1986). Bedload Management on Torrent Control by Functional Check Dams. *Contributions to Research on Torrent Erosion and Avalanches*, IUFRO Subject Group S1.04-00 Mittelungen der Forstlichen Bundesversuchsanstalt, Wien, 325-332.

Gavrilović, (1972). *Inženjering o bujičnim tokovima i eroziji.* Izgradnja, Specijalno izdanje, Beograd.

Herheulidze, I. I. (1970 a). K vaprosu o predelenom nasištenij selevih patokov tverdimi materialami Sbornik: *Dviženie nanosov v atkritih ruslah*, Moskva.

Herheulidze, (1970 b). Neka pitanja industrijske izgradnje konstrukcija radi uređenja bujica. Monografija: *Erozija, bujični tokovi i rečni nanos*, Jugoslovenski komitet za medunarodnu hidrološku deceniju, Beograd, 101-115.

Jevtić, and Kostadinov, (1992). Analiza stabilnosti ravne bujčarske pregrade u uslovima poplavnog talasa. *Glasnik Šumarskog fakulteta br. 74*, Šumarski fakultet Univerziteta u Beogradu, Beograd, 669-677.

Kitin, (1975). Rakovodstvo za upražnenia po erozia i borbata s neja, *Selskostopanska nauka*, Sofija.

Kostadinov, S. (1979). *Studija antierozionog dejstva taložnice kao specifičnog tipa pregrade prof. Rosića u zavisnosti od tipa geološke podloge.* Magistarski rad u rukopisu, Šumarski fakultet, Beograd.

Kostadinov, S. (1990). Analiza poplavnog talasa Prisjanske reke kod Pirota na dan 26. 06. 1988. godine. *Šumarski list, br. 3-5*, Zagreb, 159-169.

Kostadinov, S. (1993). Posibility of Assesment of the Slope of Siltation Based on the Some Hydraulic Characteristics of the Torrential Flows. *Journal of the Japan Society of Erosion Control Engineering, Vol. 45, No 5, Ser. No 184*, Kiodo, 28-33.

Kostadinov, S. (1995). Analisys of the Effects of Classical and Filtration Check Dams in the Torrents of Serbia. *Proceedings of the XXVI IECA Conference*, Feb. 28th - March 3rd, 1995, Atlanta, 111-124.

Kostadinov, S. (1996,a). Check Dams in the Torrent Control Practice in Serbia: Former and Present Experiences and Future Perspectives. *Proceedings of IECA Conference XXVII: Erosion Control Technology Bringing it Home.* Feb. 27-March 1, 1996; Seatle Washington, USA, 507-522.

Kostadinov, S. (1996,b). *Bujični tokovi i erozija.* Beograd, Šumarski fakultet,.

Kostadinov, S. (1996,c). Torrent Control in Mountainous Landscapes. *Proceedings of International Symposium "Stream Protection and Environment Management* ", Athens, Nov. 23-24, 1995; Published by Ministry for the Environment, Physical Planning and Public Works of Greece, Athens, 14-33.

Rosić, (1952). Novi sistem uređenja bujica. *Glasnik Šumarskog fakulteta, No 1*, Belgrade, 149-197.

Sonett, and Stovall, (1996). Small Grade Control (Check Dam) Function and Design in the Embudo Catchment, Albuqurque, New Mexico. *IECA Compilation: Methods and Techniques for Stabilizing Gullys and Using Check Dams;* Steambat Springs, IECA. 83-96.

Šamov, (1959). *Rečnie nanosi*, Leningrad.

Thiéry, (1914). *Restauration des montagnes.* Paris et Liege.

Todosijević, (2005): *Efekti poprečnih objekata u bujičnim tokovima sliva reke Drine na sektoru Loznica–Bačevci.* Magistarski rad u rukopisu, Šumarski fakultet, Beograd.

Velojić, (2000). *Proučavanje efekata poprečnih objekata u bujičnim tokovima sliva reke Nišave.* Magistarski rad u rukopisu, Šumarski fakultet, Beograd.

Zakov, (1995). Ukrepjavane na hidrografskata mreža na poroite. *Gora, 8/95,* Sofia, 6-8.

In: Check Dams, Morphological Adjustments…
Editors: C. Conesa and M. A. Lenzi
ISBN: 978-1-62618-856-3

Chapter 5

BLOCK RAMP FOR TORRENT CONTROL AND STREAM RESTORATION

Stefano Pagliara*[*] *and Michele Palermo
Department of Civil Engineering, University of Pisa,
Via Gabba 22, 56122 Pisa, Italy

ABSTRACT

One of the main problems for river engineers is the design of hydraulic structures which can control rivers natural dynamics and at the same time assure their environmental sustainability. In fact, especially in the last few decades, ecological and biological hydraulic structure's impact has become more and more important. Thus the main problem is to design an eco-friendly structure which is able to conjugate the traditional hydraulic scopes with a minimization of the environmental impact. Block ramps can be one possible solution. Block ramps are hydraulic structures which are generally used in mountain river restoration projects. They consist of either loose or fixed rocks disposed on a sloping bed. They can be inserted in stream corridors and can also be used as fish passages, as they don't break river morphology. Block ramps are structures which dissipate a big amount of energy, both on the ramp bed and in the downstream stilling basin. The dissipation is due to the localized stream slope variations and the high roughness conditions. Generally, a hydraulic jump takes place downstream of the ramp toe, thus a scour depth is formed which has to be controlled in order to assure the structure stability. The scour control can both be done by using a particular stilling basin granulometry downstream of the ramp toe or inserting structures like sills in opportune spatial positions in the stilling basin, which consent to minimize the scour lengths.

INTRODUCTION

In the last few decades, one of the main problems that river engineers had to solve is to conjugate a correct balance between the insertion of a hydraulic structure in a natural river

[*] Corresponding author: s.pagliara@ing.unipi.it

with the necessity to minimize its environmental impact as to reduce its inevitable effects on the natural biological mechanisms which take place in the river, such as fishes migration. Thus block ramps can constitute a valid solution which consents both to dissipate a large amount of energy, diminishing the solid transport remarkably, and preserving the natural biological functions of the species which live in the river. Moreover, they can also be used to convert traditional hydraulic works as to avoid the morphological discontinuity caused by them. This type of structure has to be correctly designed in order to be competitive with the other classical hydraulic structures and assure the above cited ecological advantages. A correct design of this type of structures has to take into account numerous factors and parameters. First of all, the stability of the work, meaning that although they can be considered elastic and flexible structures, the stones constituting the ramp can not be transported in normal functioning conditions or for the designed flood event. The analysis of the ramp stability has been the first factor analyzed by different researchers. In particular, Whittaker and Jäggi (1986) conducted experiments on block ramps analyzing their structural behavior and proposing design guidelines and semi-empirical equations to prevent structural failure of the ramp. The stability of such structures was also investigated by Robinson et al. (1998). Recently, the use of piles (Pagliara and Chiavaccini, 2003) or boulders placed over the base material (Pagliara and Chiavaccini, 2004) was proposed in order to increase the stability of the structure. Pagliara and Chiavaccini (2007) conducted experiments to evaluate the stabilization effect of boulders with three boulder configurations (in rows, randomly or in arc configuration). They stated that the ramp failure is characterized by three steps: initial movement, local failure and global failure. Moreover, the Authors analyzed the morphological ramp evolution. Another aspect which characterized a block ramp is the large energy dissipation which can be obtained on the same structure due to its roughness.

This process is due both to the localized variation of stream slope and the high roughness conditions exerted by the bed material (Pagliara and Dazzini 2002). The analysis of the energy dissipation mechanism for such structures was carried on by numerous researchers. In general, all of the studies put in evidence that the energy dissipation process depends both on the ramp geometric characteristics and on the flow conditions. Experiments on block ramp for a fixed slope were carried out by Platzer (1983). Studies on the energy dissipation on stepped channels were also carried out by Essery and Horner (1978), Chamani and Rajaratnam (1999a,b), Diez-Cascon et al. (1991), Peyras et al. (1992), Stephenson (1991), and Christodoulou (1993), Armanini and Gregoretti (2005), who highlighted that the dissipation process and mechanism, in case of skimming flow, is similar to the block ramp one. Chanson (1994) analyzed the energy dissipation phenomenon in the cases of nappe and skimming flow, furnishing a useful equation to estimate it. The skimming flow condition can occur both on flat sloping stepped spillways and block ramps (Pagliara and Peruginelli 2000) and it is characterized by an unstable wake beyond the gravel or in the cavity of the steps, formed of eddies which increase the flow resistance.

As other hydraulic structures, generally downstream of block ramp, a scour hole takes place. Thus, a correct design of this type of structure has to also include the analysis of the scour mechanism which occurs in the stilling basin in order to prevent their structural collapse. The scour process involves different phenomena whose comprehension is necessary. Namely, the erosive forces dissipated by vortices transporting the bed material downstream, the development of the scour hole and the stable "dynamic" geometry condition (Sumer et al. 1993, Canepa and Hager 2003, Dey and Sarkar 2006, Ferro et al. 2004) have to be foreseen.

The analysis of the scour process downstream of hydraulic structures is one of the most important topics in hydraulic engineering. Thus many studies were conducted by different researchers. One of the first among these was Veronese (1937), who analyzed the scour phenomenon downstream of a spillway and furnished useful empirical equations. Other studies on local scour were conducted by Hassan and Narayanan (1985) who proposed a theory for the prediction of the scour time rate. The analysis of the similarity in the scour process and the scour hole profiles downstream of hydraulic jump flow was proposed by Farhoudi and Smith (1985). Perhaps one of the first systematic studies which described the scour process dynamic downstream of grade control structures was conducted by Bormann and Julien (1991), and successively further developed by D'Agostino and Ferro (2004), who furnished useful design equations to foresee the equilibrium scour lengths. Both Breusers and Raudkivi (1991) and Hoffmans and Verheij (1997) furnished a comprehensive literature review and proposed guidelines to understand the phenomena related to scour in engineering practice and for different structure typologies and flow regimes.

A theoretical analysis based on momentum equation was carried on by Hoffmans (1998), both for plunging and horizontal jets. The Author proposed useful equation in order to foresee the maximum scour depth. Many researchers (among these Dey and Raikar 2005) put in evidence that the effect of the stilling basin granulometry was relevant and deeply influences the scour lengths. Systematic studies on the scour mechanism downstream of the block ramp toe and the jump typologies which can take place were also conducted in the hydraulic laboratory of The University of Pisa. Pagliara and Hager (2004) conducted experiments on block ramps in the case of uniform stilling basin material proposing useful design relationships which consent to evaluate the scour geometry. Successively, the effect of stilling basin sediment gradation on the scour mechanism (in the absence of any control structure) and the jump typologies which can occur were investigated by Pagliara (2007). He observed that two jump typologies can occur: the Free hydraulic jump in a Mobile Bed F_{MB}, and the Submerged hydraulic jump in a Mobile Bed S_{MB}.

Figure 1. Example of a block ramp built in the Cecina river (Tuscany, Italy).

In addition, he furnished simple design equations for the scour depth and length. More recently, studies on scour control inserting sills in the stilling basin in opportune spatial positions were conducted by Pagliara and Palermo (2008a, b, c). They investigated the effect of continuous, dentated and rock sills located downstream of block ramps for both uniform and non-uniform stilling basin materials, concluding that rock sills have to be preferred in order to minimize scour geometry. Figure 1 shows a picture of a block ramp built in the Cecina river (Tuscany, Italy), and Figure 2 presents an example of a block ramp simulated in laboratory.

Figure 2. Example of a block ramp model built in laboratory.

HYDRAULICS OF BLOCK RAMPS

Generally, a block ramp is characterized by a high slope bed which is preceded and followed by two mild sloped reaches. According to the ramp geometry and different hydraulic flow regimes, different possible scenarios can occur. Some are represented in Figure 3a-c. Namely, upstream of the ramp, the flow is generally in sub-critical conditions (Figure 3a-c). In the upstream section of the ramp, the critical depth occur (Figure 3a-b) if on the ramp super-critical flow condition takes place. The jump can be either localized downstream of the ramp in the stilling basin (Figure 3a) or on the ramp itself (Figure 3b). The flow can also be in sub-critical condition both downstream and upstream of the ramp, as well as on the ramp itself (Figure 3c). Thus the energy dissipation process is deeply affected by both hydraulic and geometric conditions.

Many studies were conducted to understand the hydraulics of stepped channels (Essery and Horner 1978, Pagliara and Ferro 2003). In particular Chanson (1994) developed equations to evaluate the energy dissipation both in nappe and skimming flow conditions. This latter condition is also the one that characterizes block ramps (Pagliara and Peruginelli 2000). In fact, in this case, the flow resistance increases as unstable wake beyond the gravel or micro vortex in the cavity of the steps occur. Chanson (1994) estimated the energy dissipation between the entrance and the toe of a chute and proposed the following equation:

$$\Delta E_{r1} = \frac{\Delta E_1}{E_0} = 1 - \frac{h_u \cos\alpha + k^3}{1.5k + H} \qquad (1)$$

where h_u is the uniform flow depth, k the critical depth, H the ramp height and α the bed slope angle. $\Delta E_1 = E_0 - E_1$, where $E_0 = H + 1.5k$ is the total upstream energy amount and $E_1 = h_1 + q^2/(2gh_1^2)$; h_1 is the water depth at the ramp toe, Q is the inflow discharge, q the discharge per unit width and g the acceleration due to gravity (see Fig. 4a). Moreover, Bathurst (1978) and Bathurst et al. (1981) proved that the roughness condition strongly affect the energy dissipation process on a sloping bed. A detailed analysis of the dissipative process on block ramp was conducted by Pagliara and Chiavaccini (2006a) who stated that for a base ramp the relative energy dissipation ΔE_{r1} can be expressed as follows:

$$\Delta E_{r1} = \frac{\Delta E_1}{E_0} = A + (1-A)e^{(B+CS)(k/H)} \tag{2}$$

where A, B and C are three parameters depending on the roughness conditions. Moreover, basing on Bathurst' studies, Pagliara and Chiavaccini (2006a) individuated three scale roughness condition ranges. Namely the Authors stated that large scale roughness condition (LR) occurs when the ratio $k/d_{50}<2.5$, intermediate scale roughness (IR) when $2.5<k/d_{50}<6.6$ and small scale roughness (SR) when $6.6<k/d_{50}<42$. Being this classification Pagliara and Chiavaccini (2006a) proposed the following expression of the coefficients A, B and C of the Eq. (2). A is equal to 0.33, 0.25, 0.15 if the roughness condition is LR, IR, SR, respectively; B is equal to -1.3, -1.2, -1.0 if the roughness condition is LR, IR, SR, respectively; C is equal to -14.5, -12.0, -11.5 if the roughness condition is LR, IR, SR, respectively. In Eq. (2) S is the ramp slope and d_{50} is the mean diameter of the material constituting the ramp. From Eq. (2) it can be observed that the energy dissipation is a decreasing function of the ramp slope. Moreover it increases passing from small scale roughness to large scale roughness.

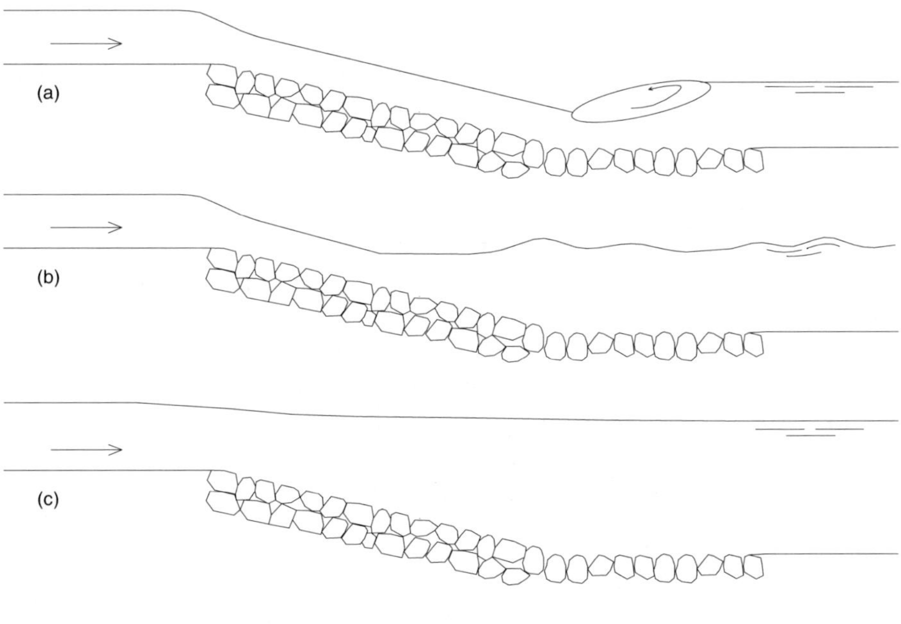

Figure 3. Flow transition in presence of a block ramp: (a) hydraulic jump located downstream of the ramp toe, (b) hydraulic jump located on the ramp, (c) sub-critical flow conditions (i.e no transition).

Generally, the ramp stability is increased locating boulders in different geometric arrangements on it. It is also evident that in this case the hydraulic and the dissipative process which takes place on it, is deeply affected by the presence of these elements. The flow results to be strongly disturbed by the boulders tending to diminish the shear stresses acting on the base material, and consequently increasing its stability. The analysis of the dissipative process in this case (presence of boulders on the block ramp) was conducted by Pagliara and Chiavaccini (2006b). The Authors stated that both the boulders' concentration and their disposition on the base ramp strongly affect the dissipative mechanism. They introduced a new parameter Γ (boulder's concentration) and modified Eq. (2) taking it into account. Thus, in the presence of boulders Eq. (2) becomes:

$$\Delta E_{r1} = \frac{\Delta E_1}{E_0} = A + (1 - A)e^{(B+CS)(k/H)}\left(1 + \frac{\Gamma}{E + F\Gamma}\right) \quad (3)$$

where E and F are two coefficients depending on the boulders' arrangements. Namely, they found that for random disposition and rounded boulders E=0.6 and F=13.3; for row disposition and rounded boulders E=0.55 and F=10.5; for random disposition and crushed boulders E=0.55 and F=9.1; and for row disposition and crushed boulders E=0.4 and F=7.7. Eq. (3) is valid for Γ<33%.

In the presence of a mobile stilling basin, Pagliara et al. (2008a) analyzed the dissipative process occurring between section 0 and section 2 (ramp entrance and downstream end of the hydraulic jump section, respectively), see Figure 4f. Furthermore, they used a tailgate to change the water depth so to obtain different submergence levels for the same flow discharge. The energy at the downstream end of the hydraulic jump (section 2) can be expressed as follows: $E_2=h_2+q^2/(2gh_2^2)$ where h_2 is the tailwater depth, Q is the inflow discharge, q the discharge per unit width and g the acceleration due to gravity. Thus the energy loss between section 0 and 2 is equal to $\Delta E_2=E_0-E_2$ and the relative energy loss is equal to $\Delta E_{r2}=\Delta E_2/E_0$. Pagliara et al. (2008a) tested ramp slopes ranging between $1/8 \leq S \leq 1/4$ and three different hydraulic jump configurations. Namely, they conducted experiments in which the jump was both located at the toe of the ramp (*free jump,* L/L_T=0) and at 1/3 or 2/3 (*submerged jump*, L/L_T=1/3 and L/L_T=2/3, respectively). L is the horizontal distance of the beginning of the jump from the ramp toe and L_T is the horizontal length of the ramp (Figure 4f).

They compared the dissipative process in the case in which the channel bed downstream of the block ramp was either fixed or mobile. Moreover, in the case of a mobile bed they also compared the energy dissipation both in presence and absence of scour at the toe of the ramp. The Authors found that no significant differences can be estimated in terms of energy dissipation in respective submergence ramp conditions both in presence and absence of scour downstream of the ramp toe. Additionally, the same occurrence (negligible differences in terms of energy dissipation) was also noted for all of the tested slopes and L/L_T=1/3 and L/L_T=2/3 in the case of either a fixed channel bed or a mobile channel bed is present downstream of the ramp.

This occurrence is mainly due to the fact that in the case of submerged hydraulic jump the main part of the dissipative process (i.e. the major energy dissipation) occurs on the ramp and the scour presence does not significantly affect it. In the case of submerged ramp condition, for a certain value of the ratio L/L_T, the relative energy dissipation depends on the following parameters: the ratio k/H and the scale roughness. Namely, the Authors proved that for the same value of the ratio k/H, the relative energy dissipation is an increasing monotonic function of the roughness condition, meaning that it increases with roughness condition.

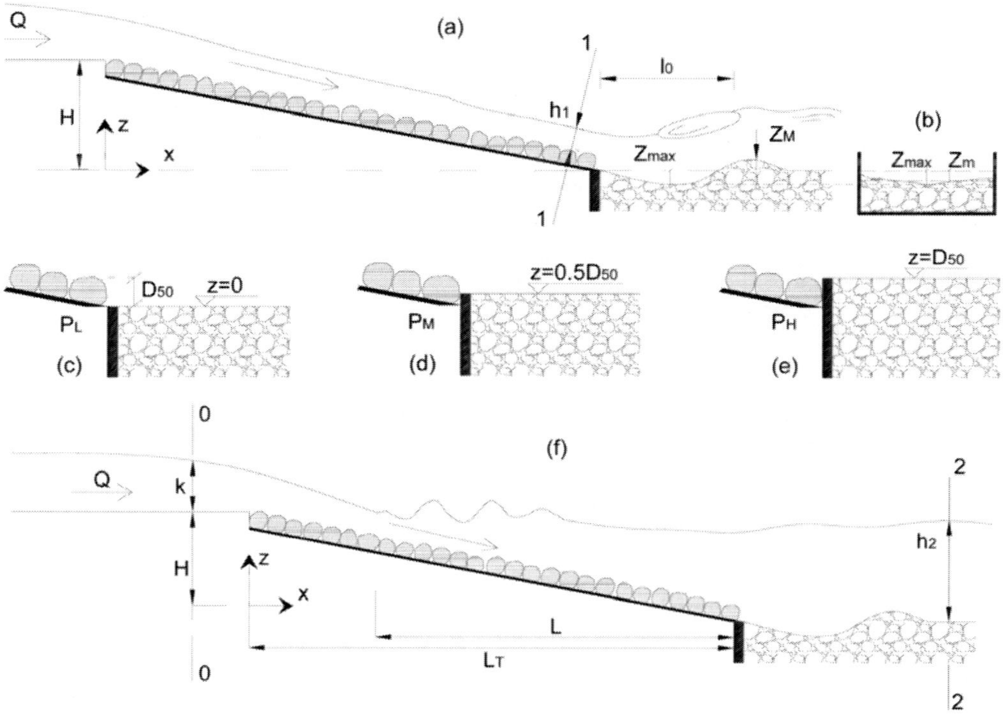

Figure 4. (a) Longitudinal profile of a block ramp and the downstream stilling basin with the indication of the main parameters in the absence of any stilling basin protecting sills and for L/L_T=0 (free jump); (b) transversal cross-section in which the maximum scour occurs; particular of the pile location at the toe of a block ramp: (c) low pile case (P_L), (d) medium pile case (P_M) and (e) high pile case (P_H); (f) longitudinal profile of a block ramp and the downstream stilling basin in the absence of any stilling basin protecting sills and for L/L_T=2/3 (submerged jump).

In the case of free jump condition and at scour equilibrium, the relative energy dissipation between section 0 and section 2 results to be only a function of the ratio k/H, as both the effects of scale roughness and ramp slope were found negligible. Based on these observations, Pagliara et al. (2008a) proposed an analytical expression which is algebraically similar to Eq. (2) in order to foresee ΔE_{r2}, which can be expressed as follows:

$$\Delta E_{r_2} = A + (1 - A)e^{(B)k/H} \tag{4}$$

For the different submergence of the ramp tested ($0<L/L_T<0.7$) and for ramp slope varying between 0.125 and 0.25, Pagliara et al. (2008a) proposed the following expression for the coefficients A and B of Eq. (4), according to different roughness conditions:

$$A = 0.239e^{-2.323\frac{L}{L_T}} \tag{4a}$$

$$B = -\left(10.7\frac{L}{L_T} + 1.729\right) \tag{4b}$$

valid for SR roughness condition;

$$A = 0.249e^{-1.618\frac{L}{L_T}} \tag{4c}$$

$$B = -\left(9.95\frac{L}{L_T} + 1.863\right) \tag{4d}$$

valid for IR roughness condition;

$$A = 0.256e^{-1.245\frac{L}{L_T}} \tag{4e}$$

$$B = -\left(8.475\frac{L}{L_T} + 1.931\right) \tag{4f}$$

valid for LR roughness condition.

As shown above, the process of energy dissipation in the case in which a stilling basin is present downstream of the ramp toe, also has to take into account the dissipation which occurs in the hydraulic jump which is present in the basin. Thus the comprehension of the different hydraulic jump typologies which can occur varying both the hydrodynamic and the geometric conditions is fundamental. The analysis of the different jump typologies which occur downstream of a block ramp was conducted by Pagliara (2007). According to different ramp slopes and hydraulic conditions Pagliara (2007) observed two different jump types. The first one is called S_{MB} (Submerged jump in Mobile Bed) where the flow at the ramp toe is characterized by the same inclination of the ramp itself. A macro-vortex occurs and it is counter clockwise, thus tending to submerge the toe of the ramp (see Figure 5a). Whereas the second jump type which can occur is called F_{MB}. This jump appears similar to a traditional free jump as a succession of waves starting from the ramp toe and proceeding downstream occurs. The roller which forms in this case is clockwise (see Figure 5b). According to the different jump typology which occurs, the sediment transport dynamic changes. In fact, if an F_{MB} occurs, the sediment transport is directed both in the upstream and downstream part of the scour hole. The reverse is true in the case in which an S_{MB} jump occurs. The sediment transport is directed only downstream. Pagliara (2007) proposed a graph which classifies the jump typologies which can occur in terms of ramp slope and densimetric Froude number (see Figure 6). This classification is valid for clear water conditions and for uniform channel bed materials. Likewise it delimitates the so called "*no scour*" and "*transport condition*" regions. The "*no scour*" region is characterized by the absence of scour. The flow does not have enough energy to remove sediments. It generally occurs for low densimetric Froude numbers. In reverse, the "*transport condition*" region is characterized by the continuous movement of the stilling basin material as the scour hole does not reach its dynamic equilibrium configuration and there is no ridge formation, which is a fundamental limiting factor for scour evolution.

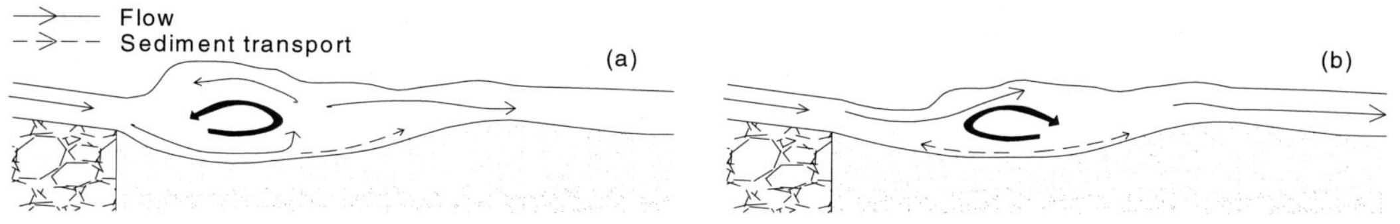

Figure 5. Diagram sketch of the different jump typologies: (a) S_{MB}; (b) F_{MB}.

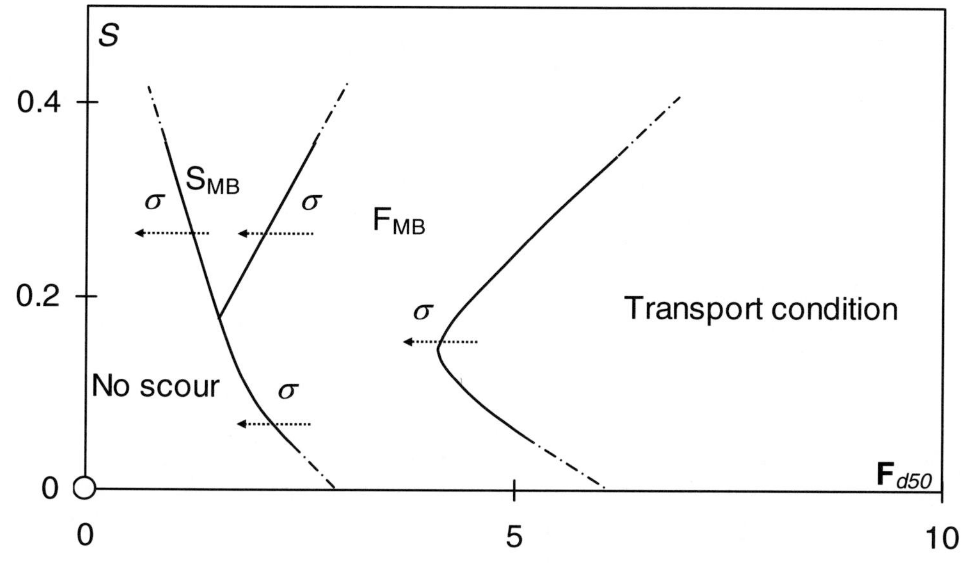

Figure 6. Hydraulic jump classification in terms of F_{d50} and S for uniform channel bed material.

More recently, Palermo et al. (2008) conducted experiments with non-uniform stilling basin materials in order to extend the jump classification. They analyzed the existing fields of the different jump types both in terms of F_{d90} and F_{d50} and proposed a classification similar to that shown in Figure 6 with F_{d90} as parameter for non-uniform material with geometric standard deviation σ up to 2.8. Note that for uniform material (σ <1.2) $F_{d90} \sim F_{d50}$, thus the existing fields of the different regions does not change significantly with respect to those individuated in Figure 6 for uniform channel bed material. Inversely, for non-uniform channel bed materials, the limiting curves of the existing fields of both jump types (F_{MB} and S_{MB}), "*no scour*" and "*transport*" regions, are qualitatively the same of those valid for uniform case. But increasing the material non-uniformity σ they result in being shifted towards lower densemetric Froude numbers, as indicated by the arrows in Figure 6. Palermo et al. (2008) proposed simple analytical equations in order to describe the limiting curves for both "*no scour*" region and "*transport condition*" region, which are valid for ramp slope varying between 1/12 and 1/4, and σ<2.8. The "*no scour*" region can be limited by different parallel lines (one for each different channel bed material tested) whose parametric equation is:

$$S = -0.35F_{d90} + 0.02\sigma^2 - 0.15\sigma + 0.8 \quad (5)$$

whereas the "transport condition" region can be limited by two straight lines whose equations are:

$$S = -0.1F_{d90} + 0.05\sigma^2 - 0.29\sigma + 0.8 \tag{6}$$

valid for S ranging between 1/12 and 1/6 and $\sigma<2.8$ and

$$S = 0.09F_{d90} - 0.05\sigma^2 + 0.28\sigma - 0.41 \tag{7}$$

valid for S ranging between 1/6 and 1/4 and $\sigma<2.8$.

Block Ramp Stability

One of the main problems related to block ramps is that their functionality is assured if the ramp is stable, meaning that the blocks must not be removed from the sloping bed. Thus the dimension of the blocks is one of the main parameters which a designer has to take into consideration in order to avoid the structural failure. A correct calculation of the blocks' dimension has to ensure that no channelization or scour holes on the face of the ramp have to occur for a certain design discharge. More simply no local failure has to take place. As this is a fundamental problem many researchers conducted experiments on block ramp stability proposing semi-empirical relationships which allow the evaluation of the critical discharge for which the ramp bed material remains stable. Although the criteria followed by the different researchers to define the critical conditions are not common, in general the most used theories to dimension ramp blocks were proposed by Whittaker and Jäggi (1986) and Robinson et al. (1998). Whittaker and Jäggi (1986) proposed the following relationship:

$$q_{c0} = 0.257S^{-1.167}\left(g\left(\frac{\gamma_s}{\gamma_w} - 1\right)D_{65}^3\right)^{0.5} \tag{8}$$

where q_{c0} is the failure unit water discharge expressed in m^2/s, S is the slope of the ramp, g the acceleration due to gravity, γ_s is the rock specific weight, γ_w is the water specific weight and D_{65} is the rock diameter for which 65% of the mixture is finer. As to be expected, the failure unit discharge is strongly dependent on the ramp slope and the granulometric characteristics of the blocks. In particular it decreases with the ramp slope and results to be a monotonic increasing function of the rock size.

Robinson et al. (1998) conducted experiments on chutes whose slope varied between 0.1 and 0.4 and proposed the following relationship to evaluate the critical discharge:

$$q_{c0} = (D_{50}S)^{(1.40+0.213/\sqrt{S})} \cdot \exp(-11.2 + 1.46/\sqrt{S}) \tag{9}$$

where S is the slope of the ramp and D_{50} (in mm) is the rock size for which the 50% of the sample is finer. In order to estimate the inception of motion on a rock chute, a different approach was introduced by Aguirre Pe and Fuentes (1991). They proposed to identify the inception of motion using the particle densimetric Froude number $F_d = U/(g(s-1)D_{50})^{0.5}$ in order to take into account both the effect of velocity and drag force on the sediments in macro-roughness condition, in which U is the mean flow velocity, $s=\rho_s/\rho$ is the relative particle density and g the acceleration due to the gravity. Based on this approach, Aguirre Pe et al. (2003) proposed the following relationship in order to evaluate the critical densimetric Froude number:

$$F_{dc} = 0.9 + 0.5\ln\frac{h}{D_{50}} + 1.3\frac{D_{50}}{h} \tag{10}$$

where h is the water depth and h/D_{50} is the relative submergence, with D_{50} as the mean diameter of the block constituting the ramp. Equation (10) is valid for the following ranges of parameters: $0.02<S<0.065$ and $0.02<h/D_{50}<30$. This approach was further developed by Pagliara and Chiavaccini (2007) who analyzed the failure mechanism of base and reinforced block ramps. They conducted experiments in order to identify the different failure phases in presence and absence of boulders on a base block ramp. Moreover they described the failure mechanism for the following boulders' dispositions: rows, random, arc and reinforced arc. The following expression was proposed by the Authors in order to evaluate the critical densimetric Froude number for all the tested boulders' arrangements:

$$F_{dc} = 1.98 S^{0.18}\left(\frac{h}{D_{84}}\right)^{0.36}(1+\Gamma)^{a_1} \tag{11}$$

In which D_{84} is the block ramp diameter for which 84% of the mixture is finer, Γ is the boulders' concentration and a_1 is a coefficient depending on the boulders' disposition. Namely, a_1 is equal to -2.2 for rows disposition, -2.0 for random disposition, -2.6 for arc disposition, -2.6 for reinforced arc (Type 1) disposition and -2.8 for reinforced arc (Type 2) disposition.

Sediment Transport on Block Ramps

Block ramps, such as other hydraulic structures, are inserted in natural contests and contribute to modify them. But also the natural contest can affect the hydraulic functionality of the structures, contributing to change both the structural and the hydraulic characteristics. In the case of a block ramp, as said above, one of the main problems is the structural stability. But another important phenomenon which is relatively not well explored is the change of bed roughness, due to the deposition of the sediments naturally transported by the river between the rocks constituting the ramp. In fact, in normal functioning conditions, these type of structures can be filled by the river sediments thus changing the morphologic characteristics, although no structural collapse can occur. This occurrence was recently analyzed by Pagliara et al. (2008b) who studied the main geometric parameters controlling the filling phenomenon and its effects in terms of energy dissipation in order to foresee the ramp morphological changes and analyze the interaction between the filling process and the geometric and hydraulic parameters.

They conducted experiments increasing the discharge with different channel bed materials and ramp geometry configuration supplying at the upstream section of the ramp sediments of various granulometries in order to reproduce the natural river sediment threshold. They also reproduced different hydraulic conditions increasing step by step the water discharge. Initially Pagliara et al. (2008b) tested very low discharges and contemporarily they supplied filling material. They noted that the flow, in this case, has not enough energy to transport the filling material downstream and an accumulation of it occurs in the upper sections of the ramp which changes its morphology. Then they increased the discharge up to Q_F, which is defined as the "filling discharge". This discharge is the minimum, having enough energy to transport the filling material up to the downstream ramp section. In this case, it was noted that if a further quantity of filling material is supplied the ramp bed morphology does not significantly change meaning that a dynamic equilibrium with the maximum level of filling has been already reached. In this dynamic equilibrium condition, the filling material is almost uniformly distributed on the ramp. Increasing the flow discharge, the filling percentage of the ramp decreases up to a limit value. The discharge corresponding to this limit condition was defined by the Authors as the "emptying discharge" Q_E. When this configuration is reached the only filling material that remains on the ramp is got trapped by ramp blocks or hidden by them. The filling percentage results to be less than 15%. A further increase of the discharge does not significantly change the reached dynamic equilibrium level and the ramp configuration remains almost the same of that described in the previous step. Figure 7a-b shows both a longitudinal and a plan view of a filled ramp in dynamic equilibrium condition, in the case in which a Q_F discharge was tested.

Pagliara et al. (2008b) proposed simple relationships in order to evaluate both the "*filling discharge*" Q_F and the "*empty discharge*" Q_E. Based on the analysis conducted by Bathurst et al. (1987), the Authors proved that the unit critical discharge q, in presence of filling material, can be expressed as a function of the bed slope, the mean diameter of the material constituting the bed D_{50} and the gravitational acceleration g, and the mean diameter d_{50} of the filling material. They obtained the following functional relationship:

$$k/D_{50}=f\,(d_{50}/D_{50},\, S\,) \tag{13}$$

where k is the critical depth relative to the unit discharge q. They proposed the following relationships in order to estimate the non dimensional groups k_F/D_{50} and k_E/D_{50}, in which k_F and k_E are the critical depths relative to the discharges Q_F and Q_E respectively.

$$\frac{k_F}{D_{50}}=\frac{\dfrac{d_{50}}{D_{50}}}{(3.54S+0.11)+0.23\dfrac{d_{50}}{D_{50}}} \tag{14a}$$

$$\frac{k_E}{D_{50}} = \frac{\dfrac{d_{50}}{D_{50}}}{0.0175 + (5.12S - 0.24)\dfrac{d_{50}}{D_{50}}} \tag{14b}$$

As it can be seen from Eqs. (14a) and (14b) both k_F/D_{50} and k_E/D_{50} in the tested range are monotonically increasing function of d_{50}/D_{50} and are valid in the following range $0.05<d_{50}/D_{50}<0.35$ and $0.125<S<0.25$.

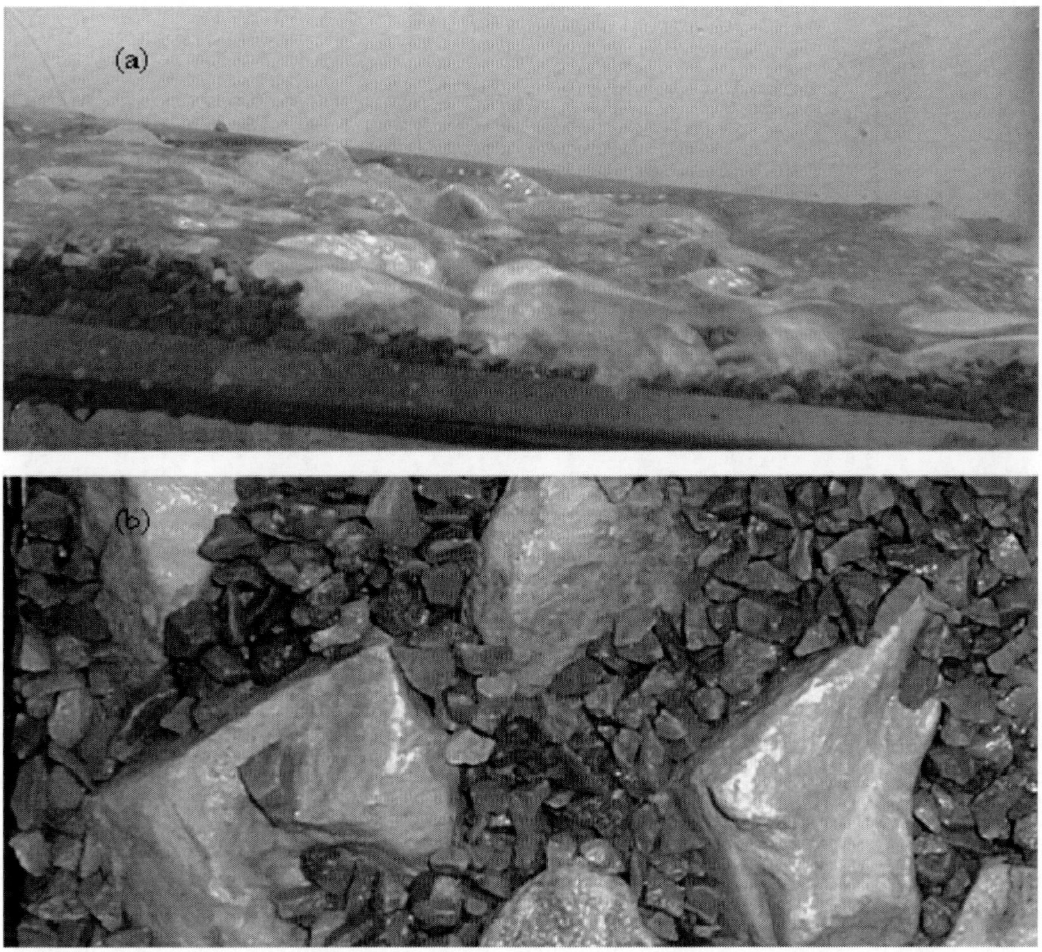

Figure 7. (a) longitudinal and (b) plan (zoomed around some blocks) view of a filled ramp in the case in which the "*filling discharge*" is tested.

STILLING BASIN DESIGN: SCOUR MECHANISM

The control of the scour process is one of the main problems related to the presence of an hydraulic structure. It is common that a scour hole is formed downstream of hydraulic structures. Thus, river engineers have to control the scour geometry and foresee its main lengths (namely maximum depth and length) in order to avoid structural problems which can also lead to the structure collapse. The stilling basin erosion can be controlled in different ways. Namely, the basin can be protected by either locating blocks (Figure 8a) or sills on it. Moreover, it is also possible to leave it without any type of protection (Figure 8b). In this last case, according to the designed flood event, stilling basin granulometry and ramp geometry, the scour profile has to be foreseen in order to avoid excessive erosion that can mine the ramp stability.

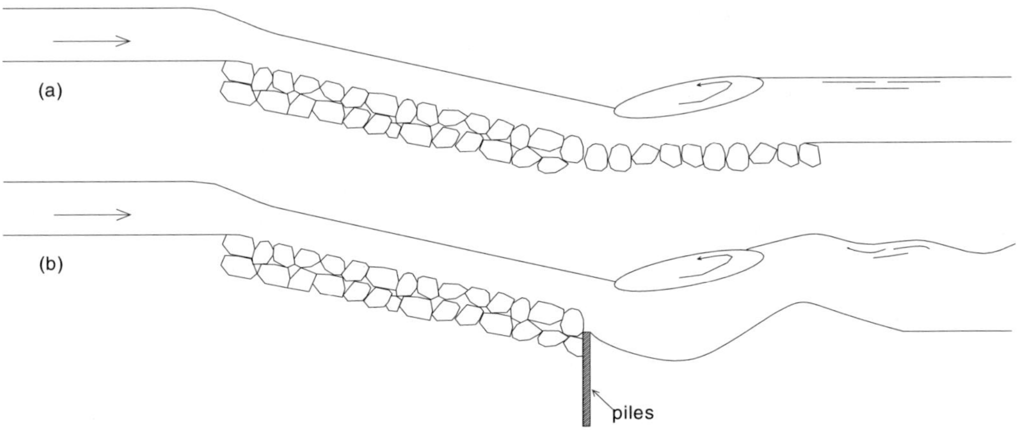

Figure 8. Stilling basin downstream of a block ramp (a) protected with blocks; (b) without any protection.

Thus it is necessary to not only correctly design the structure itself, but also investigate the impact that the structure has in the river. It means that the structure does not need to be considered an isolated element inserted in the natural contest, but a part of it. For this reason, the block ramp and the stilling basin downstream of it, can be considered a unique antropic work and the scour downstream of it is a fundamental part of the design of the structure. Many studies were conducted in order to evaluate the stilling basin geometry in various ramp configurations (ramp slope ranging between 1V:12H and 1V:4H) and different hydraulic conditions (Pagliara and Hager 2004, Pagliara 2007, Pagliara and Palermo 2006, Pagliara and Palermo 2008a-b). In particular, the analysis of the scour hole was conducted by Pagliara (2007) in clear water conditions and in the case in which an F_{MB} jump type occurs. Moreover the Author analyzed the scour geometry in presence of both uniform and non-uniform stilling basin materials. He proposed some useful practical equations in order to evaluate the main lengths of the scour hole in the absence of any stilling basin protection work. Namely, he proposed the following relationship to evaluate the medium non dimensional scour depth $Z_m=z_m/h_1$ in the section of maximum scour and the non dimensional scour length $L_0=l_0/h_1$, respectively:

$$Z_m = 0.58 \cdot \sigma^{0.55} \cdot S^{0.75} \cdot F_{d90}^{\ \ 1.8} \tag{15a}$$

$$Z_m = 0.58 \cdot \sigma^{-0.55} \cdot S^{0.75} \cdot F_{d50}^{\ \ 1.8} \tag{15b}$$

$$L_0 = 3.75 \cdot \sigma^{0.8} \cdot S^{-0.5} \cdot Z_m^{0.8} \tag{16}$$

valid in the ranges: $1/12 \leq S \leq 1/4$, $1 \leq \sigma \leq 2.8$ and $1.0 \leq F_{d90} \leq 3.75$, $1.5 \leq F_{d50} \leq 5.0$. z_m is the measured medium cross-sectional scour depth in the section of maximum scour, l_0 is the scour length and h_1 is the approach flow depth at the ramp toe. In the Eqs. (15a) and (15b) $F_{dxx} = v_1 / (g' \cdot d_{xx})^{1/2}$ is the densimetric Froude number, S the block ramp slope, σ the non-uniformity parameter; with $v_1 = Q / (Bh_1)$ as the average approach flow velocity, Q is the water discharge, B is the channel width, $g' = [(\rho_s-\rho)/\rho]g$ is the reduced gravitational acceleration, ρ_s is the sediment density, ρ is the water density, g is the gravitational acceleration and d_{xx} is the sediment diameter, in which the subscript xx is a number that indicates the percentage of finer. In Figure 4a, a diagram sketch with the indication of the main geometric parameters is reported. From both Eqs. (15a) and (15b) a significant effect of the ramp slope S is evident as having fixed it and for the same densimetric Froude number, Z_m is an increasing monotonic function of this parameter. Moreover, an increase of Z_m can also be observed for more non-uniform materials (i.e. for an increase of σ) in the case in which F_{d90} is used as parameter. Conversely, a decrease of Z_m can be observed if σ increase with F_{d50} as parameter. The different analytical behavior of Eqs. (15a) and (15b) in terms of the parameter σ was also observed by Aderibigbe and Rajaratnam (1998); who noted that an increase of σ caused a decrease of Z_m if F_{d50} is used as parameter; whereas if F_{d95} is used as parameter, Z_m increases with σ. This is due to changing the x – axis from F_{d50} to F_{d90} the data points are shifted towards lower x values and this shift is bigger with the increase of σ, thus inverting the trend valid in the case in which F_{d50} is used as parameter. Regarding the non-dimensional scour length L_0, it increases if scour depth increases. Additionally in this case, both the effect of ramp slope and sediment non-uniformity are prominent.

It has to be noted that the section of maximum scour is not perfectly two-dimensional. Thus the maximum scour depth z_{max} is slightly bigger than the average one (see Figure 4b). Pagliara (2007) proved that especially for low values of the densimetric Froude number, due to both the presence of the block on the ramp and the wall effect, the difference between z_m and z_{max} is more prominent. Thus he introduced the dimensionless variable $Z_{max}=z_{max}/h_1$, by which it is possible to estimate the value of the maximum scour depth. He proposed the following relationships both in terms of F_{d50} and F_{d90}:

$$\frac{Z_{\max}}{Z_m} = 1 + 1.75 \cdot F_{d50}^{-1.75} \tag{17a}$$

$$\frac{Z_{\max}}{Z_m} = 1 + 0.55 \cdot F_{d90}^{-0.75} \tag{17b}$$

It can be seen from Eqs. (17a) and (17b) that the ratio Z_{max}/Z_m tends to 1 for high tested densimetric Froude numbers.

All the deductions reported above are relative to a particular configuration of the ramp toe (low pile level P_L, Figure 4c). In practical application, ramp toe generally is stabilized using piles or micro-pile. It was experimentally proven that the height of the pile can affect scour profile. Let us define low pile level (P_L) as the configuration for which both the levels of the stilling basin and the upper pile edge were fixed at z=0 (Figure 4c), medium pile level (P_M) as the configuration for which both the levels of the stilling basin and the upper pile edge were fixed at $z=0.5D_{50}$ (Figure 4d) and high pile level (P_H) as the configuration for which both the levels of the stilling basin and the upper pile edge were fixed at $z=D_{50}$ (Figure 4e), where D_{50} is the mean grain diameter of the ramp material tested. Pagliara et al. (2008c) analyzed the effect of the pile level on the scour lengths. They introduced the dimensionless parameter P_i/D_{50}, where the subscript i is equal to L, M or H according to the different pile levels taken into consideration. Namely, in the case of low pile level P_L/D_{50}=0, for the medium pile level P_M/D_{50}=0.5 and for the high pile level P_H/D_{50}=1. To evaluate the scour depth in case of P_M and P_H they introduced the correction factor $(1+P_i/D_{50})^{\alpha}$ to modify Eqs. (15a) and (15b), obtaining the following relationships for Z_m:

$$Z_m = (0.58 \cdot \sigma^{0.55} \cdot S^{0.75} \cdot F_{d90}{}^{1.8}) \cdot (1 + P_i / D_{50})^{-\alpha_{90}} \tag{18a}$$

$$Z_m = (0.58 \cdot \sigma^{-0.55} \cdot S^{0.75} \cdot F_{d50}{}^{1.8}) \cdot (1 + P_i / D_{50})^{-\alpha_{50}} \tag{18b}$$

where α_{50} and α_{90} are parameters depending on the ramp slope S and can be expressed as follows:

$$\alpha_{50} = 0.27e^{4.24S} \tag{19a}$$

$$\alpha_{90} = 0.32e^{3.91S} \tag{19b}$$

It can be seen from Eqs. (18a) and (18b) that they converge to Eqs. (15a) and (15b) respectively for the case studied by Pagliara (2007) which is relative to low pile configuration P_L for which P_L/D_{50}=0. The effect of pile height is that of reducing the scour depth. Namely, an increase in pile level height cause a reduction of the scour depth. This occurrence is due to the fact that generally increasing the pile level, the jump occurs over a longer length and the energy dissipation is less localized than in the case of low pile configuration.

As said above, the erosion of the stilling basin can also be controlled inserting protection structures. In this case, the hydrodynamic of the scour phenomenon completely changes thus also influencing the main lengths, being the same hydraulic and geometric conditions. For low pile configuration, Pagliara and Palermo (2008a-b) tested three different sills (namely, rock, continuous and dentated sills) in various spatial positions and they concluded that generally rock sills perform better than the others. To put in evidence the effect of a rock sill on the scour geometry, the Authors compared the scour depth and length experiments for each combination of discharge, stilling basin material and ramp slope in absence of any protection work (*reference tests*) with those obtained in the presence of a rock sill located in different spatial positions. They tested the spatial positions indicated in Figure 9a with the symbol "+". In fact, in the same conditions, tests were carried out with the sill placed in different longitudinal ($\lambda=x_s/l_0$), and vertical positions ($Z_{op}=z_{op}/z_m$), in which x_s is the longitudinal position of the protection work measured from the ramp toe and z_{op} is the vertical position of the upper sill corner point. For both scour length and depth Pagliara and Palermo (2008b) proposed the following general relationship, in which the expression of the coefficients a_1, b_1, c_1, d_1, e_1, f_1 and δ are reported in Table 1. For non uniform material (N-U.M.) the coefficient δ is reported in brackets, while for uniform material (U.M.) is always equal to 1. ξ is equal either to Z_{ms}/Z_m, where $Z_{ms}=z_{ms}/h_1$ and $Z_m=z_m/h_1$ are the non dimensional scour depths of the test with the protection structure and the respective reference test or to

$\xi = L_s/L_0$ for scour length, where $L_s = l_s/h_1$, in which l_s is the scour length in presence of a protection sill (see Figure 9b). The following equation is valid in the λ-range specified in Table 1 and for $1/12 \leq S \leq 1/4$, $1 \leq \sigma \leq 2.8$ and $1.0 \leq F_{d90} \leq 3.75$. Z_m and L_0, as denominators of the function ξ are evaluated using Eq. (15a) and (16), respectively.

$$\xi = \left(a_1\lambda^2 + b_1\lambda + c_1\right)\cdot\left(d_1 Z_{op}^2 + e_1 Z_{op} + f_1\right)\cdot\delta \tag{20}$$

Table 1. Expression of coefficient of Eq. (20)

ξ	λ-range	a_1	b_1	c_1	d_1	e_1	f_1	δ U.M (N-U.M.)
Z_{ms}/Z_m	$0.25 \leq \lambda \leq 0.45$	0	25.3*S*-2.9	1.8-8.8*S*	0	-0.1	1.05	1 (0.7)
	$0.5 \leq \lambda \leq 1.5$	0	0.3-2*S*	3.6*S*+0.4	0	-0.1	1.05	1 (0.7)
L_s/L_0	$0.25 \leq \lambda \leq 0.45$	0	4.0-30.6*S*	10.9*S*-0.06	0.52	0.1	0.82	1 (0.4)
	$0.5 \leq \lambda \leq 1.5$	3.4-16.5*S*	39.0*S*-7.5	4.6-18.8*S*	0	-0.1	1.05	1 (0.4)

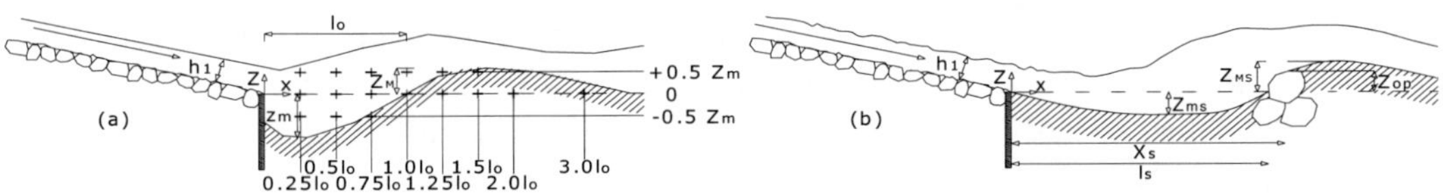

Figure 9. Diagram sketch of a block ramp with the indication of (a) the main geometric parameters in absence of protection sills and different sill spatial positions and (b) the main geometric parameters in presence of a sill.

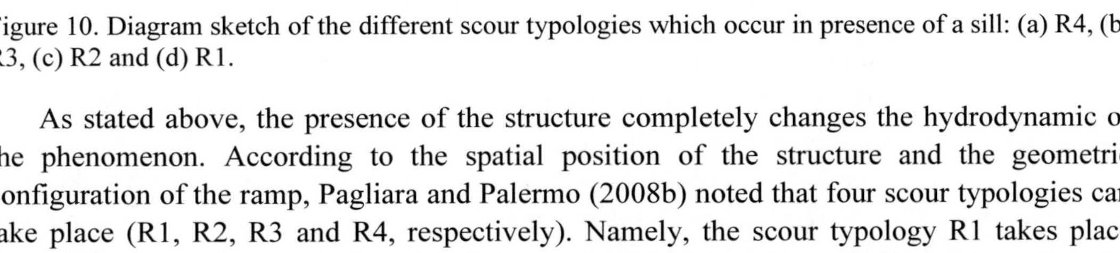

Figure 10. Diagram sketch of the different scour typologies which occur in presence of a sill: (a) R4, (b) R3, (c) R2 and (d) R1.

As stated above, the presence of the structure completely changes the hydrodynamic of the phenomenon. According to the spatial position of the structure and the geometric configuration of the ramp, Pagliara and Palermo (2008b) noted that four scour typologies can take place (R1, R2, R3 and R4, respectively). Namely, the scour typology R1 takes place when the scour hole is located downstream of the sill and it generally occurs when the protection structure is placed close to the ramp toe (see Figure 10d). Type R2 (see Figure 10c) occurs when two scour holes are formed, both upstream and downstream of the protection sill. Scour typologies R3 (see Figure 10b) and R4 (see Figure 10a) both have a scour hole which is located entirely upstream of the sill, but in the first case the protection structure is partially covered by the ridge (type R3), whereas for type R4 the deposited material totally covers the sill.

Applicative Example

Let us calculate the dimension of the blocks in a ramp in order to assure their stability and the stilling basin scour geometry in the case of low pile configuration and having fixed the following parameters: river width B=10 m; design water discharge Q=20 m^3/s; height of the ramp H= 3 m and ramp slope S=0.125. Let us suppose to use uniform rocks to build the block ramp whose density is 2650 kg/m^3. Applying Eq. (8) it is possible to evaluate the minimum D_{65} of the ramp material in order to assure its stability for the selected flood event. Substituting the values of the different parameters in Eq. (8), one obtains that the minimum D_{65} to be used is equal to 0.31m. Thus, adopting a safe coefficient equal to 1.8, the dimension of the blocks that can be used is equal to D_{65}=0.55m=(0.31·1.8)m. Being that the material is uniform, D_{65}~$D_{50,}$ thus the D_{50} of the selected ramp material is equal to 0.55m. Now suppose having a uniform material constituting the stilling basin whose σ=1.2 and the mean diameter d_{50} is equal to = 0.1m.

The relative energy dissipation between the entrance and the toe of the ramp can be estimated using Eq. (3) in which Γ=0, as no boulders are present on the base ramp. Let us calculate the non dimensional parameter k/D_{50}, in order to evaluate the scale roughness condition, and the parameter k/H, which is present in Eq. (3). In this case $k=(q^2/g)^{1/3}$=0.74m (with $q=Q/B$). Thus k/D_{50}=1.35 and the scale roughness condition is *LR*. For *LR*, the coefficients *A*, *B* and *C* of Eq. (3) are equal to 0.33, -1.3, -14.5, respectively. From Eq. (3) one can derive ΔE_{r1}=0.64. Knowing ΔE_{r1}=0.64 it is possible to estimate h_1 which results to be equal to 0.44m. Thus, assuming that the density of the stilling basin material is ρ_s=2650kg/m^3, F_{d50}=3.58. From Figure (6) it can be noted that for ramp slope S=0.125 and F_{d50}=3.58 an F_{MB} jump type takes place. Thus using Eqs.(15b) and (16) it is possible to calculate Z_m and L_0. For the adopted stilling basin material Z_m=1.09 and z_m results to be equal to 0.48m, whereas L_0=13.2 and l_0=5.8m.

Conclusion

This chapter aims to synthesize the main achievements in block ramp design. Block ramps are hydraulic structures which in the last few decades have become more and more utilized for river restoration projects as they have the unique peculiarity to sustain the ecological balance in the rivers in which they are inserted. They can be used in alternative to traditional structures (like check dams) as they do not break river morphology and allow fish passage. Moreover they dissipate more energy than the former. For a correct functioning of the ramp, it is necessary that the stability of the blocks constituting it must be assured. Thus experimental tests were conducted by many researchers in order to evaluate the critical discharge and the size of the blocks which have to be adopted. Additionally the effect of roughness and geometric configuration of the ramp on the energy dissipation process was deepened and it was found relevant. Also in this case experimental relationships have been furnished. The energy dissipation process has been analyzed also in the case in which a movable stilling basin is present at the toe of the ramp. The different jump typologies which can occur were analyzed and described. As for the traditional hydraulic structures, generally a scour process takes place in the stilling basin. Thus the analysis and the evaluation of the main lengths of the scour hole is fundamental in order to avoid structural risks. For these reasons some useful relationships which consent to evaluate both the scour depth and length in presence and absence of protection sills located in the stilling basin were recently developed. In conclusion, we can state that many aspects of block ramps have already been investigated, but further studies have to be conducted in order to solve many open questions and to find a more theoretical approach which can explain the observed phenomena.

REFERENCES

Aguirre Pe, J., and Fuentes, R. (1991) *Movement of big particles in step, macro-rough stream.* Proceeding of 24th IAHR Congress, Madrid, Spain, vol. A, 149-158.

Aguirre Pe, J., Olivero, M.L., and Mocada, A.T. (2003) Particle densimetric Froude number for estimating sediment transport. *J. Hydr. Engng. 129(6)*, 428-437.

Armanini, A., and Gregoretti, C. (2005) Incipient sediment motion at high slope in uniform flow condition. *Water resources research 41*, 1-8.

Aderibigbe, O., and Rajaratnam, N. (1998) Effect of sediment gradation on erosion by plane turbulent wall jets. *J. Hydr. Research 124(10)*, 1034-1042.

Bathurst, J.C. (1978) Flow resistance of large-scale roughness. *J. Hydraul. Div., Am. Soc. Civ. Eng., 104(12)*, 1587–1603.

Bathurst, J.C., Simons, D.B., and and Li, R.M. (1981) Resistance equation for large-scale roughness. *J. Hydraul. Div., Am. Soc. Civ. Eng., 107(12)*, 1593–1613.

Breusers, H.N.C., and Raudkivi, A.J. (1991) *IAHR Hydraulic structures design manual 2.* Balkema, Rotterdam, The Netherlands.

Bormann, E., and Julien, P.Y. (1991) Scour downstream of grade control structures. *J. Hydr. Engng. 117(5)*, 579-594.

Canepa, S., and Hager, W.H. (2003) Effect of air jet content on plunge pool scour. *J. Hydr. Engng. 128(5)*, 358-365.

Chamani, M.R., and Rajaratnam, N. (1999a) Characteristics of skimming flow over stepped spillways. *J. Hydraul. Eng., 125(4)*, 361–368.

Chamani M.R., and Rajaratnam, N. (1999b) Onset of skimming flow on stepped spillways. *J. Hydraul. Eng., 125(9)*, 969–971.

Chanson, H. (1994) *Hydraulic design of stepped cascades, channels, weirs and spillways*, Pergamon, Oxford, U.K.

Christodoulou, G.C. (1993) Energy dissipation on stepped spillways. *J. Hydraul. Eng., 119(5)*, 644–650.

D'Agostino, V., and Ferro, V. (2004) Scour on alluvional bed downstream of grade-control structures. *J Hydr. Engng, 130(1)*, 1-14.

Dey, S., and Raikar, V. (2005) Scour in long contractions. *J. Hydr. Engng. 131(12)*, 1036-1049.

Dey, S, and Sarkar, A (2006) Scour downstream of an apron due to submerged horizontal jets. *J. Hydr. Engng. 132(3)*, 246-257.

Diez-Cascon, J., Blanco, J.L., Revilla, J., and Garcia, R. (1991) Studies on the hydraulic behaviour of stepped spillways. *Int. Water Power Dam Constr., 43(9)*, 22–26.

Essery, I.T.S., and Horner, M.W. (1978) *The hydraulic design of stepped spillways*. Rep. No. 33, CIRIA, London.

Farhoudi, J., and Smith, K.V.H. (1985) Local scour profile downstream of hydraulic jump. *J. Hydr. Res. 23(4)*, 343-358.

Ferro, V., Dalla Fontana, G., Pagliara, S., Pugliesi, S., and Scotton, P. (2004) *Opere di sistemazione idraulico-forestale a basso impatto ambientale*. McGraw-Hill, Milano.

Hassan, N.M.K.N., and Narayanan, R. (1985) Local scour downstream of an apron. *J. Hydr. Engng. 111(11)*, 1371-1385.

Hoffmans, G.J.C.M., and Verheij, H.J. (1997) *Scour manual*. Balkema, Rotterdam, The Netherlands.

Hoffmans, G.J.C.M (1998) Jet scour in equilibrium phase. *J. Hydr. Engng. 124(4)*, 430-437.

Pagliara, S., and Peruginelli, A. (2000) *Energy dissipation comparison among stepped channel, drop and ramp structures*. Proc. Int. Workshop on Hydraulics of Stepped Spillways, Zurich, Switzerland, 111–118.

Pagliara, S., and Dazzini, D. (2002) *Hydraulics of block ramp for river restoration*. Proc., 2nd Int. Conf. on New Trends in Water and Environmental Engineering for Safety and Life: Eco-compatible Solution for Aquatic Environments, Capri, Italy.

Pagliara, S., and Ferro V. (2003). Contributo al dimensionamento idraulico delle rampe in pietrame. *Rivista di ingegneria agraria, vol. 1* Anno XXXIV, pp. 23-31.

Pagliara, S., and Chiavaccini, P. (2003) *The use of piles to increase block ramp stability*. Proc., XXX IAHR Congress, Thessaloniki, Greece, vol.II, 199-206.

Pagliara, S., and Hager, W.H. (2004) *Scour downstream of block ramps*. Proc., 2nd International Conference on Scour and Erosion, Singapore, vol. 2, 215-222.

Pagliara, S., and Chiavaccini, P. (2004) *Stability of reinforced block ramps*. Proc. River flow 2004 International Conference, Balkema, Rotterdam, 1291-1295.

Pagliara, S., and Chiavaccini, P. (2006a) Energy dissipation on block ramps. *J. Hydr. Engng. 132(1)*, 41-48.

Pagliara, S., and Chiavaccini, P. (2006b) Energy dissipation on reinforced block ramps. *J. Irr. And Drainage Engng. 132(3),* 293-297.

Pagliara, S., and Chiavaccini, P. (2007) Failure mechanisms of base and reinforced block ramps. *J Hyd. Res., 45(3),* 407-420.

Pagliara, S. (2007) Influence of sediment gradation on scour downstream of block ramps. *J Hyd. Engng, 133(11)*, 1241-1248.

Pagliara, S., and Palermo M. (2008a) Scour control downstream of block ramps. *J Hydr. Engng, 134(9)*, 1376-1382

Pagliara, S., and Palermo, M. (2008b) Scour control and surface sediment distribution downstream of block ramps. *J Hydr. Res., 46(3)*, 334-343.

Pagliara, S., and Palermo, M. (2008c) *Scour control downstream of block ramps in basins with continuous and dentated sill.* Proc. of 16th IAHR-APD Congress and 3rd Symposium of IAHR-ISHS, Nanjing, China, vol. VI, 2112-2117.

Pagliara, S., Das, R., and Palermo, M. (2008a) Energy dissipation on submerged block ramps. *Journal of Irr. And Drainage Eng., 134(4)*, 527-532.

Pagliara, S., Palermo, M., and Lotti, I. (2008b) Sediment transport on block ramp: filling and energy recovery *KSCE Journal of Civil Engineering, 13(2)*, 129-136.

Pagliara, S., Das, R. and Palermo, M. (2008c) *Scour depth for different pile positions at the toe of a block ramp* Proc. of 16th IAHR-APD Congress and 3rd Symposium of IAHR-ISHS, Nanjing, China, vol. VI, 2031-2036.

Palermo, M., Das, R., and Pagliara, S. (2008) *Hydraulic jump classification downstream of block ramps for non-uniform channel bed material* 2nd IJERW on Hydraulic Structures, Pisa, Italy, 129-134.

Peyras, L., Royet, P., and Degoutte, G. (1992) Flow and energy dissipation over stepped Gabion weirs. *J. Hydraul. Eng., 118(5)*, 707–717.

Platzer, G. (1983) *Die Hydraulik der breiten Blochsteinrampe, rampenbeigung 1:10*, Bundensanstalt, Wien, Austria.

Robinson, K.M., Rice, C.E., and Kadavy, K.C. (1998) *Design of rock chutes*. Trans. ASAE, 41(3), 621–626.

Stephenson, D. (1991) Energy dissipation down stepped spillways. *Int. Water Power Dam Constr., 43(9)*, 27–30.

Sumer, B.M., Christiansen, N., and Fredsøe, J. (1993) Influence of cross section on wave scour around piles. *J. Waterw., Port, Coastal, Ocean Engng. 119(5)*, 477-495.

Whittaker, W., and Jäggi, M. (1996) *Blockschwellen*. Mitteilung 91, Versuchsanstalt für Wasserbau, Hydrologie und Glaziologie. ETH Zurich, Zürich, Switzerland [in German].

Veronese, A. (1937) Erosioni di fondo a valle di uno scarico. *Ann. Lavori Pubbl. 75(9)*, 717-726 [in Italian].

In: Check Dams, Morphological Adjustments…
Editors: C. Conesa and M. A. Lenzi

ISBN: 978-1-62618-856-3

Chapter 6

EFFECTIVENESS OF CHECK DAMS IN THE RESERVOIR SEDIMENTATION CONTROL UNDER TORRENTIAL CONDITIONS

Jian Liu*[*,1], Shuhei Minami[2] and Shinji Egashira[2]

[1]College of Civil Engineering, Shenzhen University,
Nanshan District, Shenzhen 518060, China
[2]River Group, NEWJEC Inc., 3-20, Honjyo-Higashi 2-chome,
Kita-ku, Osaka, 531-0074, Japan

ABSTRACT

This chapter addresses the roles and effects of check dams in reservoir sedimentation control. Section 6.1 describes an outline of sediment management measures in the whole drainage basin. Section 6.2 gives qualitative analysis on the roles of the check dams in managing reservoir sedimentation. In Section 6.3, a numerical simulation is made in the basin of the Dashidaira dam on the Kurobe River to explore the mitigation effects of the check dams on the reservoir sedimentation issues. The numerical results show that the reservoir sedimentation problems could be mitigated even if only one check dam is employed. Section 6.4 introduces the sedimentation measures at the Sakuma and Dashidaira dams in Japan and the Sanmenxia dam in China as case studies.

Keywords: Reservoir sedimentation, check dams, storage volume, qualitative analysis, numerical simulation, Sakuma dam (Japan), Dashidaira dam (Japan), Sanmenxia dam (China).

1. INTRODUCTION

To deal with reservoir sedimentation issues, it is necessary to understand sediment behaviors comprehensively and evaluate the sediment transport rates for short and long terms

* Tel and Fax: +86-755-2673-2827; email: liujian@szu.edu.cn

at any target location. The main forms of the sediment transport in a drainage basin are bed load, suspended load and debris flow as shown in Figure 1 (Egashira 2001, Egashira et al. 2007). Slope collapses due to a storm generally occur at a mountain slope of more than 30 degrees, and parts of the deposits are transformed into the debris flows during their movement process.

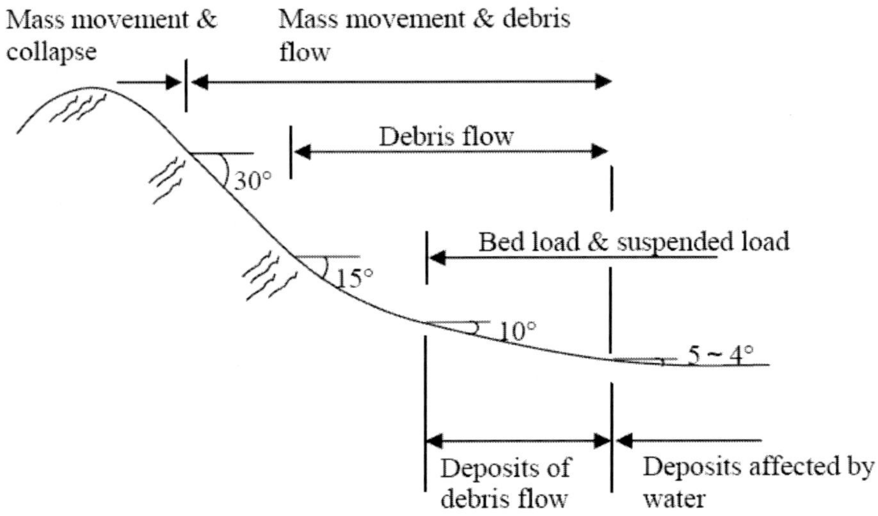

Figure 1. Sediment production in a river basin

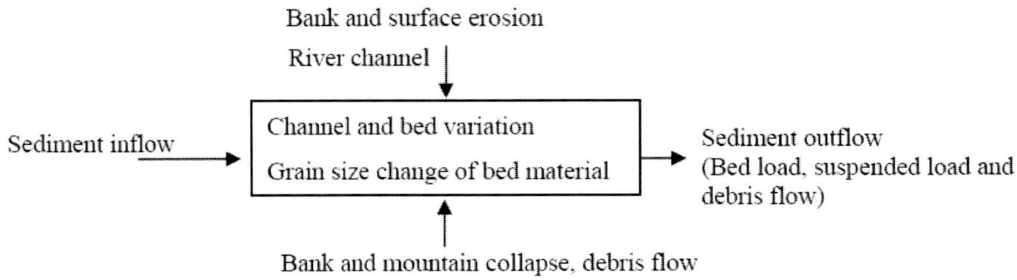

Figure 2. Basic system of sediment transport in a river.

The debris flow usually occurs in steep streams with a bed slope of over 12–15 degrees where non-cohesive materials are loosely deposited, and it ceases in the domain of around 10–4 degrees, resulting in a lot of sediment deposits. These deposits are transported downstream as suspended load and bed load when floods occur, and form the sediment source of a river. Figure 2 shows a basic system of the sediment transport in a drainage basin (Ashida, Egashira and Nakagawa 2008). Although a river basin contains numerous streams, a whole sediment transport system can be expressed by means of series and parallel connections of the basic system. Using this sediment transport system and rainfall runoff model, short- and long-term sediment transportation was evaluated for bed load, suspended load and wash load in several drainage basins (Egashira et al. 2007). When a dam is constructed on a river, the sediment will be deposited in the reservoir because the transport capacity of the river flow decreases suddenly. Figure 3 shows a typical profile of reservoir sedimentation.

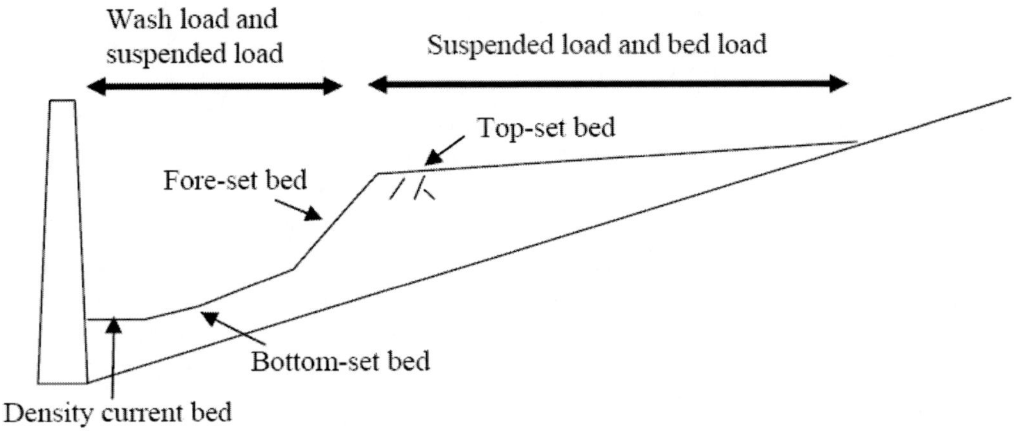

Figure 3. Typical reservoir sedimentation profile.

Bed load creates a top-set bed and fore-set bed, wash load and suspended load form a bottom-set bed, and density current creates a density current bed (Ashida, Takahashi and Michiue 1983). Such a bed profile and sediment size structures of the deposited sediment layers play an important role in planning reservoir sedimentation management. At present, several mathematical tools based on sediment hydraulics can be utilized to evaluate the reservoir sedimentation.

The reservoir sedimentation causes many environmental problems such as maintenance issues related to water withdrawal system, decrease of reservoir capacity, riverbed aggradation in the upstream, degradation in the downstream, coastal erosion and the corresponding deterioration of ecological systems. In order to reduce and recover such issues associated with reservoir sedimentation, it is necessary to properly manage sediment at the dam and in the upstream and downstream reaches. Several counter measures at the dam which are explained in the following two sections have already been operated for many years. The sediment management measures in the downstream reaches are the ones protecting the river channel and banks from erosion. In the upstream, some structural and non-structural measures such as check dams, slope surface cover works and forestation have been performed.

The check dams are common features in Europe, North America and East Asia. There are wall-type structures and open structures such as a grid dam, beam dam and split dam (Chanson 2004). The most common types of check dams are earth, rock-fill and concrete wall. They can trap sediment material during the initial period, and will control the inflowing sediment by changing the bed slope of the storage area when the reservoir is filled up with sediment. The downstream face of the concrete check dam is nearly vertical, followed by a short stilling structure. Apparently from a view of short-term sediment management, construction of the check dams in the upstream reaches is an effective and powerful countermeasure to reduce sediment inflowing into the reservoirs. This means sediment accumulates in the upstream reaches and the sediment control function of the check dams will be lost gradually with decrease of the storage capacities. Correspondingly, in order to maintain the sediment transport rate to some allowable level for long years using only the check dams, it is necessary to construct many check dams continuously. As a result, a lot of sediment will be stored in the upstream reaches and result in different sediment problems. In spite of having such issues, construction of the check dam may still be one of the effective measures. Ideally, the check dams are employed together with some drainage management performances such as forestation and surface stability measures to decrease sediment erosion and sediment yield. As a result, the sediment inflowing into the reservoir will be reduced to the allowable level without accumulating unstable sediment in the upstream reaches.

In the following sections, reservoir sedimentation management measures are described briefly with attention paid to the effects of the check dams to reduce the reservoir sedimentation. The effects of the check dams are addressed by the numerical simulation in detail because of a lack of observed data. The sedimentation measures at the Sakuma and Dashidaira dams in Japan and the Sanmenxia dam in China are introduced as the examples using the check dams to reduce the reservoir sedimentation.

2. Role of Check Dams in Managing Reservoir Sedimentation

The sedimentation measures are generally classified into decrease of inflowing sediment, artificial removal and flushing the deposits from a reservoir. Among these measures, bypass tunnel which is used for the Asahi dam and the Miwa dam in Japan and sediment flushing that is used for the Dashidaira and Unazuki dams in Japan are effective, however the facilities are very expensive and restricted by the hydrological and geography conditions. Although dredging is a usual measure, it has some restrictions with respect to possible dredging sediment volume and treatments of the removed sediment may sometimes cause different problems to be solved.

The experiences of the sediment management in the world show that the inflowing sediment is usually over the allowable rate in most of the reservoirs where the sediment management is conducted. The allowable rate is defined as a possible rate of sediment volume that can be treated economically and ecologically. Figure 4 shows the relationship between the inflowing sediment rate and allowable volume rate for sediment management. If the inflowing sediment rate is larger than the allowable rate, the excess rate of the sediment inflowing into the reservoir will accumulate in the reservoir even if some countermeasures are operated. In this case, the amount of inflowing sediment corresponding to the excess rate should be managed in the upstream reaches to decrease to zero. If the check dams are chosen only for the countermeasure in this circumstance, the inflowing sediment could be reduced to the allowable level by constructing the check dams continuously. However, a lot of unstable sediment will accumulate in the upstream reaches in this case. Such a sediment environmental condition will not be recommended. Instead, if possible, the inflowing rate of sediment should be reduced to the allowable value in view of sediment management planning, using the integrated countermeasures which may combine the check dams, slope stability works and forestation. In order to develop such an integrated countermeasure, a comprehensive understanding of sediment transport process is needed together with the tools for evaluating the sediment transport process in the whole drainage basin, as mentioned above.

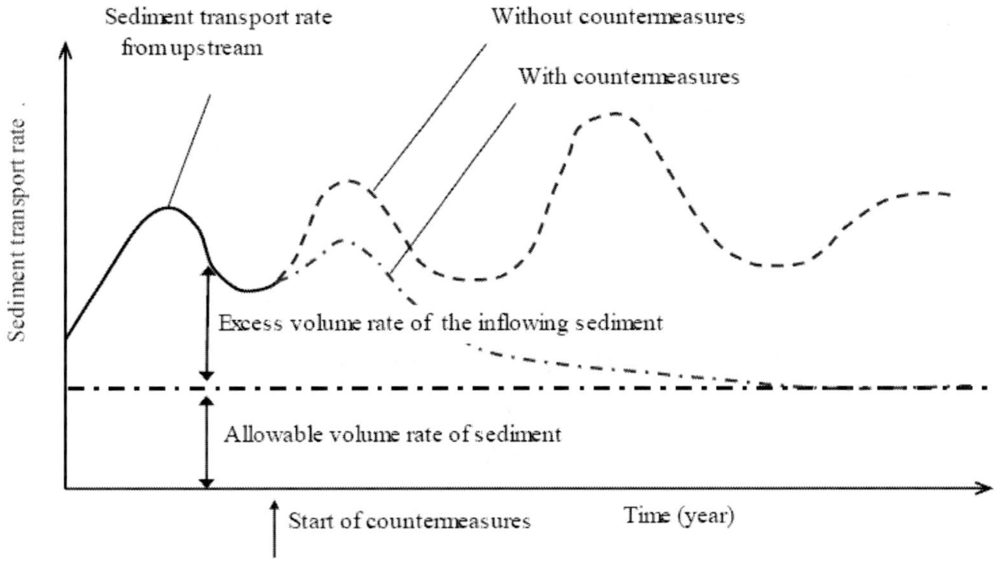

Figure 4. Relationship between inflowing sediment rate and allowable volume rate.

It is also indispensable to develop methods for evaluating the effectiveness of each measure in the integrated countermeasure.

In spite of many unsolved issues for the countermeasures against reservoir sedimentation, the check dams are still now one of the useful measures to manage the inflowing sediment. A wall type of concrete check dam is taken as an example for discussion. It is recognized that the check dam of this type has the functions of sediment trapping and sediment control. Generally, the sediment trapping function appears during initial operation stage immediately after completion, and will lose when the reservoir is filled up of sediment. After finishing the trapping stage, the check dam controls the sediment transportation quantitatively and qualitatively by changing the bed slope (storage volume) corresponding to a hydrograph of the flood laden with sediment. In other words, sediment hydrograph, sediment sizes and sediment transport formations, all of which result from the differences of the sediment carrying capacity of the flow between the reservoir and the upstream reaches, are controlled as follows.

1) The peak discharge of sediment is decreasing and it appears late in comparison with the flood peak.
2) Sediment sorting or selective transportation takes place, thus, the out-flowing sediment sizes tend to be finer in comparison to the inflowing ones.
3) Debris flow changes to a usual flow laden with sediment since the sediment deposits into the check dam' reservoir. Moreover, some parts of the suspended sediment changes to contact loads.

In contrast to such useful functions, it should be noted that the sediment control is not always directed to the expected effects, but to negative effects in certain conditions; e.g. the deposited sediment in the check dam may be a main source of debris flow although there are rare cases.

Taking the above-mentioned control functions into considerations, and evaluating the locations, sizes and the number of the check dams, the inflowing sediment into reservoirs could be controlled to be the allowable level. There are neither data for testing effects of check dams on reservoir sedimentation nor the predicted results although specific cases are chosen for investigating the effects of the check dams. Correspondingly, a single check dam will be chosen in the following section to investigate numerically the effect of check dam on reservoir sedimentation.

3. Evaluation of the Effects of a Check Dam by Numerical Simulation

3.1. Research Object

The Dashidaira reservoir on the Kurobe River in Japan is selected as a research object. The reservoir bed 2500 - 3000m upstream of the dam still rises although the sediment flushing at the dam is performed when a large flood occurs. The outlet facility of a power plant is operated at the place located near the high water level as shown in Figure 5 and used to be damaged often due to the river bed rising. Therefore, the deposited sediment, parts of which cannot be drained by the sediment flushing because of sediment composed of coarse size particles, is dredged artificially to prevent from sediment induced damage. To investigate how to mitigate such an issue using a wall type check dam, a computation is conducted under the following conditions. The check dam with a height of 10m is assumed to be constructed 5km upstream of the Dashidaira dam. The calculated reach is 6.4km. The location of the check dam and initial longitudinal profile of the reach are shown in Figure 5.

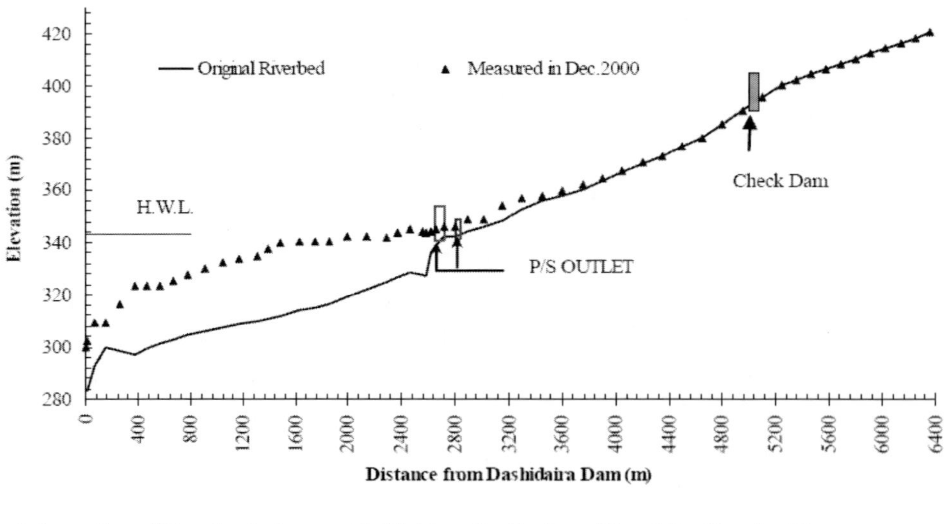

Figure 5. Location of the check dam and initial longitudinal profile of the Kurobe River.

3.2. Governing Equations

Flow Equations

The flow is governed by the following continuity and momentum conservation equations.

$$\frac{\partial Q}{\partial x} = q' \tag{1}$$

$$\frac{1}{gA}\frac{\partial}{\partial x}\left(\frac{Q^2}{A}\right) + \frac{\partial}{\partial x}(Z_b + h) + I_e = 0 \tag{2}$$

in which Q = discharge in cross section, q' = tributary discharge per unit length, A = cross-sectional area, Z_b = elevation of riverbed, h = water depth, g = gravitational acceleration, x = streamwise direction, and I_e = energy slope that can be specified by the following Manning's resistance formula:

$$I_e = \frac{n^2 V^2}{R^{4/3}} \tag{3}$$

where

n = Manning's roughness, V = mean cross-sectional velocity and R = hydraulic radius.

Sediment Transport Equation

The riverbed evolution is described by the following equation (Liu et al. 2004):

$$\frac{\partial Z_b}{\partial t} + \frac{1}{B_b}\left[\sum_{k=1}^{Nb}\left(\frac{B_s}{1-\lambda_b}\frac{\partial q_{bk}}{\partial x} + \frac{B_s(E_{sk} - D_{sk})}{1-\lambda_s}\right) + \sum_{k=Nb+1}^{Nt}\frac{B_s(E_{wk} - D_{wk})}{1-\lambda_w}\right] = 0 \tag{4}$$

where t = time, B_b = width of river bed, B_s = width of sediment transport, q_{bk} = bed load discharge of grain size d_k per unit width, E_{sk} and D_{sk} = rising and falling fluxes of the suspended load with grain size d_k per unit area near bed, E_{wk} and D_{wk} = rising and falling fluxes of the wash load with grain size d_k per unit area near bed, N_b and N_t = numbers of the grain sizes of bed material load and total load, and λ_b, λ_s, and λ_w = porosities of the deposits by bed load, suspended load, and wash load, respectively.

E_{sk} and D_{sk} can be expressed as

$$E_{sk} = W_{sk} C_{ek} \tag{5}$$

$$D_{sk} = W_{sk} C_{ak} \tag{6}$$

where

W_{sk} = fall velocity of kth grain, C_{ek} = concentration of kth grain at an equilibrium datum plane, and C_{ak} = concentration of kth grain at 0.05 h above riverbed.

E_{wk} and D_{wk} can be computed by the following equations:

$$E_{wk} = \beta u_{*wk} p_{bk} \left(\frac{u_*^2}{u_{*wk}^2} - 1 \right)^{\chi}, \quad u_* \le u_{*wk}; \qquad E_{wk} = 0, \quad u_* > u_{*wk} \tag{7}$$

$$D_{wk} = W_{sk} C_{ak}, \quad u_* \le u_{*dk}; \qquad D_{wk} = 0, \quad u_* > u_{*dk} \tag{8}$$

where u_* = shear velocity, u_{*dk} = critical fall shear velocity of kth wash load, u_{*wk} = critical suspension shear velocity of kth wash load, p_{bk} = percentage of kth grain in bed material, $\sum p_{bk}$ = 1, parameters β = 0.00004-0.1 and χ = 1.0-1.5 were proposed by Iwashita et al. (1988), and β = 0.0002 and χ = 1.5 were used in this study.

The concentration parameters C_{ek} and C_{ak} can be determined by the following equations (Ashida and Michiue 1970):

$$C_{ek} = 0.25 P_{ek} \left(g(\zeta_0) / \zeta_0 - G(\zeta_0) \right) \tag{9}$$

$$g(\zeta_0) = \frac{1}{\sqrt{2\pi}} \exp\left(-\frac{1}{2} \zeta_0^2 \right) \tag{10}$$

$$G(\zeta_0) = \frac{1}{\sqrt{2\pi}} \int_{\zeta_0}^{\infty} \exp\left(-\frac{1}{2} \zeta^2 \right) d\zeta \tag{11}$$

$$C_{ak} = F C_k (1 - e^{-F}) \tag{12}$$

where p_{ek} = percentage of kth grain in suspended load, $\zeta_0 = W_{sk}/\sigma_p$, $\zeta = W_p/\sigma_p$, W_p = fluctuating velocity of particle in the vertical direction, σ_p = standard variation of W_p ($=0.75u_*$), $F = 14.25 W_{sk}/u_*$, and C_k = mean cross-sectional concentration of grain size d_k, which is determined by the following diffusion equation.

$$\frac{\partial}{\partial t}(A C_k) = \frac{\partial}{\partial x}\left\{ A \left(\varepsilon_{sx} \frac{\partial C_k}{\partial x} - v C_k \right) \right\} + B_s F_{ed} \tag{13}$$

where $F_{ed} = E_{sk} - D_{sk}$ for suspended load and $F_{ed} = E_{wk} - D_{wk}$ for wash load, ε_{sx} = diffusion coefficient in the flow direction($= 0.5\kappa h u_*$), and κ = Karman constant.

In consideration of the sediment characteristics of the Japanese rivers, the bed load discharge q_{bk} is computed by the following formula developed by Ashida and Michiue (1972):

$$\frac{q_{bk}}{\sqrt{(\sigma/\rho - 1) g d_k^3}} = 17 p_{bk} \tau_{*ek}^{3/2} \left(1 - \frac{u_{*ck}}{u_*} \right) \left(1 - \frac{\tau_{*ck}}{\tau_{*k}} \right) \tag{14}$$

where σ and ρ= densities of sediment and water, u_{*ck} = critical shear velocity of kth grain, τ_{*k},τ_{*ck}, and τ_{*ek} = non-dimensional shear stress, critical shear stress, and effective shear stress of kth grain, respectively, and these variables are defined as

$$\tau_{*k} = u_*^2 / F_k \tag{15}$$

$$\tau_{*ck} = u_{*ck}^2 / F_k \tag{16}$$

$$\tau_{*ek} = u_{*e}^2 / F_k \tag{17}$$

in which F_k = $(\sigma/\rho\text{-}1)gd_k$, $\boldsymbol{u_{*e}}$ = effective shear velocity, and it can be determined by the following equations (Engelund 1967, Ashida and Michiue 1972).

$$V / u_{*e} = 6.0 + 5.75 \log_{10} \frac{R}{d_m(1 + 2\tau_{*m})} \tag{18}$$

$$\tau_{*m} = u_{*e}^2 / F_m \tag{19}$$

where d_m = mean grain size of sediment mixture, $\boldsymbol{\tau_{*m}}$ = nondimensional shear stress of sediment with d_m, $F_m = (\sigma/\rho\text{-}1)\, gd_m$, and Fr = Froude number.

$\boldsymbol{u_{*ck}}$ can be calculated by the following equations (Ashida and Michiue 1972):

$$u_{*ck}^2 / u_{*cm}^2 = 0.85, \quad for\ d_k / d_m < 0.4 \tag{20}$$

$$u_{*ck}^2 / u_{*cm}^2 = \frac{1.64}{(\log_{10}(19 d_k / d_m))^2}, \quad for\ d_k / d_m \geq 0.4 \tag{21}$$

in which $\boldsymbol{u_{*cm}}$ = critical shear velocity of sediment with d_m, and it can be determined by the following formula proposed by Iwagaki (1956):

$$u_{*cm} = \sqrt{a(d_m / 10)^b} \tag{22}$$

where a, b = constants, a = 0.0226, b=1 for d_m<0.065 mm, a= 0.000841, b=11/32 for 0.065<d_m<0.565 mm, a = 0.0055, b=1 for 0.565<d_m<1.18 mm, a = 0.01346, b=31/22 for 1.18<d_m<3.03 mm, and a = 0.00809, b=1 for d_m>3.03 mm.

The percentage of each grain size in the bed material p_{bk} is computed by the following equations, assuming that the riverbed is composed of a mixing layer, transition layer, and deposition strata consisting of the deposition layers with the same thickness from the surface to a reference plane (Liu et al. 2004).

$$\frac{\partial p_{bk}}{\partial t} + \frac{1}{T_m}\left(\frac{\partial z_b}{\partial t}(\delta P_{bk} + (1-\delta) P_{bk0}) - \frac{\partial z_{bk}}{\partial t} \right) = 0 \tag{23}$$

$$p_{bk0} = p_{tk}, \quad if\ N_{dl} \geq N_{dl0}; \quad p_{bk0} = p_{Ndl}, \quad if\ N_{dl} < N_{dl0} \tag{24}$$

$$Z_b = Z_0 + T_m + T_t + N_{dl} T_{1dl} \tag{25}$$

$$\frac{\partial T_t}{\partial t} = \frac{\partial Z_b}{\partial t}, subject\ to\ 0 < T_t \leq T_{1dl} \tag{26}$$

$$T_t = T_{t0} + T_{1dl}, N_{dl} = N_{dl0} - 1, \quad if\ T_t < 0 \tag{27}$$

$$T_t = T_{t0} - T_{1dl}, N_{dl} = N_{dl0} + 1, \quad if\ T_t > T_{1dl} \tag{28}$$

$$\frac{\partial p_{tk}}{\partial t} = \frac{\partial Z_b}{\partial t}\frac{p_{bk}}{T_t}, \quad if\ \frac{\partial Z_b}{\partial t} > 0\ and\ N_{dl} \le N_{dl0} \tag{29}$$

$$p_{tk} = p_{bk}, \quad if\ \frac{\partial Z_b}{\partial t} > 0\ and\ N_{dl} > N_{dl0} \tag{30}$$

$$\frac{\partial p_{tk}}{\partial t} = 0, \quad if\ \frac{\partial Z_b}{\partial t} < 0\ and\ N_{dl} \ge N_{dl0} \tag{31}$$

$$p_{tk} = p_{Ndl}, \quad if\ \frac{\partial Z_b}{\partial t} < 0\ and\ N_{dl} < N_{dl0} \tag{32}$$

$$\frac{\partial p_{Ndl}}{\partial t} = 0, \quad if\ N_{dl} \ge N_{dl0} \tag{33}$$

$$p_{Ndl} = \frac{T_t}{T_{dl1}} p_{tk} + (1 - \frac{T_t}{T_{dl1}}) p_{bk}, \quad if\ \frac{\partial Z_b}{\partial t} > 0\ and\ N_{dl} < N_{dl0} \tag{34}$$

$$p_{Ndl} = p_{Ndl-1}, \quad if\ \frac{\partial Z_b}{\partial t} < 0\ and\ N_{dl} < N_{dl0} \tag{35}$$

where Z_{bk} = bed elevation of kth grain eroded or deposited after time interval Δt, $\sum Z_{bk} = Z_b$, p_{bk0} = percentage of kth grain in transition layer or top layer of deposit strata when the bed surface is eroded ($\partial Z_b/\partial t < 0$), δ = identification factor, $\delta = 1$ if $\partial Z_b/\partial t \ge 0$, $\delta = 0$ if $\partial Z_b/\partial t < 0$, Z_0 = reference bed datum, T_m, T_t, and T_{1dl} = thickness of mixing layer, transition layer, and one deposition layer, respectively, T_{t0} = thickness of transition layer at previous time, N_{dl} = number of the deposition layers, N_{dl0} = number of the deposition layers at previous time, and p_{tk}, p_{Ndl}, and p_{Ndl-1} = percentages of kth grain, in the transition layer, top deposition layer and second deposition layer, respectively. It should be noted that the bed load transport is connected to the mixing layer and transition layer through the changing T_m and p_{bk} during deposition and linked to the mixing layer, transition layer, and deposition strata through T_m, p_{bk}, and p_{bk0} during erosion.

3.3. Initial and Boundary Conditions

The governing equations are discretized by the finite volume method and the sediment transport equations are solved by a tridiagonal matrix algorithm method after the flow equations are computed by making use of a standard step-by-step procedure to determine the flow profiles in the subcritical and supercritical flow reaches. In order to secure the numerical stability, a short Δt was used to make the maximum depth degraded or aggraded in each step being less than 0.25m. The mixing layer, transition layer, and deposition strata are set to be equal to 0.5m. The measured discharges at the Dashidaira dam were used for the input conditions for the flow equations. The main parameters such as the Manning roughness and incipient motion discharges and formulas of the inflowing wash, suspended, and bed loads were calibrated by means of the field measurement data from 1985 to 2001. The wash loads are defined as the particles with a grain diameter less than 0.2 mm in this study. The suspended sediment concentrations at the upstream sections are given by the Lane and Kalinske (1941) formula:

$$C_k / C_{ek} = F_{cs} \exp(-15(z/h - 0.05) \mathrm{W_{sk}} / \mathrm{u_*}) \tag{36}$$

where F_{cs} = correction factor that is determined by the measurement data and flood scale, and z = distance from river bed.

The sediment concentration distribution at the downstream section is as follows:

$$\frac{\partial C_k}{\partial x} = 0 \tag{37}$$

The initial grain size distributions of the bed materials are calculated values in December 2000. The initial riverbed elevation from the Dashidaira dam to the section 3150m upstream of the dam is based on the measured value in December 2000. The initial riverbed elevation from the section 3150m upstream of the dam to the upstream section is calculated values in December 2000 because of no measured data in the reach is based on the measured value in December 2000. The intervals of cross section of simulation are set to be 100 to 200m.

The sediment supply at the inlet section of the Dashidaira reservoir is obtained by the uniform flow method and the sediment transport formulae where the bed slopes are 1/55 from the measured values, roughness coefficient is 0.03, and the grain size distributions of the bed materials are determined on the basis of the survey results. The bed load is computed by the Ashida and Michiue formula, and the suspended load is computed by the Lane-Kalinske formula. The concentration near the riverbed is computed by the Ashida and Michiue formula. The wash load at the inlet section of the Dashidaira reservoir is determined on the basis of the measured the suspended sediment (SS) value of turbid water from 2001 to 2006 by the function of discharge. The bed material is divided into 10 classes by size, the representative grain sizes of the bed load and suspended load are 361.2mm, 118.3mm, 37.42mm, 11.83mm, 3.742mm, 1.183mm and 0.374mm. The representative grain sizes of the wash load are 0.1mm, 0.0158mm and 0.0022mm (Minami and Fujita 2008).

The initial riverbed upstream of the check dam is set as the state without deposits. The water level at the check dam is determined by the critical water level.

3.4. Calculated Results and Discussions

Figure 6 shows the measured and calculated bed profiles in the Dashidaira reservoir in December 2006 without the check dam (Case 1) and with the check dam (Case 2). The calculated profile of Case 1 corresponds well with the measurement in December 2006. The reservoir of the check dam is fully filled in two years from 2001 to 2002. The bed elevation of the Dashidaira reservoir in the reach from 1600m to 3200m upstream of the Dashidaira dam of Case 2 is lower than the one of Case 1, and it is favorable to the power station whose intake is located in the reach.

Figure 7 shows the variations of the bed elevation at the section 2710m upstream of the Dashidaira dam from December 2000 to December 2006 for Cases 1 and 2. It can be found that construction of the check dam reduces the bed elevation rising with a maximum value of 3m. Figure 8 indicates the calculated sediment deposited in the reach from 2580m to 3150m upstream of the Dashidaira dam for Case 1 and 2. The deposits of Case 2 are much less than the ones of Case 1, and the deposits in six years of Case 2 are only 40% of the ones of Case 1.

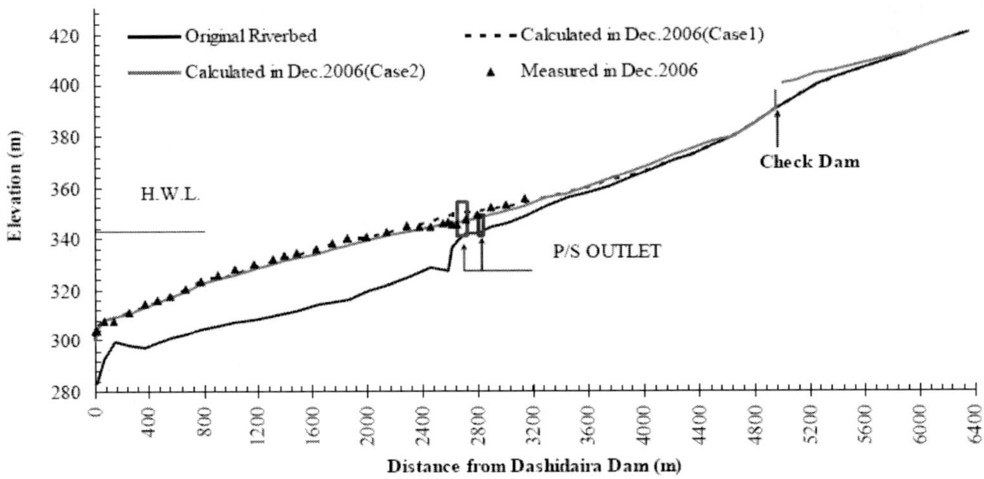

Figure 6. Measured and calculated bed profiles in the Dashidaira in December 2006.

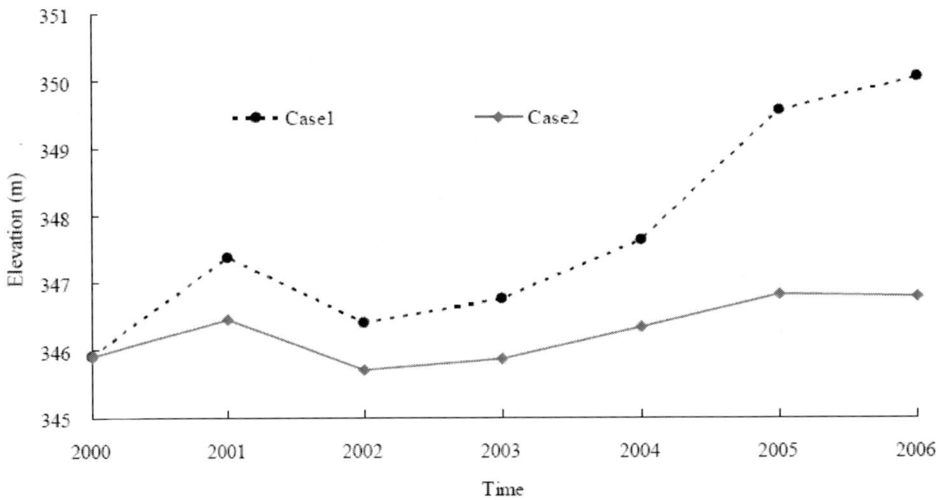

Figure 7. Variations of the bed elevation at the section 2710m upstream of the Dashidaira dam.

Figure 9 shows sediment amount of each rain size flowing through the section 3150m upstream of the Dashidaira dam from December 2000 to December 2006. The coarse sediment more than 11.83mm is controlled by the check dam, therefore, the coarse sediment of Case 2 is smaller than the one of Case 1. Figure 10 shows the water discharge, inflow and outflow sediment discharges of the grain size of more than 11.83mm at the check dam in July 2006.

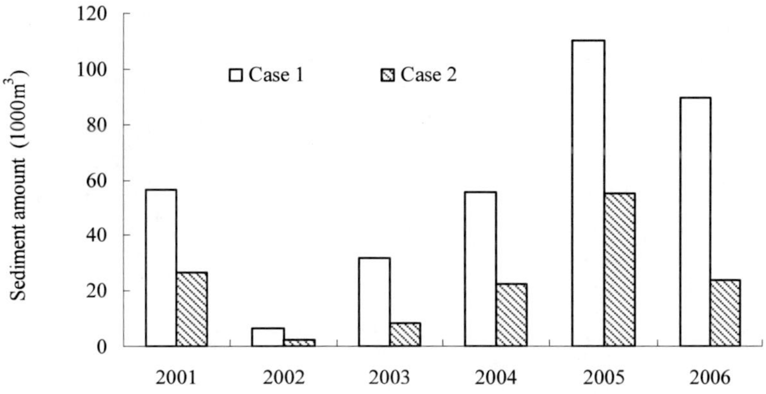

Figure 8. Deposits in the reach from 2580m to 3150m upstream of the Dashidaira dam.

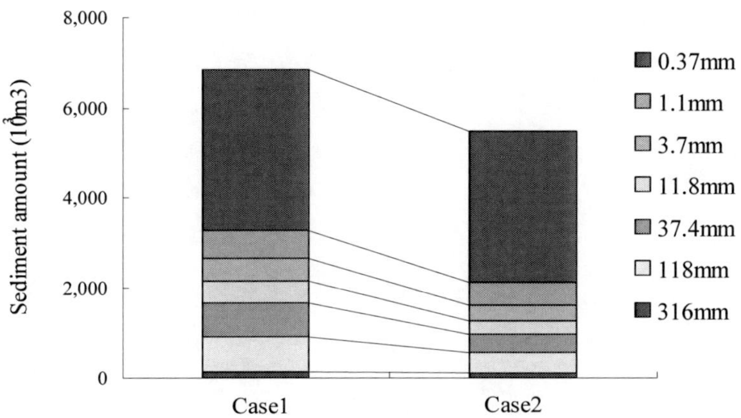

Figure 9. Sediment amount of each rain size flowing through the section 3150m upstream of the Dashidaira dam.

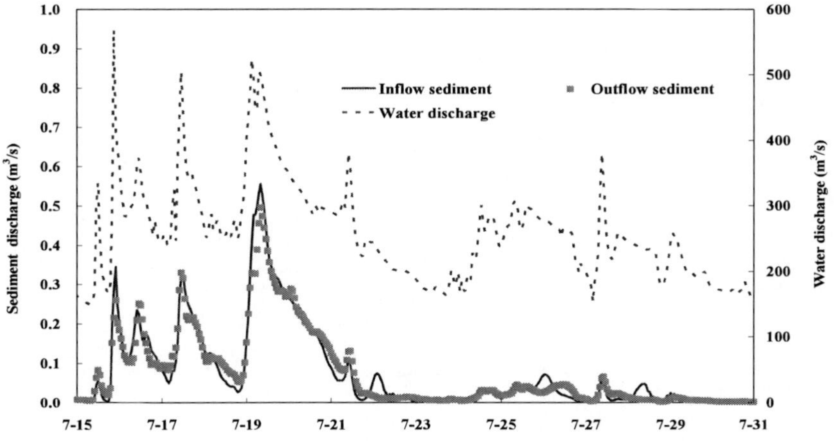

Figure 10. Inflow and outflow sediment discharges of the grain size of more than 11.83mm at the check dam in July 2006.

As compared with the sediment hydrographs, the peaks of the outflow sediment hydrograph is smaller than the ones of the inflow sediment, it means the check dam regulates the sediment flowing down to the downstream.

The above-mentioned simulation results indicate the check dam could refrain from the bed elevation rising at the end of the Dashidaira dam. If the reservoir capacity of the check dam is large or a lot of the check dams are built in the upper reaches, the check dams would have large influences on the decrease of the sedimentation in the downstream channel.

4. CASE STUDY

4.1. Sakuma Dam

The Sakuma dam is located at the site 71km upstream of the estuary of the Tenryu River on which seven large dams were constructed as show Figure 11.

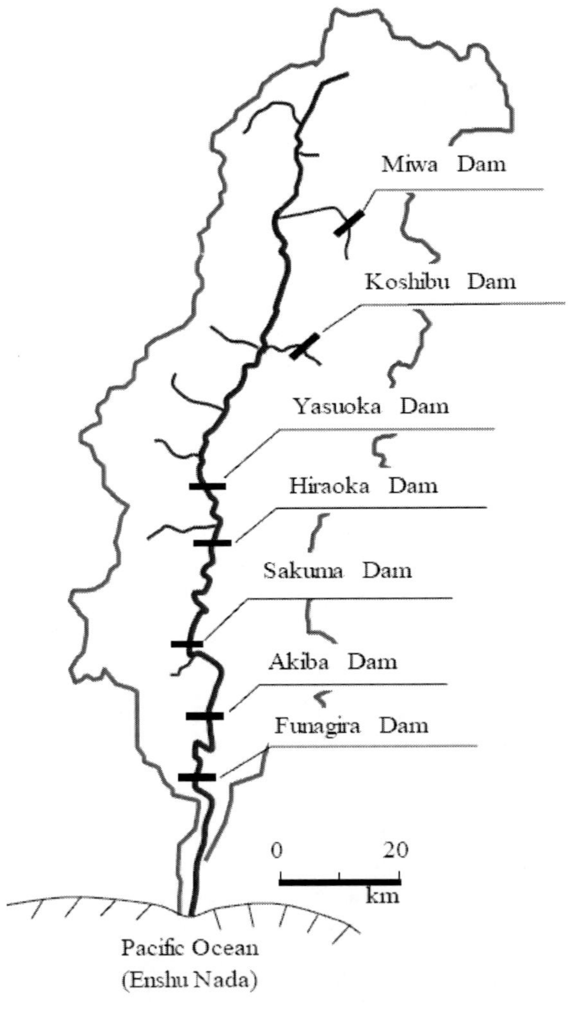

Figure 11. Location of the Sakuma dam.

The Sakuma dam was completed in 1956 and managed by the Electric Power Development Co., Ltd. The reservoir capacity is 320million m^3. Figure 12 shows the variation of the sedimentation amount in the Sakuma reservoir, and Figure 13 shows the sedimentation variations in the longitudinal profile of the Sakuma reservoir. The total sedimentation amount was 120million m^3 in 2006 and the sedimentation rate was 35%. Most of annual inflowing sediment of 2.5 millionm^3 is deposited into the reservoir. The sedimentation has affected the water withdrawal and outflow, and made the riverbed rising in the upper river channel and costal shore erosion.

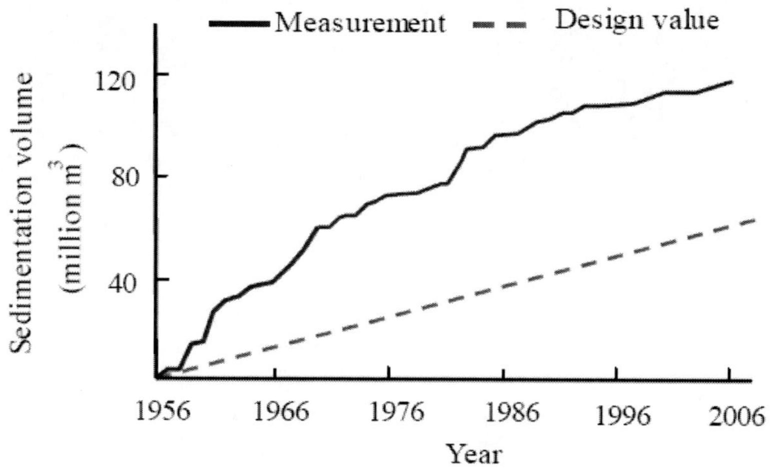

Figure 12. Variation of the sedimentation amount of the Sakuma reservoir.

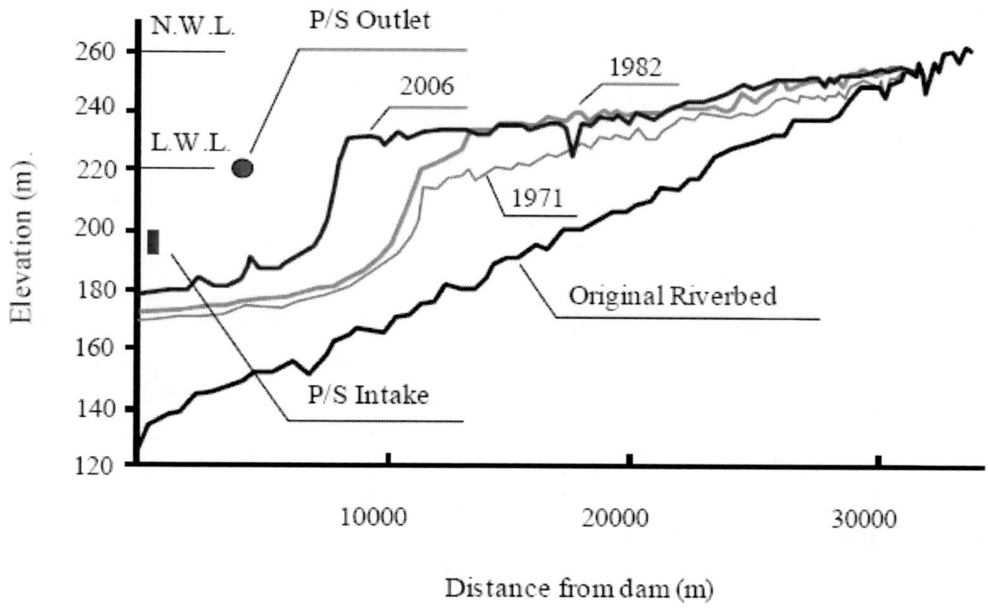

Figure 13. Profile of the Sakuma reservoir sedimentation.

Figure 14. No.6 Iijima check dam on the Yotagiri River.

Some measures such as construction of check dams in upper reaches, dredging and low water level operation have been performed to reduce influence of sedimentation. A lot of the check dams have been constructed in the tributary basins of the Tenryu River. The main types of the check dams include concrete, split, stone masonry and steel cell dams. Figure 14 shows the No.6 Iijima check dam on the Yotagiri River which is a tributary of the Upper Tenryu River (Upper Tenryu River Office of MLIT 2006). The No.6 Iijima check dam located at the site 10km upstream of the confluence of the Yotagiri River with the Tenryu River is 40m high concrete dam with a crest length of 114m. The check dam completed in February 2006 is used to reduce the sediment disaster and sedimentation problems of the reservoirs on the Tenryu River such as the Sakuma and Funagira dams.

4.2. Dashidaira Dam

The Dashidaira dam on the Kurobe River located in the Hokuriku region in central Japan as shown in Figure 15. It was completed in 1985 and managed by the Kansai Electric Power Co., Inc. The reservoir has a gross capacity of 9.01 million m^3. Two large flushing outlets with a cross section of 5m×5m are equipped in the 76.7m high gravity dam and each flushing outlet has a capacity of half of annual sediment inflow of 0.5 million m^3. The catchment area of the Kurobe River is 682km^2, and the length of the main river is 85km.The gradient of the river bed is 1/5 to 1/100 which means one of the steepest in Japan. The geological condition in the river basin is characterized by weathered granite, and the sediment production potential is very high. There are a lot of slope collapses in the river basin. The annual rainfall in the basin is about 4,000mm, and there is a great amount of sediment flowing into the reservoirs. About 10 times of annual sediment flowed into the Dashidaira reservoir when a large-scale flood occurred in the Kurobe River in July 1995. The sediment of more than 3 million m^3 was deposited in the reservoir as shown in Figure 16.

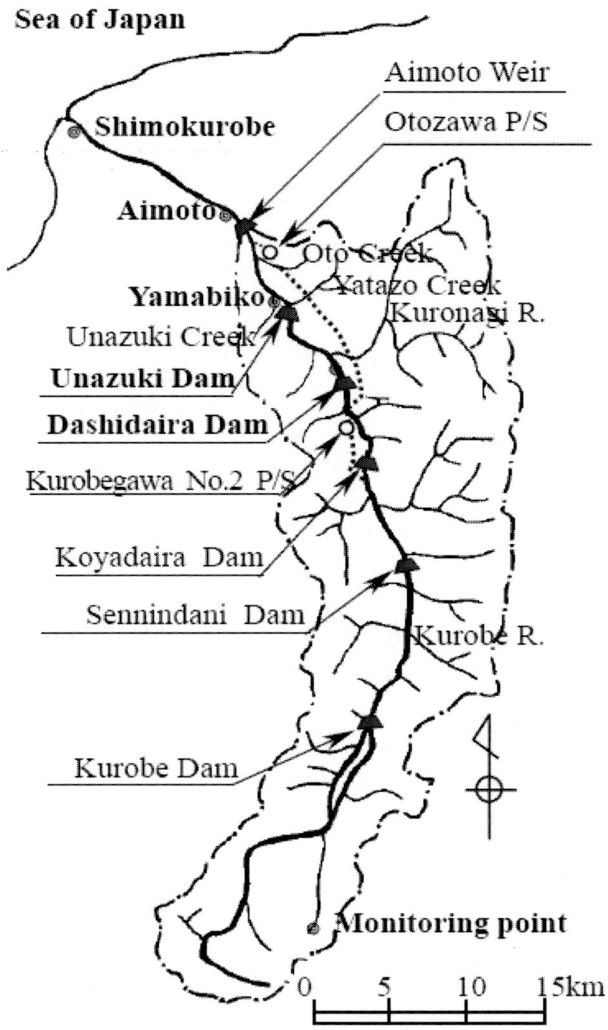

Figure 15. Location of the Dashidaira dam.

The riverbed rose 10m, and caused serious damage to the power station, apartments and railways (Sumi and Kanazawa 2006). Sediment flushing at the Dashidaira dam has been carried out since 1991, seven years after completion, at the time the accumulated sedimentation reached 3 million m^3. Since 2001, the sediment flushing has been performed in concert with the Unazuki dam 7km downstream of the Dashidaira dam. The flushed sediment and measured data on the SS concentrations and dissolved oxygen (DO) contents at the Yamabiko Bridge just downstream of the Unazuki dam are given in Table 1. The storage capacity is maintained at 5 million m^3 by means of flushing operation. The maximum SS values are relatively high but the minimum DO values are allowable as compared with the standard value of 4mg/l at the Yamabiko Bridge. In addition, on the basis of the measurements in July 1999, the abrasion on the bottom of the flushing outlet near the left bank is much severer than that in the flushing outlet near the right bank, and the maximum abrasion depth on the bottom of the flushing outlet near the left bank is 9mm and average depth is at 4.9mm (Liu et al. 2002).

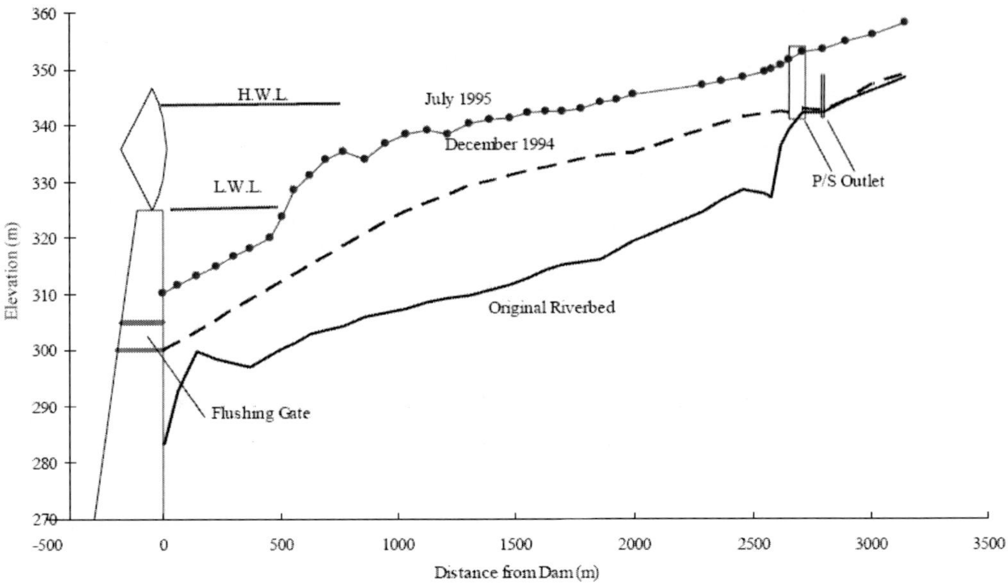

Figure 16. Sedimentation profile of the Dashidaira reservoir in 1995.

Table 1. Sediment flushing at Dashidaira and Unazuki reservoirs

Date	Flushed sediment (10^6m^3)	Free flow time (hr)	Peak inflow during flushing (m^3/s)		Min. DO at Yamabiko Bridge (mg/l)	Max. SS at Yamabiko Bridge (mg/l)
Dec. 11-22, 1991	0.46	30	50-90	Water from Kurobe Dam		
Feb.27-28, 1994	0.08	1	40	Water from Kurobe Dam		
July 8-10, 1995	0.02	0	-	Flood	11.3	3700
Qct.27-31, 1995	1.72	12	200	Water from Kurobe Dam	9.7	29400
June 27- July 1, 1996	0.80	29	1052.5	Flood	10.3	9470
July 9-13, 1997	0.46	48	304.6	Flood	9.2	28900
June 28-30, 1998	0.34	24	318.4	Flood	7.0	9400
Sep. 15-17,1999	0.70	24	341	Flood	5.8	52100
June 19-21, 2001	0.59	26 (12 at Unazuki)	369 (477 at Unazuki)	Flood	11.4	2500
June 30-July 2, 2001	-	12 (12 at Unazuki)	491 (500 at Unazuki)	Flood	10.6	3700

In addition to the sediment flushing at the dam, a lot of check dams, most of which are split type as shown in Figure 17, have been built in the basin. However, only 22% of the planned check dams have been constructed until 2006. Therefore, it is necessary to construct more check dams to reduce sediment amount flowing into the Dashidaira reservoir (Nishikawa and Hashinoki 2006).

Figure 17. No. 2 Nobosetani check dam in Kurobe River basin (The height of four steel pipes is 4m, and completed in 1979).

4.3. The Sanmenxia Dam

The Sanmenxia dam is located in the lower part of the middle reach of the Yellow River. The construction of the Sanmenxia dam was initiated in 1957, and water impoundment commenced in September 1960. The crest elevation of the dam is 353m, and the original capacity of the reservoir is 9.705 billion m^3 with a pool level of 335m. The reservoir area extends upstream a distance of 246km to the Longmen Hydrological Station. The Yellow River has a drainage area of 795,000km^2 and a length of 5464km making it the second longest river in China. The long-term annual sediment load at the Sanmenxia Station was 1.6 billion tons before 1980, ranking first of all the world's rivers, although the sediment load has reduced greatly in the past 20 years. The huge amount of sediment is mainly from the loess plateau in the middle reaches of the river.

Severe sedimentation problem became evident immediately after impoundment in 1960. During the first 18 months, 93% of the incoming sediment load was trapped in the reservoir and caused 17% of the reservoir capacity loss below El. 335m or 26% of the capacity loss below El. 330m. The river bed rose about 5.5m at the Tongguan Station in October 1961. To mitigate the sedimentation, the operation scheme was changed to detain flood water in flood season. However, the flood-releasing capacity of the outlet structures was limited. In order to increase the capacity to discharge sediment from the reservoir, reconstruction of outlet structures was carried out in two stages during the period from 1964 to 1973. Ten bottom outlets were reopened, thus, sediment deposits can be flushed from the reservoir through these outlets with the lowest elevation at 280m. Three reservoir operation modes at the Sanmenxia reservoir were developed and used, i.e., 1) storage from September 1960 to March 1962, the reservoir was operated at a high storage level in the whole year; 2) detaining flood water and sluicing sediment from March 1962 to October 1973, the reservoir was operated at a low storage level throughout the year, detaining floods only during flood season, and sluicing sediment with the largest possible discharges; and 3) storing clear water and releasing turbid water from November 1973 to the present, the reservoir has been operated at a high level (315–320m) to store relatively clear water in nonflood season (November–June), and at a low level (302–305m) to release high sediment concentrations in flood seasons (July–October). Moreover, the bottom outlets have also been used to discharge high concentration density currents. As the concentration is higher than 200kg/m^3, density current occurs in the reservoir. The density current may be directly released from the reservoir though the bottom outlets even if the pool level remains high (Xu and Cheng 2002). The third operation scheme is called as mode of storing clear water and sluicing muddy water in China, which has been used for many dams such as the Gezhouban dam and the Three Gorges Project on the Yangtze River.

During the flood in August 2001, the Sanmenxia reservoir flushed sediment of 165 million m^3, in which inflow sediment was 115 million m^3 and the deposited silt was 50 million m^3. The maximum flushing discharge and sediment concentration are 2890m^3/s and 492kg/m^3 at the damsite in response to the maximum inflowing discharge of 2750m^3/s and sediment concentration of 432kg/m^3 at the Tongguan gauging station. The riverbed at the gauging station was eroded about 1m. The flushing operation caused a density current at the downstream Xiaolangdi reservoir, which was timely passed through the flushing outlets (Liu, et al. 2002). 43% of the Yellow River's sediment load is from the Loess plateau region in the middle reaches with forest cover of lower than 5%, which only accounts for 7% of the whole basin. The Yellow River basin is arid and semi-arid and, therefore, reforestation alone is difficult to achieve the objective of erosion control. Over 100,000 check-dams have been built on the Loess Plateau since 1960s to reduce the sediment flowing into the Sanmenxia reservoir and create productive warped farmland. Approximately 100,000 check-dams in the Yellow River basin were built in the late 1960s and the 1970s. The progress to construct the check dams was slow in 1980s and 1990s, and only 1118 check dams had been built from 1986 to 1999 in the Loess Plateau (Feng 2000). However, the progress to construct the check dams and reforestation after 2000 is fast, and over 1,800 check dams were constructed on the Loess Plateau in 2004 (MWR 2006). Until the present, there still are 6758 gullies in the Loess Plateau that are suitable for building check dam, and 242,000 check-dams are planning to be built in the future (Huang 2000).

The survey results of 4877 check dams in the drainage basin of the Huangpuchuan, Kuyehe, Jialuhe,Tuweihe and Dalihe rivers which are parts of tributaries of the Yellow River upstream of the Sanmenxia dam in 1992 show that the sedimentation velocity, sediment blocking index and sediment blocking effect are related to the scale of the check dam and the basin erosion characteristics. The annual average sedimentation volume per check dam was 9000 to 24000 m^3, and the annual average silting land per check dam was 0.14 to 0.45 ha. They are positively related to the height and control area of the check dam, erosive sediment modulus and the grain size of sediment in the control area. The sediment blocking index varied from 50 to 76 million ton/km^2, and it has positive relationship with the height of check dam, and inverse relationship with the grain size. The sediment blocking effect of the check dam ranged from 23.3% to 52.9%, and has positive relationship with check dam height, inverse relationship with the control area and the modulus of coarse sediment with mean diameter of more than 0.05 mm (Jiao et al. 2003).

Figure 18. Typical check dam in the Sanmenxia reservoir basin.

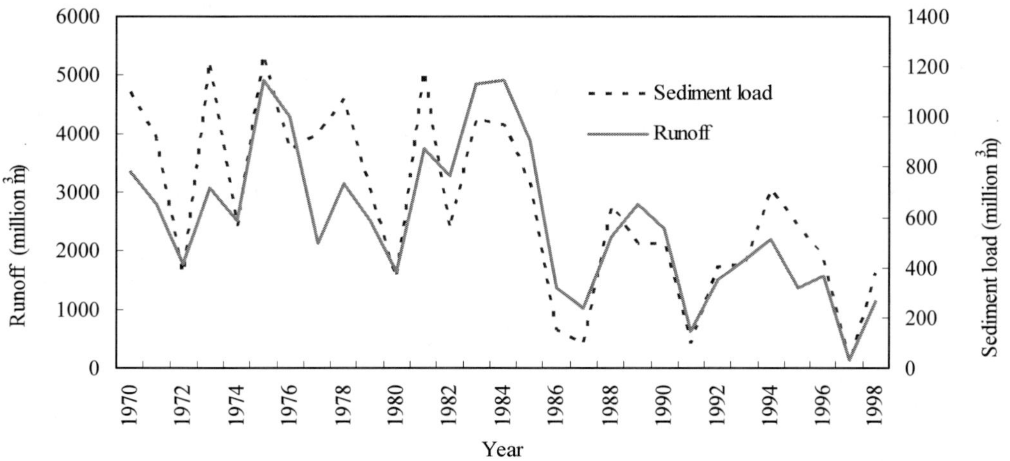

Figure 19. Variation of annual runoff and sediment load of the Yellow River at the Lijin hydrological station.

It should be noted that the most of the check dams in the Yellow River basin are earth or rockfill dams. Figure 18 shows a check dam in the Sanmenxia reservoir basin located in Pianguan County, Xinzhou City, Shanxi province. The check dam was built and managed by local farmers who plant the wheat and corn crops on the deposits in the reservoir in the spring and autumn.

As a result, the sediment fed into the Yellow River by erosion from the loess plateau has been greatly reduced since 1984. Figure 19 shows the annual water and load of the Yellow River in the period from the 1950s to 1990s measured at the Lijin hydrological station (Wang and Liang 2000). The reduction in sediment load in the 1980s and 1990s is due mainly to sediment trapping by the reservoirs on the Yellow river and the check dams in the basin.

Conclusion

First of all, the effects of the check dams to reduce the reservoir sedimentation are addressed in qualitative analysis. Secondly, the effects of the check dams are analyzed in detail by numerical simulation, taking as an example the Dashidaira reservoir on the Kurobe River in Japan. The calculated results show the check dam can decrease the sedimentation amount and keep the bed elevation from rising at the end of the reservoir. It can be found that the numerical simulation provides a useful tool to evaluate the effects of the check dams, and this method can also be used for solving the associated problems. Finally, the sedimentation measures at the Sakuma and Dashidaira dams in Japan and the Sanmenxia dam in China are introduced. Because there are few measured data on the effects of the check dam to reduce the reservoir sedimentation, it is further necessary to collect measurements of the check dams to understand their functions in the reservoir sedimentation control.

Acknowledgments

The project was supported by the Shenzhen Science and Technology Program (No. 200739) and Fund of Land Investment and Development Centre of Shenzhen Works Bureau for Rainfall Usage Study of the Longhua No. 2 Line Project (Contract No. 2007139). The authors wish to acknowledge Prof. Carmelo Conesa-García of the University of Murcia for review.

References

Ashida, K., Egashira, S. and Nakagawa, H. (2008). *River morphodynamics for the 21st century*. Kyoto, Japan: Kyoto University Press, 169-188 (in Japanese).

Ashida, K.and Michiue, M. (1970). Study on the suspended sediment (1): Concentration of the suspended sediment near the bed surface. *Disaster Prevention Research Institute Annuals, Kyoto University*, 13B, 233-241 (in Japanese).

Ashida, K. and Michiue, M. (1972). Study on hydraulic resistance and bedload transport rate in alluvial streams. *Proceedings of JSCE*, 206, 59-69 (in Japanese).

Ashida, K., Takahashi, T. and Michiue, M.(1983). *Sediment disaster and measures*. Tokyo, Japan: Morikita Press, 157.

Chanson, H. (2004). Sabo check dams - Mountain protection systems in Japan. *Journal of River Basin Management, 2(4)*, 301-307.

Egashira, S. (2001). Method to predict sediment transport process. *Lecture note for summer seminar, Hydraulics and Coastal Engineering Committee, JSCE*, A2, 1-14. (in Japanese)

Egashira, S., Itoh, T., Horie, K. and Nishimoto, N. (2007). A method to predict sediment transport process in drainage basins with dams. *River, Coastal and Estuarine Morphodynamics, RCEM 2007*, London, UK: Taylor and Francis, 1193-1201.

Engelund, F. (1967) Closure to discussion on hydraulic resistance of alluvial streams. *Journal of Hydraulic Division, ASCE*, 93(4), 287-296.

Feng, G. (2000). The key to bring the Yellow River under control is to speed up construction of check-dams in the sandy and grit areas. *Science and Technology Review*, 7, 53–57 (in Chinese).

Huang, Z. (2000). Practice and Plan for Construction of Dam System Ecological Project in Gullies in the Loess Plateau. *Soil and Water Conservancy in China* 10, 1–4 (in Chinese).

Iwagaki, Y. (1956). Study on critical tractive force. *Proceedings of JSCE*, 41, 1-21 (in Japanese).

Iwashita, O., Shioda, K. and Kitamura, Y. (1988). The mechanism of sediment erosion and turbidity in reservoir. *Electric Power Civil Engineering, Japan Electric Power Civil Engineering Association*, 212, 93-103 (in Japanese).

Jiao, J-Y., Wang, W-Z., Li, J. and Zheng B-M. (2003). Silting land and sediment blocking benefit of check dam in hilly and gully region on the Loess Plateau. *Transactions of the Chinese Society of Agricultural Engineering*, 6, 302-306 (in Chinese).

Lane, E. W. and Kalinske, A. A. (1941). Engineering calculations of suspended sediment. *Transaction of American Geophysics Union*, 22, 603-607.

Liu, J. Liu, B-Y. and Ashida, K. (2002). Reservoir Sedimentation Management in Asia. *Advances in Hydro-Science and Engineering (CD-ROM), the 5th International Conference on Hydro-Science and Engineering (ICHE-2002)*, Warsaw, Poland, September 18-21.

Liu, J., Minami, S., Otsuki, H., Liu, B-Y. and Ashida, K. (2004). Prediction of concerted sediment flushing. *Journal of Hydraulic Engineering, ASCE, 130(11)*, 1089-1096.

Minami, S. and Fujita, M.(2008). The Sediment Responses in the Downstream by Sediment Flushing of the Reservoirs. *Proceedings of River Engineering, JSCE,Vol.14*, 259-264.

Ministry of Water Resources (MWR). (2006). *2004-2005 Annual Report*. General Office of MWR, April 4, http://www.chinawater.gov.cn/english1/20060404/69725.asp

Nishikawa, H. and Hashinoki, T. (2006). The sediment system of the Kurobe River. *Journal of the Japan Society of Erosion Control Engineering, Vol.59 No.2*, 64-72.

Sumi, T. and Kanazawa, H. (2006).Environmental Study on sediment Flushing in the Kurobe River. *The 22nd International Congress on Large Dams,* Q85-R16, Barcelona, 219-242.

Upper Tenryu River Office of MLIT. (2006). Performance of the check dams to protect the lives in the Yotagiri River. *Construction Club*, 178, 11-11(in Japanese).

Wang Z-Y. and Liang Z-Y. (2000). Dynamic characteristics of the Yellow River mouth. *Earth Surface Process and Landforms, 25*, 765-782.

Xu, J and Cheng, D. (2002). Relation between the erosion and sedimentation zones in the Yellow River. *Geomorphology, 48 (4)*, 365-382.

In: Check Dams, Morphological Adjustments…
Editors: C. C. Garcia and M. A. Lenzi

ISBN: 978-1-62618-856-3

Chapter 7

MANAGING STREAM MORPHOLOGY WITH CHECK DAMS

Ellen Wohl[1], Anne Chin[2], Jeffrey P. Haltiner[3] and G. Mathias Kondolf[4]

[1]Department of Geosciences, Colorado State University,
Ft. Collins, CO 80523-1482 USA
[2]Department of Geography, University of Oregon,
Eugene, OR 97403-1251 USA
[3]Phillip Williams and Associates, Ltd., 550 Kearny Street,
Suite 900, San Francisco, CA 94108, USA
[4]Department of Landscape Architecture and Environmental Planning,
University of California, Berkeley, CA 94720, USA

ABSTRACT

Check dams in steep streams create local base levels that limit channel erosion, store sediment, and dissipate flow energy. When designed to mimic naturally formed step-pool sequences with an appropriate ratio of *H/L/S* (dam height/downstream spacing/streambed gradient) and natural cobble and boulder clasts, check dams can enhance channel stability as well as diversity of aquatic habitat.

Conversely, poorly designed check dams can exacerbate erosion below the dam, in extreme cases resulting in failure of the check dam and excess sediment yields downstream. Inappropriate spacing and height of the check dams can also restrict the longitudinal movement of aquatic organisms.

This chapter describes the dynamics of steep streams and reviews the literature that pertains to the morphological effects of check dams. Following a brief summary of the design process, it illustrates a case study of the restoration of Martin Canyon Creek in northern California, USA, completed in 1999, that utilized check dams. The chapter concludes with recommendations for using check dams as management tools in steep stream networks.

INTRODUCTION

High-gradient streams in the world's mountainous environments are particularly prone to episodic channel instability. Steep hill slopes that are readily destabilized by precipitation inputs, short transport paths from hill slopes to valley bottoms for water and sediment, and narrow valley bottoms with limited ability to store or attenuate increased water and sediment volumes, all facilitate rapid and large-magnitude fluctuations in water and sediment supplied to stream channels. The resulting erosion and deposition along mountain streams can create hazards to people living in the vicinity, as well as in the communities downstream. As a result, people in Europe and Asia have been engineering mountain streams to reduce these hazards for centuries (Japanese Ministry of Construction, 1993; Jaeggi and Zarn, 1999). In addition to the climatic and morphological causes described above, human activities are a major factor initiating channel instability in mountainous regions of the western United States and elsewhere (Wohl, 2000). Agriculture (grazing and farming), logging, and urbanization are all linked to rapid channel incision, and the need for channel stabilization (Haltiner et al., 1996).

Check dams are among the structures most widely used to stabilize mountain streams and reduce flood hazards. These structures can be designed to: create a local base level that limits upstream migration of bed incision and associated bank destabilization; store sediment and limit deposition and overbank flooding or channel avulsion downstream; or enhance profile irregularity and associated dissipation of flow energy (Lenzi, 2002; Castillo et al., 2007; Shieh et al., 2007). Open check dams are designed to trap large sediment clasts while permitting water and finer sediment to pass downstream (Figure 1a). Closed check dams, which are typically made of concrete, rock (grouted or loose), or gabions (wire cages filled with rock), are traditionally arranged in a cascade, analogous to a natural step-pool sequence (Figure 1b) (Wohl, 2000).

More recently, construction of closed check dams in the European Alps have utilized large boulders, with some concrete dams partially buried beneath the boulders to enhance the ecological and esthetic value of the dams (Lenzi, 2002). Although the selection of the location and design of check dams has mostly focused on hazard reduction, ecological and esthetic considerations (hereafter referred to as environmental value) are becoming increasingly important. Thus, small-scale check dams designed to mimic natural step-pool sequences are increasingly being used for stream restoration in steep channels (Chin et al., in press a).

This chapter reviews the morphological effects of check dams on steep channels, including empirical relations that describe stable sequences of check dams, derived from analogies with natural step-pool sequences. We then present examples of successful design of stream restoration projects incorporating small check dams, focusing on the case of Martin Canyon Creek in the city of Dublin in northern California, USA.

We conclude with recommendations for managing process and form along steep channels using check dams. We focus on regions other than the European Alps and arid rangelands to provide geographic representation in light of the material presented in other chapters in this volume.

Figure 1a. Example of an open check dam designed to allow finer sediment to pass through the structure. This dam, along the Hopfgartnergraben, a small mountain stream in Austria, is full of sediment.

Figure 1b. A sequence of three closed check dams along the Alptal in Switzerland. Note the small plunge pool at the base of the lowest dam, and the gravel bar just downstream.

Morphological Effects of Check Dams on Steep Channels

At the most fundamental geomorphic level, check dams create a local base level that promotes sediment deposition upstream from the structure, and a vertical drop that induces scour immediately downstream. Transverse depositional features such as boulder berms are typically included at the downstream end of the scour hole and upstream of the next check dam. The installation of check dams can also produce changes in channel cross-sectional geometry, grain-size distribution, bedforms, and bed slope (Martín-Vide and Andreatta, 2009). The details of these changes vary with each site as a result of the interactions among multiple control variables. These control variables include height and longitudinal spacing of successive check dams, water and sediment supply in the channel, and substrate resistance of the channel (Lai et al., 1998). Sediment retention behind check dams can facilitate channel narrowing, armouring, and/or incision downstream (Boix-Fayos et al., 2007). Decreased flow velocity upstream from the check dam can cause substrate fining and associated changes in stream habitat and trophic structure (Shieh et al., 2007) (Figure 2). By creating unnaturally high barriers, check dams can also potentially limit longitudinal movement of stream organisms (Jaeggi and Zarn, 1999; Shieh et al., 2007).

Where check dams are widely spaced along channels with relatively low gradient, local scouring can form a pronounced pool below each dam while deposition slightly raises the bed downstream (Lenzi and Comiti, 2003). The intervening stream segment between successive check dams will maintain a slope that facilitates subcritical to critical uniform flow and equilibrium conditions for the particular sediment supply (Lenzi et al., 2003a). In steeper streams, check dams can be closely spaced such that each check dam is at the end of the pool created by the structure immediately upstream. Flow undergoes repeated transitions from sub- to supercritical in these cases (Comiti et al., 2005).

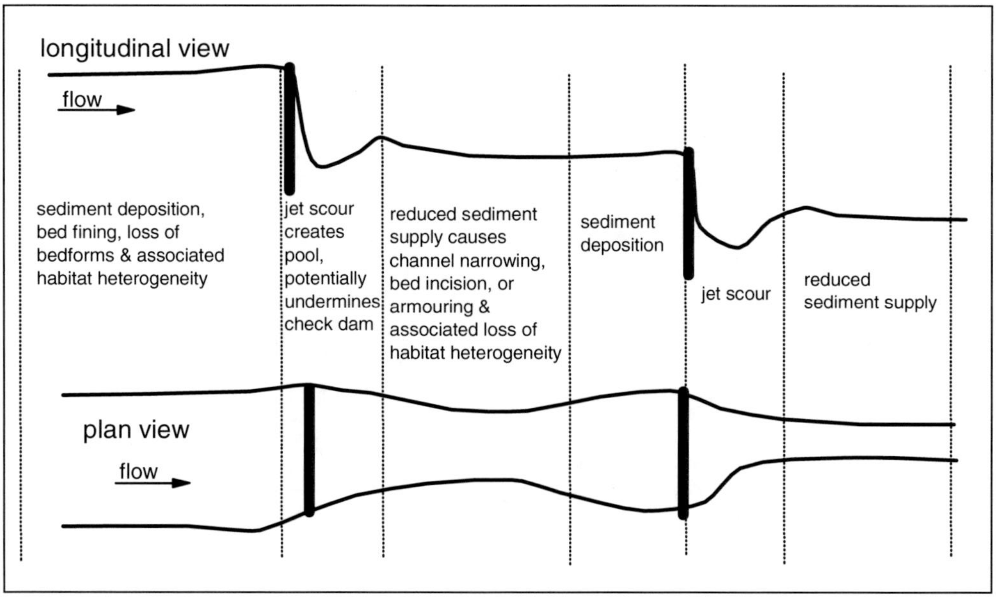

Figure 2. Schematic longitudinal and plan view illustration of morphological effects of check dams.

Closely spaced dams, where the ratio of scour length to downstream distance between dams (l_s/L) is between 0.4 and 0.67, can interfere with pool scour and reduce the maximum depths and scour lengths of pools (Lenzi and Comiti, 2003). The height and longitudinal spacing of check dams thus exert a primary control on morphological adjustments in steep streams. Variability in flow and sediment supply exerts a secondary control, as do streambed substrate and original channel morphology.

As a first approximation, the greater the difference between the "natural" streambed gradient and the slope created by a check dam, the greater will be the morphological effects of the check dam. If the check dam is being installed primarily to store sediment, substantial reductions in gradient may be desirable to facilitate sediment deposition. If the check dam is part of a stream restoration designed to maintain or enhance channel stability and environmental value, then a dam geometry that minimizes changes in gradient is likely preferable. This geometry is most likely achieved through a series of lower dams rather than a single tall structure.

Self-Formed Step-Pool Sequences as Analogs for Stable Check Dams

Naturally formed step-pool sequences provide an analog for designing stable check dams along steep streams because step-pool sequences create the same type of stepped longitudinal profile as check dams, yet are self-maintaining. Step-pool sequences in which the ratio of step height (*H*) to longitudinal step spacing (*L*) and streambed gradient (*S*) is between 1 and 2 create maximum energy dissipation, and most natural, stable step-pool bedforms exhibit this ratio (Abrahams et al., 1995):

$$1 \leq H/L/S \leq 2 \qquad (1)$$

Step geometry that maximizes energy dissipation likely reflects the conditions within steep channels between 3 and 20% gradient along which step-pool sequences typically form. Although the sediment transport capacity in these channels can be episodically overwhelmed by abundant sediment supply from flash floods or debris flows, the channels are typically supply-limited with respect to finer sediments (Chin and Wohl, 2005). High streambed gradient and coarse-grained substrate of limited mobility combine to create highly turbulent flow (Wohl and Thompson, 2000; Wilcox and Wohl, 2007). Relatively frequent flows effectively remove exposed finer sediment from most portions of the channel boundary, leaving only the coarsest material and relatively well-defined steps and pools. Infrequent larger flows that mobilize the entire stream substrate, or debris flows, can result in such high sediment transport that pools either partially or completely fill, creating a hydraulically smoother bed. Subsequent lower flows then selectively remove the finer sediment and re-create the step-pool sequence (Lenzi et al., 1999). Warburton (1992) designated three phases of sediment transport in step-pool channels: low-flow flushing of fines; frequent high-flow mobilization of pool-filling gravels; and less-frequent higher discharge mobilization of step-forming grains. The frequency and magnitude of discharges capable of mobilizing step-forming grains ranges from an estimated >50 years in Oregon (Grant et al., 1990) to 30-50 years in the Italian Alps (Lenzi, 2001) and annual events in Japan (Ashida et al., 1976; Sawada et al., 1983).

Different phases of sediment transport within step-pool channels reflect not only differences in sediment supply, but also variation through time in available flow energy in relation to changes in discharge. Peterson and Mohanty (1960) described tumbling flow in which a supercritical jet flows over the step and goes through a hydraulic jump to become subcritical flow at some distance across the pool immediately below the step. A flow regime with this type of downstream alternation between super- and subcritical flow is now more commonly known as nappe flow (Chanson, 1994). Wohl and Thompson (2000) noted the importance of wake turbulence from mid-profile shear layers immediately downstream from steps and bed-generated turbulence at sites upstream from steps, at step lips and in runs. Turbulence intensities are greatest in pools and at high discharges (Wilcox and Wohl, 2007), up to the point that the tumbling flow transitions to skimming flow. Skimming flow occurs above a specific discharge threshold when the entire flow becomes supercritical (Chanson and Toombes, 2002). Flow resistance in skimming flow is dominated by form losses and cavity recirculation (Chanson et al., 2002), although skin friction may also be important (Gonzalez et al., 2008). Skimming flow occurs only at high, infrequent discharges close to step-forming events. Distinctly higher reach-averaged velocities present at the onset of skimming flow (Comiti et al., in press) can lead to step instability because larger jets flowing over steps have greater scouring efficiency (Comiti and Lenzi, 2006), and the undermining of pools is the most frequent cause of the collapse of steps (Lenzi, 2001; Crowe, 2002).

Constructed check dams will thus remain stable or, if comprised of boulders, adapt and re-form as flow energy and sediment transport changes through time along steep channels. Events during which particularly large flow energy is directed against the channel boundaries, or unusually large volumes of sediment are mobile, will likely destabilize check dams. Consistent with processes in natural step-pool streams, the undermining of pools is the most common cause of collapse for artificial steps (Giacometti, 2000) because of the impinging jet created by the vertical drop (Bormann and Julien, 1991; Stein and Julien, 1994). It therefore becomes important to understand the controls on the dimensions and processes relating to pools. Lenzi et al. (2003b) found that the dimensions of a plunge pool below a vertical drop depends on jet thickness (or critical flow depth) and drop height, although substrate grain size also exerts some influence. They developed an empirical equation for stable channels:

$$0.6 \leq y_s/(z + H_s) \leq 1.4, \quad (2)$$

where y_s is maximum scour depth with respect to crest elevation (m), z is measured drop height (m), and H_s is critical specific flow energy (m). Lenzi (2002) also suggested a range for the design height of the structure H:

$$H/D_{90} = 1\text{-}4, \quad (3)$$

where D_{90} represents a large grain-size fraction of the streambed substrate.

Equations 1-3 provide guidelines for designing check dams that should remain stable, but they do not necessarily facilitate prediction of the morphological effects of constructing check dams. Busnelli et al. (2001) described one-dimensional numerical models of hydraulics and mobile-bed morphology downstream from open check dams. Shieh et al. (2007) further used the one-dimensional step-backwater program HEC-RAS to simulate changes in hydraulic variables associated with check dams. Use of numerical models, such as these, can facilitate evaluation of the potential morphological effects of check dams before the construction of such dams. Such modeling efforts can also aid in the structure design.

The Design Process for Check Dams

The success of check dam structures is measured by their function and long-term stability, which are dependent on several components (Haltiner and Beeman, 2003):

- an in-depth understanding of the geomorphic and hydraulic regimes of the site,
- a comprehensive design process,
- competent installation by an experienced contractor, and
- a long-term monitoring and maintenance program (critical during the initial establishment period).

Although the specific scale and type of analysis for each of the factors noted above is site-dependent, several general steps are common to most projects involving check dams:

1. The expected long-term equilibrium slope (recognized to be dynamic) of the channel is the most important parameter in the design process. This slope can be estimated using a combination of empirical, geomorphic, and engineering procedures. If possible, all three approaches should be used to capture the uncertainty inherent in predicting channel slope. The difference in slope between existing and future conditions reflects the expected changes in the hydraulic and sediment regime at the site, compared with historic, undisturbed conditions.
2. Next, the pattern of the channel planform is identified. For steep mountain channels, this pattern may be unchanged from existing conditions. For intensive restorations involving significant grading, however, increased channel meandering maybe be an effective tool to reduce channel gradient.
3. The effective height (H) of the individual check dams is determined. In general, a greater number of structures with lower height is preferred over fewer, larger structures. Experience has shown that multiple low structures are more stable, provide improved habitat, and have more environmental value than higher structures (as discussed above). This approach involves greater site disturbance, however, and may be more expensive.
4. The design process proceeds from a selected starting location downstream where the channel is stable (stable reach, bedrock control, culvert, crest of an existing check dam, etc). The design (stable) channel slope is extended upstream from this point to an approximate location where it intersects the existing channel thalweg, or is below the existing channel by a depth, h. This location specifies the crest elevation for the next upstream structure.
5. The specific location of the structure is then selected based on local conditions (straight reach versus a meander; presence of stabilizing bank conditions (trees, bedrock, other structures)).

6. Once the specific locations for the check dams are specified, a detailed design can be developed for each structure. This includes:
    (a) Determining the width of the check dam based on hydraulic analysis; the structure must extend laterally a sufficient distance to be anchored firmly in the side banks, and vertically such that large floods do not "flank" the structure and result in "end erosion";
    (b) Extending the base of the check dam sufficiently deep to prevent structure failure by local scour or upstream migration of headcuts; where possible, the base of the upstream structure is set at, or below the crest of the next downstream structure;
    (c) Selecting the type of materials for the check dam (e.g., loose or grouted boulders, concrete, sheet piles, gabions, soil cement);
    (d) Calculating the sizes of rocks comprising the dam (if boulders are used); methods to ensure "plugging" of the voids between boulders are identified;
    (e) A scour pool is designed below the check dam to force the location of the expected hydraulic jump, and provide an erosion-resistant pool to dissipate excess energy from the drop.

In addition, the empirical design parameters obtained from natural step-pool sequences (equations 1-3 above), used in conjunction with these steps, will improve the long-term stability and function of the constructed check dams.

Examples of Successful Restoration of Steep Channels Using Check Dams

Several examples are now available in the published literature of using boulder structures to restore or stabilize steep channels. As noted by Lenzi (2002), artificially emplacing vertical drop structures of any kind will most likely work under conditions conducive to forming steps and pools, which include bed gradients in the range of 3 to 20% and large boulders present in the channel. Consideration must also be given to catchment-scale conditions, such as upstream progressing headcuts or changes in sediment supply, which might affect bed stability (Chanson, 2004). Lenzi (2002) and Lenzi and Comiti (2003) described bouldery check-dam sequences that have functioned successfully along the Maso di Spinelle River in northern Italy. Boulders in these structures were partially buried and used to anchor each other, with lateral keying to limit bank erosion. A slight upstream arch in the step lip helped to evenly distribute stresses across the vertical face. Lenzi (2002) found that a stepped series of low boulder structures was more functionally stable and of higher environmental quality than a single large check dam along this river.

Chin et al. (in press a; in press b) reviewed five examples of restoration of steep channels in the United States using step sequences. In each case, regional hydraulic geometry relationships were used to estimate stable dimensions of severely degraded steep channels in urban or agricultural lands. Bank stabilization used planting of native riparian species and temporary structures such as willow bundles. Boulder and log steps were emplaced to stabilize the streambed and to create maximum flow resistance, as per equation 1. Some of these restoration measures were undertaken in 1996; since that time, individual boulders and entire steps have been redistributed during high discharges. Although the initial step spacing was too large in some cases, subsequent self-organization has created a stable channel with high environmental value. Below, we outline in greater detail the additional case of the restoration of Martin Canyon Creek in California, USA, including the design of the check dams and their performance since installation in 1999.

The Martin Canyon Creek Restoration

Martin Canyon is a small tributary draining 2.8 km^2 of the Alameda Creek watershed in the city of Dublin in northern California. In 1999, Phillip William and Associates, Ltd. (PWA) completed the restoration of a 1.3 km segment of the creek using a series of boulder check dam sequences. Martin Canyon Creek flows into a flood control channel (the J-3), which was experiencing problems of sedimentation at that time. Similar to other streams in the area subjected to the pressures of urbanization, signs of incision and channel widening within the creek were also evident. Because of a large housing development being planned adjacent to Martin Canyon Creek, the City of Dublin and Alameda County Flood Control and Water Conservation District (the manager of the J-3 flood channel) required the developer to restore the creek. The requirements were to prevent over-sedimentation into downstream reaches and threats to adjacent structures through channel erosion (Haltiner, 1997). An initial engineering proposal outlined a series of 20 check dams made of interlocking vinyl sheet piles and the use of rip-rap to inhibit channel erosion, but it failed to gain approval during the permit process because of its overly structural design and lack of aesthetic appeal. Thus, PWA was consulted to formulate a new design with the following goals: 1) minimize sedimentation in the flood canal downstream; 2) protect structures near stream meanders; 3) maintain the natural character and ecology of the creek; and 4) minimize costs of long-term maintenance (PWA, 1999).

The design was based on a geomorphic approach and included several key components. First, a sediment budget estimated the source and quantity of the sediment derived from the project reach (~4970 m^3 or 13%) relative to other portions of the watershed. Second, the dynamic equilibrium slope was determined (1.15% within a range of 0.75-1.5). Third, field surveys focused on assessing the rates and extent of the erosion near meanders. Fourth, a hydraulic model that incorporated estimated roughness characteristics enabled a less conservative design than a purely engineering approach that emphasized safety concerns. Finally, the design focused on the dominant discharge (2-5 year flow), rather than the 100-year event. These procedures produced a final design comprised of a series of nine check dams, each providing 0.6-0.9 m drops. The specific drop height was developed using the design approach described above, in which the predicted stable channel slope (1.15%) was projected upstream from the crest elevation of the downstream check dam to identify an approximate location for the structure (where the future slope would be about 0.6- to 0.9-m below existing grade). Local site conditions were then used to select the specific location. The check dams were designed to capture sediment upstream, thereby decreasing the average gradient over time, as well as to arrest the upstream migration of incision (via headcuts). Boulders from the local area were used to construct the check dams, replacing the original vinyl sheet pile design, and resulting in sequences that resemble natural steps and pools (Figure 3). Additional erosion control measures, including erosion fabric and rock armoring, were installed to protect meander bends that came within ~9 m of residential development.

Figure 3 (a-f). Boulder check dams installed in Martin Canyon Creek. 3a. Channel prior to restoration, with large head cut (~2.5 m) migrating upstream (1997). Photo by J. Haltiner.3b. Rock material delivered to the site for construction of check dams, 1998. Photo by J. Haltiner. 3c. Small excavator used to construct the check dams. Also visible: gravel bedding material and filter fabric used to enhance trapping of fine sediment and "sealing" of the boulder structures, 1998. Photo by J. Haltiner. 3d. A check dam following construction, Dec.1999. Photo by J. Haltiner. 3e. A photograph taken in Fall of 2006 (approx seven years following construction) showing aggradation upstream and scour downstream of a check dam; flow from right to left, channel width ~ 2 m and height from top of dam to bottom of pool is ~ 1.2 m. Photo by Mark Tompkins. 3f. The boulder check dams resemble natural step-pool sequences, photo taken in 2006. Photo by W. Wagner.

A post-project evaluation was conducted in Fall of 2006, approximately seven years after the restoration, to assess the adjustments in the restored channel and overall performance of the check dams (Wagner and Roseman, 2006). A comparison of the longitudinal profile of 2006 to one surveyed immediately following construction (the "as-built" profile) shows the changes that have occurred (Figure 4).

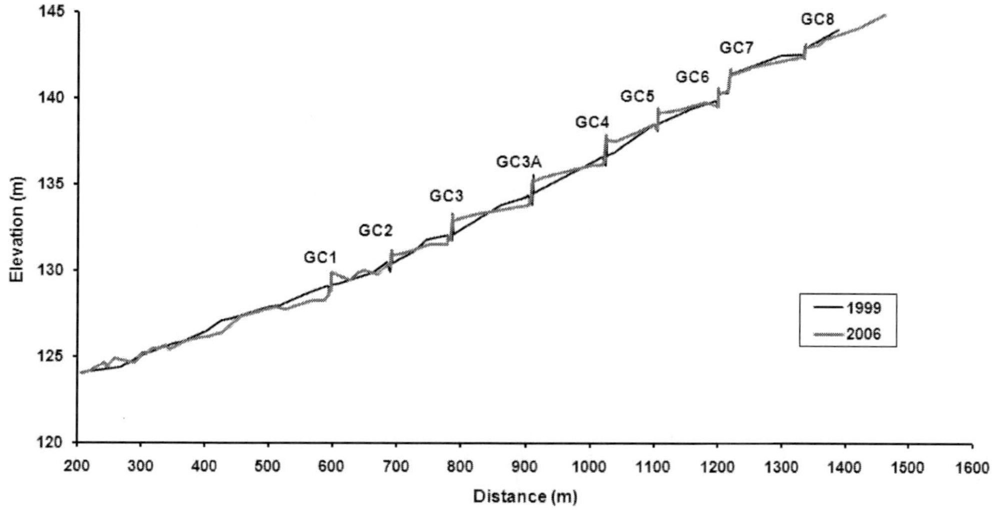

Figure 4. Longitudinal profile surveyed in 2006 compared to the profile constructed following restoration (the "as-built"), showing the locations of the nine check dams (e.g., GC1, GC2).

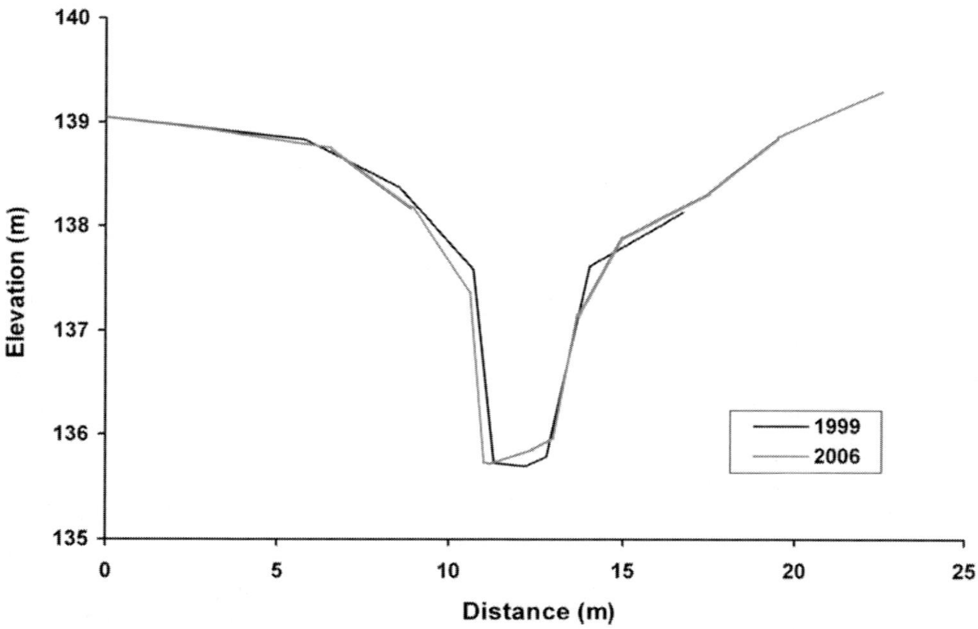

Figure 5. Cross section of Martin Canyon Creek at a location between check dams GC4 and GC3A (see Figure 4): 1999 and 2006.

In 1999, the as-built profile exhibited a relatively smooth gradient that approximated the profile of the original channel (PWA, 1999). In the years since, sedimentation had occurred behind the check dams, as intended, such that a stepped profile was clearly evident by 2006. This stepped profile is characterized by a reduction in gradient in the segments between the check dams, which was further produced by scour in the pools immediately downstream of the check dams (Figure 3e). This decrease in slope from scouring in pools is similar to the adjustments documented in other cases of stream restoration using artificial steps (Chin et al., in press a). Comparison of the slopes of the individual segments between check dams in 2006 to those in 1999 (Table 1) shows that a reduction in gradient had been achieved in all but one segment. The overall values are within the range estimated for equilibrium slopes in the restoration design (PWA, 1999). A cross-section surveyed at one location (Figure 5), as well as visual field inspection, shows that the channel has remained generally stable in the years following restoration (Figures 3e, f). The check dams have remained intact, except for one near a meander bend that is exhibiting signs of rock movement and erosion.

Table 1. Comparison of gradients of segments between check dams, 1999 and 2006

Location	1999 (%)	2006 (%)	% Change in slope
Culvert	1.31	1.16	-12
GC1	1.55	1.18	-31
GC2	1.91	0.75	-155
GC3	1.99	0.72	-176
GC 3A	2.03	0.94	-116
GC 4	2.53	0.70	-261
GC 5	1.47	1.41	-4
GC 6	0.87	0.71	-22
GC 7	1.02	1.08	6
GC 8 (estimated)	2.08	1.58	-32

CONCLUSION

Based on the current state of knowledge of the morphological effects of check dams, as well as our experience in the design/implementation and observations of numerous check dams, we offer the following suggestions:

- As noted by Chin et al. (in press, a), specific site constraints may dictate whether it is feasible to allow the height and spacing of vertical drop structures to adjust to varying discharge and sediment supply, or whether drop structures must be fixed in place. This may be dictated by the availability of upstream cobbles and boulders to resupply the step-pools and check dam structures, and the risk of structure failure to adjacent infrastructure.

- In either case, use of equations 1-3 and numerical modeling of flow, sediment movement, and bed morphology can facilitate prediction of stable geometries for check dams of any type.
- Use of multiple lower check dams rather than a few very tall structures facilitates adjustment of potentially mobile structures to changing flow and sediment supply; limits changes in bed gradient and stream morphology associated with each structure; limits the damage if any single structure fails; and minimizes disruption of longitudinal connectivity for stream organisms.
- Successful designs integrate a geomorphic understanding of the site with a comprehensive engineering design.
- Competent installation by experienced contractors is crucial to realizing the intended design.
- A monitoring and maintenance program is essential to the long-term functioning of check dam structures. Particularly with loose-rock, boulder structures, some movement and adjustment of structures is likely during the initial years following implementation, or following large flows. In conjunction with this, clarity on the ownership, and responsibility for the above monitoring and maintenance, are essential to the long-term project success.

ACKNOWLEDGMENTS

We are grateful to Wayne Wagner and Jesse Roseman for providing the original data from their post-project assessment of Martin Canyon Creek, which was used to generate Figures 4 and 5, and Table 1. Shannah Anderson also provided valuable assistance with field reconnaissance and data collection.

REFERENCES

Abrahams, A.D., Li, G., and Atkinson, J.F. (1995). Step-pool streams: adjustment to maximum flow resistance. *Water Resources Research, 31*, 2593-2602.

Ashida, K., Takahashi, T., and Sawada, T. (1976). Sediment yield and transport on a mountainous small watershed (Japanese with English abstract). *Bulletin of the Disaster Prevention Research Institute, Kyoto University, 26*, 119-144.

Boix-Fayos, C., Barbera, G.G., Lopez-Bermudez, F. and Castillo, V.M. (2007). Effects of check dams, reforestation and land-use changes on river channel morphology; case study of the Rogativa Catchment (Murcia, Spain). *Geomorphology, 91*, 103-123.

Bormann, N. and Julien, P.Y. (1991). Scour downstream of grade-control structures. *J. of Hydraulic Engineering, 117*, 579-594.

Busnelli, M.M., Stelling, G.S., and Larcher, M. (2001). Numerical morphological modeling of open-check dams. *ASCE Journal of Hydraulic Engineering, 127*, 105-114.

Castillo, V.M., Mosch, W.M., Garcia, C.C., Barbera, G.G., Cano, J.A.N., and Lopez-Bermudez, F. (2007). Effectiveness and geomorphological impacts of check dams for soil erosion control in a semiarid Mediterranean catchment; El Carcavo (Murcia, Spain). *Catena, 70*, 416-427.

Chanson, H. (1994). Hydraulics of nappe flow regime above stepped chutes and spillways. *Australian Civil Engineering Transactions CE, 36(1)*, 69-76.

Chanson, H. (2004). Sabo check dams; mountain protection systems in Japan. *International Journal of River Basin Management, 2*, 301-307.

Chanson, H. and Toombes, L. (2002). Energy dissipation and air entrainment in a stepped storm waterway: an experimental study. *ASCE J. Irrigation and Drainage Engineering, 128*, 305-315.

Chanson, H., Yasuda, Y., and Ohtsu, I. (2002). Flow resistance in skimming flows and its modeling. *Canadian J. Civil Engineering, 29*, 809-819.

Chin, A., Anderson, S., Collison, A., Ellis-Sugai, B.J., Haltiner, J.P., Hogervorst, J.B., Kondolf, G.M., O'Hirok, L.S., Purcell, A.H., Riley, A.L., and Wohl, E. in press, a. Linking theory and practice for restoration of step-pool streams. *Environmental Management.*

Chin, A., Purcell, A.H., Quan, J.W., and Resh, V.H. in press, b. Assessing geomorphological and ecological responses in restored step-pool systems. Geological Society of America Special Paper 451.

Chin, A. and Wohl, E. (2005). Toward a theory for step pools in stream channels. *Progress in Physical Geography, 29*, 275-296.

Comiti, F., Andreoli, A., and Lenzi, M.A. (2005). Morphological effects of local scouring in step-pool streams. *Earth Surface Processes and Landforms, 30*, 1567-1581.

Comiti, F. and Lenzi, M.A. (2006). Dimensions of standing waves at steps in mountain rivers. *Water Resources Research, 42*, W03411.

Comiti, F., Cadol, D., and Wohl, E. in press. Flow regimes, bed morphology and flow resistance in self-formed step-pool channels. *Water Resources Research.*

Crowe, J.C. (2002). An experimental study of the step pool bedform. Unpublished PhD dissertation, Johns Hopkins University, Baltimore, MD, 147 pp.

Giacometti, G. (2000). Local scour downstream bed sills and check dams in high gradient streams. PhD thesis, University of Padova, 190 pp. (in Italian).

Gonzalez, C.A., Takahashi, M. and Chanson, H. (2008). An experimental study of effects of step roughness in skimming flow on stepped chutes. *IAHR J. Hydraulic Research, 46*, Extra Issue 1, 24-35.

Grant, G.E., Swanson, F.J., and Wolman, M.G. (1990). Pattern and origin of stepped-bed morphology in high-gradient streams, western Cascades, Oregon. *Geological Society of America Bulletin, 102*, 340-352.

Haltiner, J.P., G.M. Kondolf and P.B. Williams (1996). Restoration Approaches in California, Chapter 11. *In* A. Brooks and D. Shields (eds.) *River Channel Restoration*, John Wiley and Sons, 433 pp.

Haltiner, J. (1997). Martin Canyon Stream Stabilization: Phase I: Problem Reconnaissance and Conceptual Plan Approach. Phillip Williams and Associates, Ltd., July 22.

Haltiner, J.P. and C. Beeman, (2003). "Restoring Floodplain and Channel Functions to Incised and Leveed Stream Systems". In: Faber, P.M.(ed.) 2003. California Riparian Systems: Processes and Floodplain Management, Ecology, and Restoration. 2001 Riparian Habitat and Floodplains Conference Proceedings, Riparian Habitat Joint Venture, Sacramento, CA.

Jaeggi, M. and Zarn, B. (1999). Stream channel restoration and erosion control for incised channels in alpine environments. In, S.E. Darby and A. Simon, eds., Incised river channels: processes, forms, engineering and management. Wiley, Chichester, pp. 343-369.

Japanese Ministry of Construction. (1993). Sabo. Ministry of Construction, Kobe, 36 pp.

Lai, J.-S., Tsay, T.-C., and Lin, G.-F. (1998). A study on equilibrium slope upstream of check dams. In, A.W. Jayawardena, J.H. Lee, and Z.Y. Wang, eds., River sedimentation; theory and applications. A.A. Balkema, Rotterdam, Netherlands, pp. 561-564.

Lenzi, M.A. (2001). Step-pool evolution in the Rio Cordon, northeastern Italy. *Earth Surface Processes and Landforms, 26*, 991-1008.

Lenzi, M.A. (2002). Stream bed stabilization using boulder check dams that mimic step-pool morphology features in Northern Italy. *Geomorphology, 45*, 243-260.

Lenzi, M.A. and Comiti, F. (2003). Local scouring and morphological adjustments in steep channels with check-dam sequences. *Geomorphology, 55*, 97-109.

Lenzi, M.A., D'Agostino, V., Billi, P. (1999). Bedload transport in the instrumented catchment of the Rio Cordon: Part 1. Analysis of bedload records, conditions and threshold of bedload entrainment. *Catena, 36*, 171-190.

Lenzi, M.A., Marion, A., and Comiti, F. (2003a). Interference processes on scouring at bed sills. *Earth Surface Processes and Landforms, 28*, 99-110.

Lenzi, M.A., Marion, A., and Comiti, F. (2003b). Local scouring at grade-control structures in alluvial mountain rivers. *Water Resources Research, 39 (7)*, 2002WR001815, 12 pp.

Martín-Vide, J.P. and A. Andreatta. (2009). Channel degradation and slope adjustment in steep streams controlled through bed sills. *Earth Surface Processes and Landforms, 34,* 38-47.

Peterson, D.F. and Mohanty, P.K. (1960). Flume studies of flow in steep, rough channels. *ASCE J. Hydraulics Division, 86*, 55-76.

Phillip Williams and Associates, Ltd., (1999). Martin Canyon Creek Stream Restoration Owner's Manual: Final Report and Recommended Monitoring Program. November 12.

Sawada, T., Ashida, K., and Takahashi, T. (1983). Relationship between channel pattern and sediment transport in a steep gravel bed river. *Zeitschrift fur Geomorphologie, Supplement band 46*, 55-66.

Shieh, C.-L., Guh, Y.-R., and Wang, S.-Q. (2007). The application of range of variability approach to the assessment of a check dam on riverine habitat alteration. *Environmental Geology, 52*, 427-435.

Stein, O.R. and Julien, P.Y. (1994). Sediment concentration below free overfall. *ASCE Journal of Hydraulic Engineering, 120*, 1044-1053.

Wagner, W. and Roseman, J. (2006). Post-project appraisal of Martin Canyon Creek Restoration. Unpublished report, Department of Landscape Architecture and Environmental Planning, University of California, Berkeley, 27 pp.

Warburton, J. (1992). Observations of bed load transport and channel bed changes in a proglacial mountain stream. *Arctic Alpine Research, 24*, 195-203.

Wilcox, A.C. and Wohl, E.E. (2007). Field measurements of three-dimensional hydraulics in a step-pool channel. *Geomorphology, 83*, 215-231.

Wohl, E. (2000*). Mountain rivers*. Am. Geophysical Union Press, Washington, D.C., 320 pp.

Wohl, E.E. and Thompson, D.M. (2000). Velocity characteristics along a small step-pool channel. *Earth Surface Processes and Landforms, 25*, 353-367.

In: Check Dams, Morphological Adjustments…
Editors: C. Conesa and M. A. Lenzi
ISBN: 978-1-62618-856-3

Chapter 8

SEDIMENT TRANSPORTATION FROM BED-LOAD TO DEBRIS-FLOW AND ITS CONTROL BY CHECK DAMS IN TORRENTIAL STREAMS

***Rabindra Osti*[*1] *and Shinji Egashira*[2]**
[1]International Centre for Water Hazard and Risk Management,
Public Work Research Institute, Minamihara 1-6,
Tsukuba 305-8516, Japan
[2]NEWJEC Inc, 2-3-20 Honjo-Higashi,
Kita-ku, Osaka 531-0074, Japan

ABSTRACT

Among many other uses, check dams are useful for controlling debris-flows quantitatively and qualitatively. In order to increase the effectiveness of debris-flow control using check dams, improved methods of planning and designing such systems are prerequisite. As a basis for these evaluations, sediment-flow characteristics need to be considered. Knowing the flow characteristics will help to propose techniques for evaluating and improving the mitigative effectiveness of check dams against debris-flows. In this chapter, a review of independent sediment transport formulas for bed-load and debris-flow as well as the unified view of mechanics of debris flow and bed-load is made available. The control function of closed-type check dams is also discussed with special attention to the potential storage volumes of check dams and inflowing sediment volumes. Results produced by using experimental and numerical models are described for investigating changes in the characteristics of a debris-flow passing through a check dam. Basically, three governing non-dimensional parameters, a non-dimensional potential storage volume, relative outflow sediment volume and normalized inflow sediment volume, are specified in terms of potential storage volume of the check dam. Additionally, this chapter explains the criteria for the selection of sizes, numbers and locations of check dams as a basis for check dam planning. As an example, the 1999 debris-flow event in the San Julian River in Venezuela is chosen for discussion.

1. INTRODUCTION

Present chapter aims to introduce a brief but critical review on mechanics of sediment transportation from bed load to debris flow and to describe mainly sediment control by check dam and associated issues. There are many debris-flow countermeasures (Figure 1) especially different kinds of check dams which are designed and constructed to control sediment transport process quantitatively and qualitatively. Although the main function of check dam is to control the debris flow, the debris flow control mechanism of each type of check dam may differ from the view points of effective storage volume, peak load reduction as well as the temporal control of outflow characteristics.

The functions of check dam can be measured in terms of magnitude for changes of sediment run out volume, sediment size and hydrograph. The control of such flow characteristics are supposed to be achieved by changing the debris flow mechanism such as by changing the regime of sediment movement from debris flow or from suspended load to contact load, by changing sediment transport rate and by sorting sediment size (Ashida and Egashira, 1989). Several types of check dams and their general functions are reported in several studies e.g., Armanini and Larcher (2001), Ashida et al. (1988), Busnelli et al. (2001), Takahashi et al. (2001), Honda and Egashira (1997), Mizuyama et al. (1984), Takahashi (1991) etc.

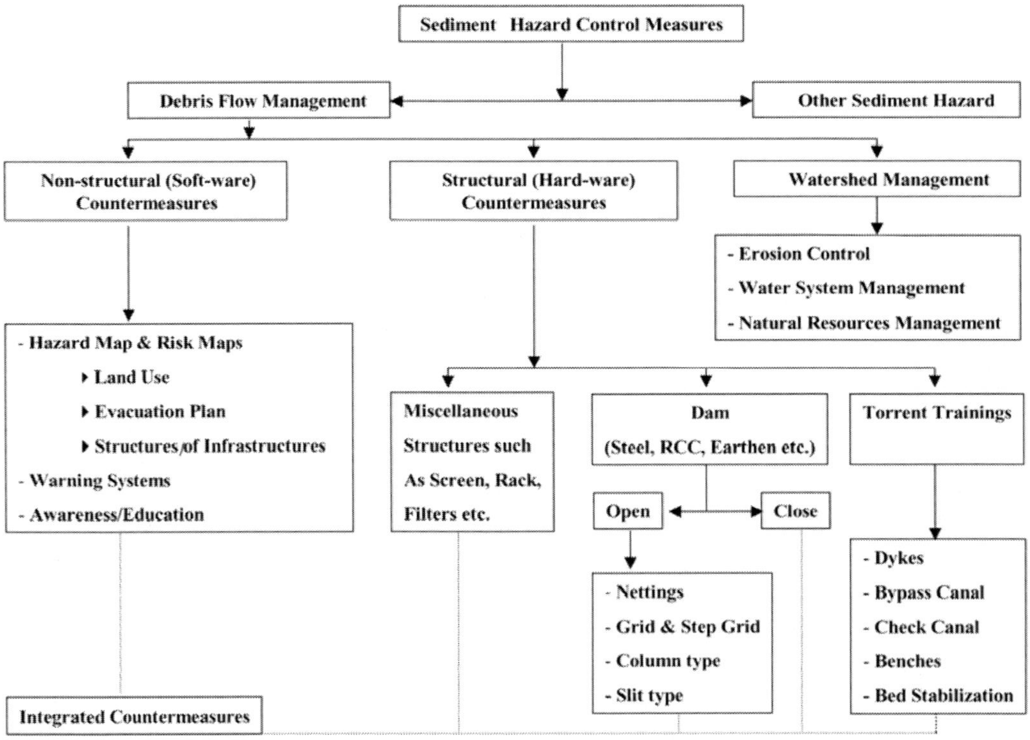

Figure 1. Classification of countermeasures against sediment hazard.

* E-mail: osti55@pwri.go.jp

Studies on specific sediment control mechanism of several types of check dams can also be found in literatures, for examples, the impact of dynamic forces induced by debris flow on the check dam was studied by Lin et al. (1999), sediment trap rate based on specific grain size was considered by Mizuyama et al., (2000), comparative study on the effects of dam's shapes such as column, gravity, slit etc. were studied by Chen and Ho (1999) and Bovolin and Mizuno (2000), general design criteria of check dam are discussed by Johnson *et al.* (1989) and design against scouring of bed material downstream to the check dam structures as well as bed sills are described by Marion *et al.* (2004). Usually, most researchers have tried to evaluate these control functions focusing on changes of sediment hydrograph of out-flowing floods with respect to the inflow ones. However, it is not easy to formulate the sediment control functions in a general term, because mechanics of sediment transportation in steep areas are very complex.

Figure 2 shows schematics of sediment transport modes occurring in a uniform flow (Egashira and Ashida, 1992). Sediment transportation takes place in various manners from debris flow (Figure 2a) to flow with bed load (Figure 2c) even if sediment particle sizes are large enough to maintain laminar motion. In a region ranging from milder than some upper threshold gradient and steeper than the lower threshold gradient, bed material can be carried in dispersion up to the free surface (Figure 2a). In case of non-cohesive sediment, these threshold gradients can be evaluated using a stability analysis for an infinitely long slope, which are estimated at about 20-23 degrees for the upper and 12-15 degrees for the lower threshold, respectively. In the region milder than the lower threshold gradient, the thickness of sediment transport layer decreases with decrease of bed slope, resulting in a clear water layer over the sediment layer (Figure 2b), and thus bed load layer is formed as shown in Figure 2(c).

Many studies have been conducted for bed-load problems based on flume tests, dimensional analyses and related tools of fluid dynamics. Although there are still many problems that yet to be solved from a view of dynamics principles, researchers and engineers could choose any of the existing bed load formulas, whenever they need to predict sediment transport rate and associated sediment transport phenomena (eg: Meyer-Peter and Müller, 1948; Einstein, 1950; Bagnold, 1966; Engelund and Hansen, 1967; Graf, 1971; Graf and Suszka, 1987; Ashida and Michiue, 1972; Nakagawa and Tsujimoto, 1980; Parker et al., 1982). However, it would be worth to reconsider why such many formulas have been proposed and what are their limitations.

In order to make it understandable, some of well known bed load formulas will be reviewed in this chapter. The review will be based on the principles of flow dynamics as well as on the methods newly described in the following section, showing a possible formula applicable for bed load to debris flow.

Debris flows have been studied in many academic fields ranging from engineering to science and valuable results have been presented in many cases. Results clarify that heavy rainfalls, earthquakes, and volcanic eruptions can trigger debris flows such as debris flow induced by channel erosion, landslides, pyroclastic flow, open slope fire or dam or levee breached flow. Such debris flows can be characterized by the properties of sediment, sediment concentration in the flow and topography (Pierson and Scott, 1985; Scott et al., 1992; Costa, 1994; Cannon et al., 2001; Cannon, 2001; Nsom, 2002; Kershaw et al., 2005; Itoh and Egashira, 2007; Osti et al., 2007; etc.). Studies on the constitutive relations of debris flows have provided many formulas. With attention given on the structure of their expression, these formulas can be grouped as follows (Egashira, 2007);

(a) Coulomb type shear stress (eg: Iverson and Denlinger, 2001; Savage and Hutter, 1989, 1991)
(b) Bingham type (eg: Chen, 1988; Julien and Lan, 1991; O'Brien and Julien, 1988)
(c) Fluid type (eg: Takahashi, 1980, 1991; Tsubaki et al., 1982)
(d) Coulomb type shear plus fluid type (eg: Egashira and Ashida, 1992; Egashira et al. 1997a,b)

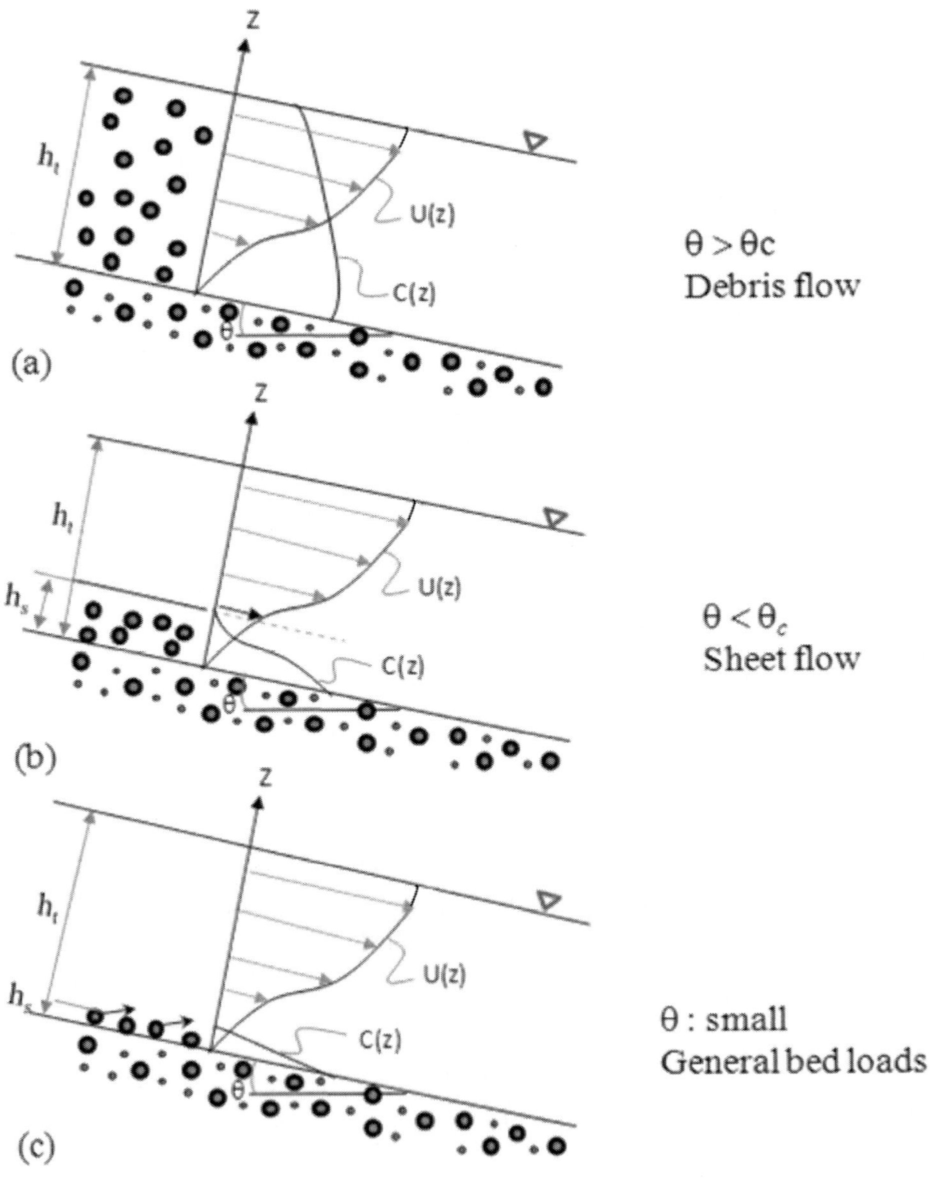

Figure 2. Regimes of sediment transportation.

There are advantages and disadvantages of each type. For instance, a steady state solution which is an important nature of flow cannot be described by the type- a), but by the type d). The Bingham type hardly distinguishes the differences between the flows over a rigid bed and over an erodible bed. The fluid type solutions cannot explain how the flow body stops, while the types -a) and -d) are able to do so. These issues suggest that further researches are desired to develop the reliable formulas.

In addition to the constitutive relations, the debris flow behaviors such as developing and diminishing process in torrents and debris fans are influenced very much by sediment erosion-deposition processes as well as by the corresponding topographic changes. In other words, a formula for the erosion-deposition rate plays an important role in the predictions of such a debris flow process. However, there are few researches on erosion-deposition relation in spite of its importance. Egashira (1993) and Egashira et al. (2001) proposed an erosion-deposition rate formula which was developed supposing local equilibrium of sediment concentration in the debris flow body and is useful for non-cohesive, coarse sediment. In present chapter, this formula is employed as shown in section 3 of this chapter.

Check dams are widely used to control the debris-flows. However, this type of structure usually filled with sediment either by the general river sediment load in normal time or by small scale debris-flows prior to any large scale event and therefore, there is always a question that whether the filled check dam has any role to control anticipated debris-flow effectively. To answer that question, Osti et al. (2008) introduced the criteria for the evaluation of control function of close type check dam that filled up with sediment up to the crest level. In the proposed criteria, a relative run-out sediment volume, which is normalized by an inflow sediment volume at a check-dam section, can be evaluated using two non-dimensional parameters; a) relative inflow sediment volume, and b) potential storage volume that can be defined by the equilibrium and initial bed slopes of deposition.

2. Relation of Bed Load and Debris Flow

2.1. Bed Load and Bed Load Formulas

It is recognized that bed loads are carried by flow of water in a manner of rolling, sliding and jumping along the bed. Generally bed load transport rate is defined as absolute volume discharge in unit length and unit time. Bed load problems have been considered in relation to various sediment issues such as river changes, channels stability and water uses. Many formulas have been proposed through researchers and one of these is Meyer-Peter and Müller's bed load formula (1948) which was developed based on flume tests.

$$q_{b*} = 8(\tau_* - 0.047)^{3/2} \tag{1}$$

in which τ_*, and q_{b*} are non-dimensional bed shear stress and non-dimensional bed load transport rate defined as

$$\tau_* = u_*^{\,2} /(\sigma / \rho - 1) g d$$

$$q_{b*} = q_b / \sqrt{(\sigma / \rho - 1) g d^3}$$

in which u_* is the shear velocity defined as ($\sqrt{\tau_b / \rho} = \sqrt{ghi}$), where τ_b is bed shear stress, h is flow depth, i is bed slope or energy slope, σ is the mass density of sediment particle, ρ is the mass density of water, g is the acceleration due to gravity and d is the reference particle size.

In formulating a complex process, we often can resolve it into some simple elements, either from a view points of phenomenology or of methodology. Although there are many differences in the methods employed, studies on bed load transport formulas could be classified roughly into two, referring to the target elements.

q_b = (pick-up rate of sediment in unit area)*(step-length) (2)

q_b = (sediment volume of bed load layer in unit area)*(average velocity) (3)

Studies on the basis of equation (2) suggested that bed load movement was a stochastic process, for which Einstein's bed load formula (1950) is well known. Einstein formulated pick-up rate and step-length taking stochastic properties of fluid force and sediment movement into account, and proposed a formula. The formula, in uniform sediment case, is as follows.

$$\frac{A_* q_{b*}}{1 - A_* q_{b*}} = 1 - \frac{1}{\sqrt{\pi}} \int_{-B^*/\tau_* - 1/\eta_*}^{B^*/\tau_* - 1/\eta_*} e^{-t^2} dt \tag{4}$$

in which η_*, A_* and B_* are universal constants determined by data fitting, which are equal to 0.5, 27.0 and 0.156, respectively.

Based on the Einstein's bed-load formula, many researches were carried out during a period from late 1960's to 1970's and produced many results (e.g. Yang and Sayre, 1971; Nakagawa and Tsujimoto, 1980). In this approach, it is very important to know how to evaluate the pick-up rate or rest-period of bed sediment as well as the step-length of sediment particle. In other words, basic issues are how to specify reasonably the probability density functions corresponding to pick-up rate and step length, and how to connect these density functions and hydraulic parameters such as bed shear stress, physical properties of sediment particles etc.

Bagnold's studies, which could be classified into the studies corresponding to Equation (3), have still influenced many researchers. His idea (Bagnold, 1966) is to correlate bed load transport rate and stream power that defined by bed shear stress times flow velocity (or shear velocity). Although it is complex and we can hardly understand his treatments for the dynamic principles, some of his ideas have attracted researches, which lead to the further proposals of many bed load formulas (eg. Engelund and Hansen, 1967; Ashida and Michue, 1972). For instance, Ashida and Michiue evaluated the sediment volume of bed load layer designated in Equation (3) using the following Bagnold's suggestion,

$$\tau_G = \sigma_n \tan \alpha \tag{5}$$

in which τ_G is the grain shear stress, σ_n is the normal stress and α is particle-particle collision angle. In addition, they evaluated sediment particle velocity using equation of motion for mass point system, and proposed a formula.

$$q_{b*} = 17\tau_{*_e}^{3/2}(1-\frac{\tau_{*c}}{\tau_*})(1-\frac{u_{*c}}{u_*}) \tag{6}$$

in which τ_{*e} is the non-dimensional effective bed shear stress, τ_{*c} is the non-dimensional critical bed shear stress for initiating sediment movement, which can be evaluated by using Shields diagram, and u_{*c} is the critical shear velocity. The effective bed shear stress is specified as

$$\tau_{*_e} = u_{*_e}^{\ 2} / [\sigma / \rho - 1)gd]$$

in which u_{*e} should be evaluated by using

$$\frac{v}{u_{*_e}} = 6.0 + 2.5\ln\frac{h}{d(1+2\tau_*)} \tag{7}$$

in which v is the spatial average velocity. Although Equation (6) was proposed through data fitting and has been widely used, it was not necessarily derived from reasonable dynamic principles.

If Bagnold's and other related researches applied the energy conservation equation in a clearly defined control volume of the bed load layer, their result might have concluded with a quite different equation from their originals. In order to understand the bed load process, it is recommended that we had better restart the study by specifying clearly the field as well as by using the corresponding field equations even if issues associated with applicability put aside.

2.2. Bed Load Formula Derived from Constitutive Relations of Water-Sediment Mixture Flow

Bed load transport rate defined by Equation (3) can be expressed in several different ways referring to Figure 2.

$$q_b = NV_s\overline{v}_s = \overline{c}_s h_s \overline{v}_s = \int_0^{h_t} c\, v_s dz \tag{8}$$

in which N is the number of sediment particle in unit area of the bed load layer, V_s is the volume of a reference sediment particle, $\overline{v}_s$ is the average velocity of bed load layer, v_s is the local velocity in bed load layer, $\overline{c}_s$ is the average sediment concentration of bed load layer, c

is the local sediment concentration in bed load layer, h_s is the thickness of bed load layer and h_t is the total flow depth.

Figure 2 suggests that thickness of sediment layer, h_s, from bed load to debris flow changes continuously and thus, sediment transport rate changes monotonically. Correspondingly, if local velocity and sediment concentration are evaluated on the basis of dynamics principle different from those employed in evaluating bed load formulas, a bed load equation can be deduced. Egashira and Ashida (1992) and Egashira et al. (1997a,b, 2001b, 2005) proposed constitutive relations for the flow of coarse sediment and water mixture and solved local velocity and sediment concentration in the sediment layer from bed load to debris flow using the proposed constitutive relations as well as mass and momentum conservation equations. Main part of their study is described as follows.

In the case of steady, uniform, unidirectional flow of sediment-water mixture, the momentum conservation equations are simplified as

$$\tau(z) = \int_z^{h_t} \rho_m g \sin\theta dz \tag{9}$$

$$p(z) = \int_z^{h_t} \rho_m g \cos\theta dz \tag{10}$$

in which h_t is the flow depth, g is the acceleration due to gravity, z is the bed-normal coordinate, θ is the inclination from the horizontal, τ is the shear stress in planes parallel to the bed, p is the isotropic component of the stress i.e. pressure, ρ_m is the mass density of sediment mixture (Eq. 11).

$$\rho_m = (\sigma - \rho)c + \rho \tag{11}$$

In which σ is the mass density of sediment particles, ρ is the mass density of water and c is the sediment concentration by volume in the mixture.

Egashira et al. (1997a,b) formulated the energy dissipation function for the flow of water and sediment mixture, supposing that energy dissipation takes place in three ways; a) inter particle contact, b) inelastic particle-particle collisions and c) turbulence of interstitial water;

$$\Phi = p_s \tan\phi_s (\frac{\partial u}{\partial z}) + k_d(1-e^2)\sigma c^{1/3} d^2 (\frac{\partial u}{\partial z})^3 + kf\rho \frac{(1-c)^{5/3}}{c^{2/3}} (\frac{\partial u}{\partial z})^3 \tag{12}$$

in which each term of right hand side is corresponding to a), b) and c) category, respectively, Φ is the dissipation function defined as $\tau(\partial u / \partial z)^3$ that is the energy dissipation rate in unit time and unit volume, and other symbols will be explained after equation (16). They derived the shear stress formula directly from the energy dissipation function. It constitutes of three parts as follows.

$$\tau = \tau_y + \tau_d + \tau_f \tag{13}$$

in which

$$\tau_y = p_s \tan\phi_s \tag{14}$$

$$\tau_d = k_d(1-e^2)\sigma c^{1/3} d^2 (\frac{\partial u}{\partial z})^2 \tag{15}$$

$$\tau_f = k_f \rho \frac{(1-c)^{5/3}}{c^{2/3}} d^2 (\frac{\partial u}{\partial z})^2 \tag{16}$$

in which τ_y is the yield shear stress resulting from particle-particle static contact, p_s is the static granular pressure resulting from particle-particle contacts, ϕ_s is the internal friction angle of sediment particles which takes 34-38 degrees in natural sands and cobbles, τ_d is the shear stress due to inelastic collisions of particle to particle, e is the restitution coefficient which is 0.85 in case of natural sands and cobbles, τ_f is the shear stress supported by interstitial water, k_d and k_f are expected to be universal constants and specified as 0.0828 and 0.16, respectively.

In addition to the shear stress formula, they derived the pressure expression composed of three components as follows.

$$p = p_s + p_d + p_w \tag{17}$$

in which

$$\frac{p_s}{p_s + p_d} = (\frac{c}{c_*})^{1/5} \tag{18}$$

$$p_d = k_d e^2 \sigma c^{1/3} d^2 (\frac{\partial u}{\partial z})^2 \tag{19}$$

$$p_w = \rho g(h_t - z)\cos\theta \tag{20}$$

in which c_* is the sediment concentration of stationary layer underlying the bed load layer, P_s denotes the static granular pressure, p_d is the dynamic pressure which is the conservative part of kinetic energy, p_w is the hydrostatic pressure of interstitial water, supposing that sediment suspension does not take place by turbulence and thus, sediment particles behave as laminar motion, θ is the bed slope, and σ, ρ, c, d and others are the same as before .

Substituting equation (13) with (14)-(16) into (9) and substituting equation (17) with (18)-(20) into (10), they derived two equations for velocity and sediment concentration, respectively. The two equations were solved numerically for velocity and sediment concentration profiles of the flows over erodible bed as well as over rigid bed, the validity of which were tested using many flume data (Egashira, et al., 1997a, 2001, 2005).

In Figure 3, numerical prediction (exact solution) and flume data are shown. We can see that both of the results change monotonically from bed load to debris flow, which suggests constitutive relations of the sediment-water mixture (debris flow) could be extend to solve the flow of bed load layer even if the thickness, h_s, is thin. In this figure, flux sediment concentration is defined as.

$$c_f = \int cudz / \int udz = q_b / (q_w + q_b) \quad (21)$$

in which q_w is the flow discharge of water only.

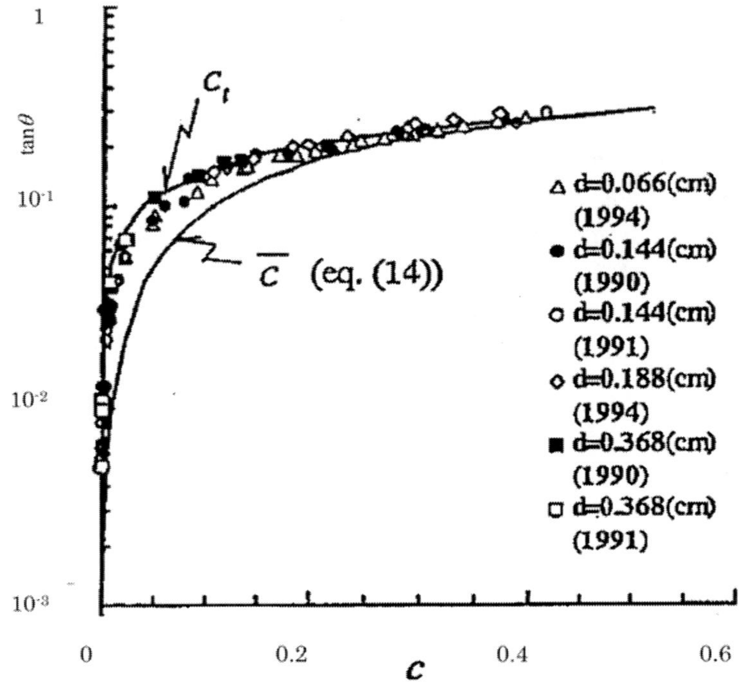

Figure 3. Flux sediment concentration with different particle size diameters.

To compute the profiles of velocity and sediment concentration of bed load layer, two equations must be solved in terms of an iterative method, because the velocity and the sediment concentration profile interact with each other. Therefore, it is somewhat complex to predict flux sediment concentration. Contrarily, the corresponding experimental data can be obtained easily by flume tests. On the other hand, it is very difficult to obtain the depth-average sediment concentration from the flume tests due to the difficulty of the measurement. In contrast to the difficulty in obtaining flume data, the corresponding equation can be deduced easily as follows.

Integrating equations (9) and (10) from $z = 0$ to free surface, and referring to $\tau = \tau_y$ and $p = p_s$ at $z = 0$ ($\partial u/\partial z = 0$ and $c = c_*$ at $z = 0$), the equation is;

$$\bar{c} = \frac{\tan\theta}{(\sigma/\rho - 1)(\tan\phi_s - \tan\theta)} \quad (22)$$

The predicted curve in terms of equation (22) is also shown in Figure 3. Herein, the authors should point out one of the serious problems related to equation (22). It is the fact that there are some research-papers in which they compared the result predicted by equation (22) to the experimental data for flux sediment concentration and give a negative evaluation to this equation for the reason why experimental data cannot be predicted well by this formula. The experimental data should not be compared to the results predicted by equation (22), but by equation (21).

Let us come back to the discussion on a bed load equation because the introduced constitutive relations predict well the flow of sediment-water mixture in a wide rage from bed load to debris flow. The formulas for the thickness of bed load layer and the bed load rate, which can be evaluated exactly by numerical computation as shown in Figure 3, are deduced analytically supposing that the average sediment concentration of the layer takes a constant value of $\bar{c}_S$. Integrating right hand sides of equations (9) and (10) from 0 to h_s and then from h_s to h_t (free surface), and referring to $\tau(0) = \tau_y(0)$ and $p(0) = p_s(0)$ the equation for the thickness of bed load layer is obtained (Egashira et al., 1997b, 2005).

$$\frac{h_s}{h_t} = \frac{1}{(\sigma/\rho - 1)\bar{c}_s} \frac{\tan\theta}{\tan\phi_s - \tan\theta} \tag{23}$$

Equation (23) multiplied by h_t/d leads to;

$$\frac{h_s}{d} = \frac{1}{\bar{c}_s \cos\theta} \frac{1}{\tan\phi_s - \tan\theta} \tau_* \tag{24}$$

in which τ_* is the same definition shown in bed load equations;

$$\tau_* = \frac{u_*^2}{(\sigma/\rho - 1)gd} = \frac{h_t \sin\theta}{(\sigma/\rho - 1)d}$$

Similarly, integrating Equations (9) and (10) from z to h_s and then, from h_s to the free surface, and introducing the constitutive relations into the left hand sides of Equations (9) and (10), Egashira and Ashida (1992) and Egashira et al. (1997b, 2005) derived the velocity profile and the average velocity of the bed load layer. The average velocity is given by

$$\frac{\bar{v}_s}{u_*} = \frac{4}{15} \frac{K_1 K_2}{\sqrt{f_d + f_f}} \tau_* \tag{25}$$

In addition, they obtained a bed load equation directly by substituting either the equations (23) and (25) or (24) and (25) into the definition equation for bed load as shown in equation (8).

$$q_{b*} = \frac{4}{15}\frac{K_1 K_2}{\sqrt{f_d + f_f}}\tau_*^{5/2} \ ,(\tau_* \geq \tau_{*c}) \tag{26}$$

in which

$$K_1 = \frac{1}{\cos\theta}\frac{1}{\tan\phi_s - \tan\theta}$$

$$K_2 = \frac{1}{\bar{c}_s}\left[1 - \frac{h_s}{h_t}\right]^{1/2}$$

$$f_d = k_d(1-e^2)(\sigma/\rho)\bar{c}_s^{\ 1/3}$$

$$f_f = k_f(1-\bar{c}_s)^{5/3}\bar{c}_s^{\ -2/3}$$

Values of universal and empirical constants which were calibrated in terms of flume data on uniform debris flows are as follows; $k_d = 0.0828$, $k_f = 0.16$ and $e = 0.85$. Moreover, $\bar{c}_s$ is supposed to be $0.5c_*$ considering that sediment concentration takes a value of c_* at the bed and decreases monotonically to 0 at the upper boundary of the bed load layer, ϕ_s is 34 to 38 degrees obtained from both of dried and submerged tests for natural and artificial sands and σ is estimated to be about 2.65 g/cm^3. Hereafter, c_*=0.52 and ϕ_s=34 degrees, which were specified by the corresponding tests, will be employed if there are no special reasons.

In Figure 4, curves predicted by Eq. (24) are shown and it is compared with flume data and following equation.

$$\frac{h_s}{d} = \frac{1}{\bar{c}_s \cos\theta}\frac{1}{\tan\phi_s - \tan\theta}(\tau_* - \tau_{*c}) \tag{27}$$

This equation, which was introduced by Egashira et al. (2005), was obtained easily by modifying a little the bed load treatments by Bagnold (1966), Engelund et al. (1967) and Ashida and Michiue (1972). We can see that the flume data are evaluated satisfactory by the both equations.

In Figure 5, a curve predicted by Equation (26) is shown and it is compared with flume data and different bed-load formulas. We can see Equation (26) predicts the flume data fairly well notwithstanding that data fitting were not performed at all, although the predicted curve is not better than those with the empirically developed formulas (Graf and Suszka, 1987; Shinohara and Tsubaki, 1959).

The results shown in Figures 3, 4 and 5 suggest that the sets of constitutive equations of debris flow can be extended to solving for bed-load transport rate as well as for the flow resistance. Accordingly, governing equations, the corresponding friction law and erosion-deposition formula proposed for the debris flow will be applied to evaluate the sediment control by check dams and associated issues in steep areas where sediment transportation takes place ranging from debris flow to bed load.

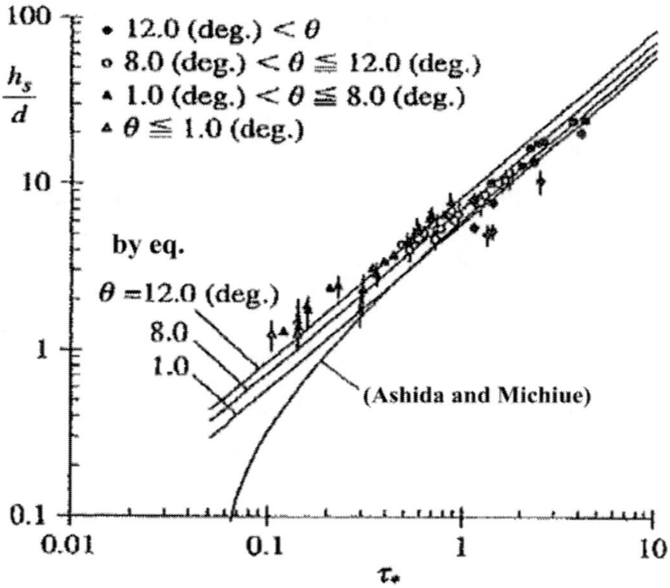

Figure 4. Relationship between non-dimensional shear stress and thickness of bed-load layer.

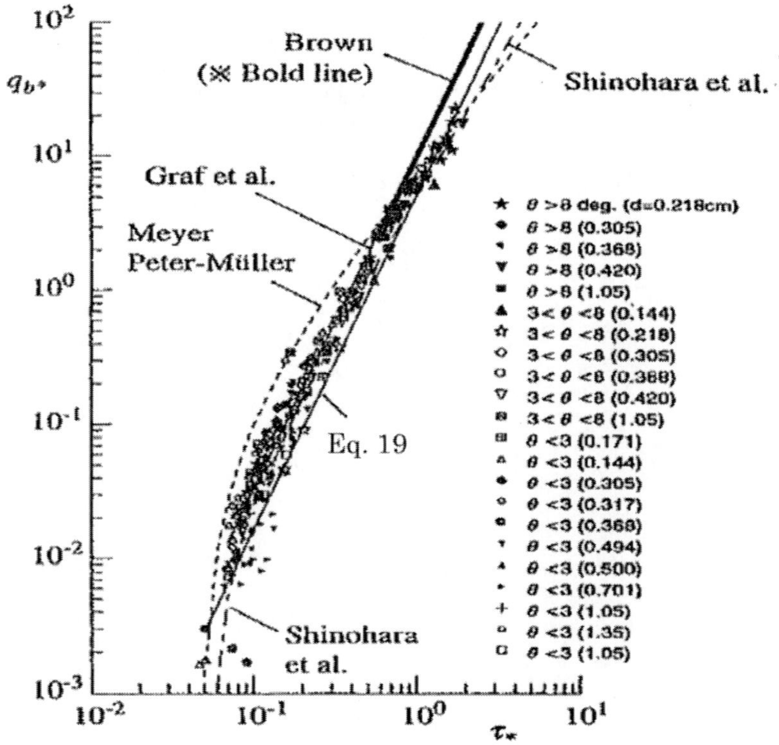

Figure 5. Bed-load formulas comparison.

3. Sediment Control by Check Dams

3.1. Review

Several empirical tools have been conventionally used to design the check dam. In most occasions, total capacity of check dams are determined based on the concept that the deposited sediment makes a horizontal surface at the top of the check dam. In fact, the total sediment volume that can be trapped by the check dam varies with the slope of deposited area and therefore greatly governed by the supplied condition, check dam height and the extended riverbed slope.

There are few studies that have considered the depositional profile of the debris-flow to estimate the capacity of the check dam (e.g. Woolbiser and Arno (1965); Arai and Takahashi (1988); Ashida and Egashira (1989); Takahashi (1991); MLIT (1984); Nedelcu, 2001). The study conducted by Woolbiser and Arno (1965) described the statistical solution to determine the bed-slope profile over check dam by improving fundamental Laursen's equation; which is

$$\frac{d}{dt}[f(B)] = q(B) - q_s b \tag{28}$$

where, $f(B)$ is the sediment volume stored in the dam and is an arbitrary function of channel boundary, $q(B)$ is the sediment discharge capacity of channel and $q_s b$ is the out-flowing rate of sediment, b is the width of check dam. It is assumed that for the equilibrium states i.e. the condition at which the sediment carrying capacity of channel is equal to rate of sediment supply. If $\frac{d}{dt}[f(B)] = 0$, Eq. (20) can be written as

$$q(B) = q_s b \tag{29}$$

By employing sediment transport equation which is

$$\frac{q(B)}{b} = \psi\tau_0(\tau_0 - \tau_c) \tag{30}$$

where ψ is the experimentally determined coefficient, $\tau_0 = \rho g R S_e$ is average shear stress explained by ρ, g, R and S_e, which are density, gravity constant, hydraulic radius and energy slope respectively, τ_c is critical shear stress. By employing the friction factor from Manning's equation, they finally developed and compared the relationship between old and present bed-slope over the check dam section, which is

$$\frac{S_{up}}{S_{u0}} = \left(\frac{G_{sp}}{G_{s0}}\right)^{5/7}\left(\frac{\psi_0}{\psi_p}\right)^{5/7}\left(\frac{Q_0}{Q_p}\right)^{6/7}\left(\frac{n_0}{n_p}\right)^{6/7}\left(\frac{b_p}{b_0}\right)^{1/7} \tag{31}$$

In this equation, S_{up} and S_{u0} are average slopes above the structure at present and for old case, $G_{sp} = q_s b$. This equation was employed to determine the values of S_{up}/S_{u0} ratio in about 50 different fields. Later, after conducting regression analysis they developed a general expression in the following form, which was validated for those specified regions

$$\frac{S_{up}}{S_{uo}} = 0.591 + 0.00135b - 0.0527H \tag{32}$$

where H is the check dam height.

If the surface slope of the deposit is milder than the critical slope, deposition will continue on the pile of newly deposited sediment as long as the debris-flow is continuously supplied from upstream. However some amounts of outflow exist from beginning that was because of squeezed overland flow (Takahashi, 1991) and due to the difference in depositional speed and supply rate (Osti et al., 2005). The deposition or erosion process over the check dam can be determined according to the probable equilibrium bed slope that supposed to be formed during infinitively long debris flow event. VanDine (1996) estimated the relationship between storage angle (α) and storage capacity (V) behind a check dam. As far as the relation is exponential, he approximated the design value of storage capacity at which the change of storage volume (ΔV) is maximum for minimum change in storage angle ($\Delta\alpha$). For the determination of storage angle, he employed the Takahashi (1991) formula which is;

$$\frac{\tan\theta_d}{\tan\theta_u} = \frac{\tan\alpha}{\tan\phi} \tag{33}$$

where θ_d is the downstream channel gradient, θ_u is the upstream channel gradient, α is the internal kinetic friction angle of debris flow and ϕ is the internal static friction angle of sediment. Moreover, Ministry of Land Infrastructure Transport and Tourism, Japan (MLIT, 1984) has proposed its own empirical concept on determining the final bed profile over check dam. It has been approximated that the final bed profile will be in-between 1/2-2/3 of original bed slope of the reach at the check dam section. There are no such reliable proofs made for the validation of this concept so far, however it has been used as a design standard in Japan.

3.2. Potential Storage Volume

It is supposed that the debris flow can be controlled due to the change of stored sediment volume, although the storage area is initially filled up with sediment to the level of the dam crest. This idea was previously discussed by us (Osti et al., 2008) in terms of experimental and numerical results and the general criterion for the control function of the check dam was also discussed. The proposed criteria are verified for different check dam heights in the laboratory and later tested in the field. According to these results, the relative outflow sediment volume, which is normalized by inflow sediment volume, can be evaluated well using two non-dimensional parameters: relative inflow sediment volume and a non-dimensional potential storage volume. The potential storage volume can be defined by the height of check dam, initial bed slope, original bed slope, and equilibrium bed slope corresponding to inflowing debris. The results also help optimize the sizes and locations of check dams in steep mountain torrents.

We can consider a case of sediment control by a close type check dam having a storage area, which is filled up with sediment as shown in Figure 6(a). If debris-flow takes place and continues to flow, the dam will gradually lose its control function. However, it is expected that the dam can control debris-flow and change its characteristics when the slope of the storage area is less than the corresponding equilibrium bed slope.

In order to evaluate sediment control as general as possible, we introduced a concept of potential storage volume, V_p, which is defined as:

$$V_p = \frac{(H_d \cos\theta)^2 B}{2}\left\{\frac{1}{\tan(\theta-\theta_e)} - \frac{1}{\tan(\theta-\theta_0)}\right\} \tag{34}$$

in which H_d is the height of dam, B is the flow width, θ is the slope of original torrent bed, θ_0 is the initial bed slope of the storage area prior to a debris-flow, and θ_e is the equilibrium bed slope corresponding to the sediment concentration of debris-flow. The equilibrium bed slope can be explained by Eq. 35,

$$\tan\theta_e = \frac{(\sigma/\rho - 1)\bar{c}}{(\sigma/\rho - 1)\bar{c} + 1}\tan\phi_s \tag{35}$$

From a theoretical view point, the potential storage volume V_p, will be infinitely large if the equilibrium bed slope, θ_e, is steeper than the original slope, θ. However, such condition is unrealistic and unlikely to happen in nature because a debris-flow deposits part of its sediment load upstream where θ is shallower than θ_e. Accordingly, it is considered that the original bed slope, θ, is always larger than the corresponding equilibrium bed slope, θ_e.

It is convenient to obtain a non-dimensional form for potential storage volume in discussing problems associated with sediment control. Eq. 34 yields:

$$V_{p*} = \frac{V_p}{\dfrac{H_d^2 B}{2}\dfrac{\cos^2\theta}{\tan(\theta-\theta_e)}} = 1 - \frac{\tan(\theta-\theta_e)}{\tan(\theta-\theta_0)} \tag{36}$$

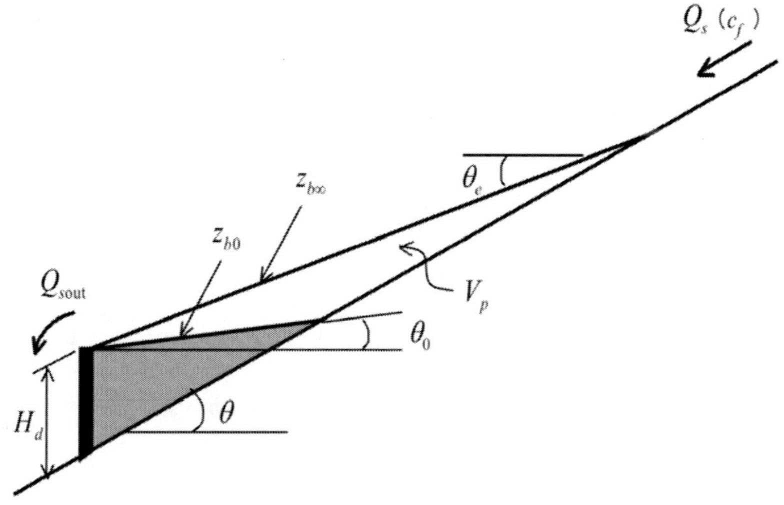

(a) Schematics of close type check dam.

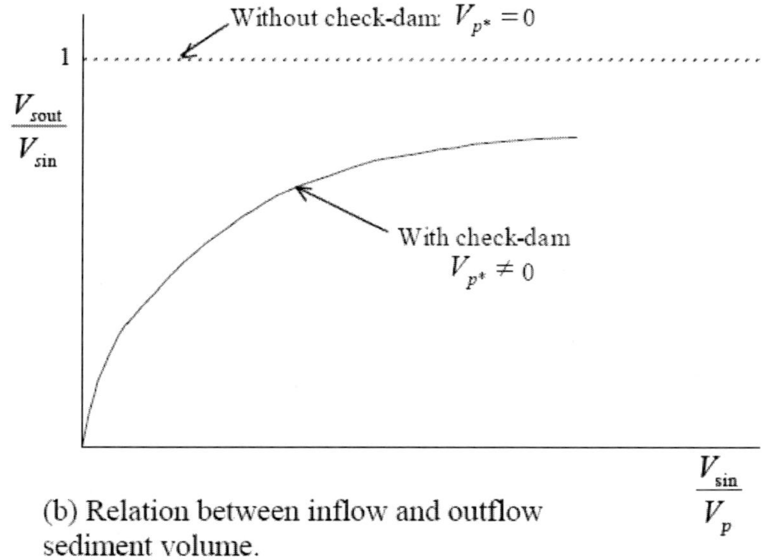

(b) Relation between inflow and outflow sediment volume.

Figure 6. Sediment control function using potential storage volume of a check dam.

Generally, non-dimensional potential storage volume, V_{p*}, takes a value from negative infinity to unity. However, negative values of V_{p*} have no physical meanings and should be considered zero. Therefore, the value of V_{p*} effectively ranges from zero to unity ($0 \leq V_{p*} \leq 1$). In addition, we suggest that a check dam will be more effective if V_{p*} increases toward unity and, in contrast, will lose its control function if V_{p*} tends toward zero. The condition that there is no dam can be expressed as $H_d = 0$ and $\theta_0 = \theta$.

In Figure 6(b), a relation between relative outflow sediment volume, V_{sout}/V_{sin}, and relative inflow sediment volume, V_{sin}/V_p, is shown schematically. Herein, V_{sin} is the total amount of sediment volume transported from upstream and V_{sout} is the total outflow sediment volume from the dam. We consider that a debris-flow can be controlled well if the relative inflow sediment volume is small. In contrast, the dam will lose its control function if the relative inflow sediment volume increases infinitely. It should be noted that debris-flow size will be further developed rather than controlled if the initial storage area's bed slope, θ_0, is steeper than θ_e. In the following sections, we assume sediment control conditions such that $\theta_0<\theta_e$. The criteria will be.

3.3. Debris Flow Model

In order to simulate the debris flow along a torrential reach, one-dimensional governing equations for the flow of the sediment-water mixture and the constitutive equations developed by Egashira et al. (1997a) are employed. The equations of mass conservation describing the water-sediment mixture and sediment-only are respectively expressed as:

$$\frac{\partial h}{\partial t}+\frac{1}{B}\frac{\partial \bar{u}hB}{\partial x}=\frac{E}{c_*} \tag{37}$$

$$\frac{\partial \bar{c}\, h}{\partial t}+\frac{1}{B}\frac{\partial \gamma\, \bar{c}\bar{u}hB}{\partial x}=E \tag{38}$$

The equations for momentum conservation and bed elevation are as follows:

$$\frac{\partial h\bar{u}}{\partial t}+\frac{\partial \beta h\bar{u}\bar{u}}{\partial x}=gh\sin\theta-gh\cos\theta\frac{\partial h}{\partial x}-\frac{\tau_b}{\bar{\rho}_m} \tag{39}$$

$$\frac{\partial z_b}{\partial t}=-\frac{E}{c_*\cos\theta} \tag{40}$$

where h is the flow depth, t is the time, x is the coordinate toward the flow direction, E is the erosion rate of bed sediment (its negative value denotes the deposition), $\bar{c}$ is the depth-averaged and volumetric sediment concentration, c_* is the sediment concentration of underlying sediment bed, $\bar{u}$ is the depth-averaged velocity of mixture, B is the river bed width, g is the acceleration due to gravity , and ρ_m is the depth-averaged mass density of sediment-water mixture defined as:

$$\rho_m=(\sigma-\rho)\bar{c}+\rho \tag{41}$$

where σ is the mass density of sediment particle, ρ is the mass density of water including fine sediment, τ_b is the bed shear stress, z_b is the bed elevation from a reference level, and θ is the bed slope, which is expressed as:

$$\theta=\sin^{-1}\left[-\frac{\partial z_b}{\partial x}\right] \tag{42}$$

$\gamma = (1/A)\int_A \{(cu)/(\bar{c}\,\bar{u})\}dA$ in Eq. (38) is a correction factor whose value ranges from 0 to 1 (Egashira et al., 2002) , depending on the bed-slope and inter-particle friction angle. This correction factor is introduced into Eq. (38), because the spatially averaged sediment concentration is generally used instead of the flux one. β in Eq. (39) is the momentum-correction factor and takes a value from 1.10 to 1.40 (Egashira et al., 2002) for a debris-flow. Equations for the bed shear stress and erosion rate from Egashira *et al.* (1997a,b, 2002; Egashira, 2007) and erosion rate (Egashira, 1993, Egashira et al., 2001) are employed:

$$\tau_b = \tau_y + \rho f \bar{u}^2 \tag{43}$$

$$\frac{E}{\bar{u}} = c_* \tan(\theta - \theta_e) \tag{44}$$

where τ_y is the yield stress caused by particle-to-particle contacts, f is the friction factor, θ is the bed slope, and θ_e is the equilibrium bed-slope (Eq. 35) corresponding to sediment concentration $\bar{c}$ of a debris-flow. Parameters τ_y and f are expressed as follows:

$$\tau_y = \left(\frac{\bar{c}}{c_*}\right)^{1/5} (\sigma - \rho)\bar{c}\, g\, h \cos\theta \tan\phi_s \tag{45}$$

$$f = \frac{25}{4}\left[k_f \frac{(1-\bar{c})^{5/3}}{\bar{c}^{2/3}} + k_d\left(\frac{\sigma}{\rho}\right)(1-e^2)\bar{c}^{1/3}\right]\left(\frac{h}{d}\right)^{-2} \tag{46}$$

where d is the particle diameter, $k_f = 0.16$, $k_d = 0.0828$, e and ϕ_s are the restitution coefficient (e=0.85) and the inter-particle friction angle of sediment particles, respectively. This model has been successfully tested in several river basins including San Julian in Venezuela (Osti and Egashira, 2008, Osti et al., 2004; Egashira et al., 2003).

3.4. Evaluation of Sediment Control Function of Check Dam

Let us try to show how the expression of sediment control can be simplified by using three non-dimensional values: relative outflow sediment volume; relative inflow sediment volume; and non-dimensional potential storage volume. In this section, we will also introduce some of our numerical and experimental results (Osti and Egashira, 2008; Osti et al., 2005; Osti et al., 2004). The experimental conditions are shown in Table 1. In Figures 7(a) and 7(b), the relative outflow sediment volumes are shown with respect to the relative inflow sediment volume and non-dimensional storage volume respectively, which are obtained from flume data for each check dam height as well as from the numerical computations based on the method proposed by Egashira et al. (2002) as described under debris flow model in this chapter. Herein, V_{sout} and V_{sin} are defined as:

$$V_{sout} = \int_0^{\infty} q_{sout}\, dt \tag{47}$$

$$V_{s\,in} = \int_0^T q_{s\,in}\,dt \tag{48}$$

in which q_{sout} is the out-flowing sediment discharge, q_{sin} is the inflowing sediment discharge rate, and T is the duration of sediment supplied by a debris-flow. In this chapter, uniform debris-flow supply for a specified time period is referred to a particular size of debris-flow.

Table 1. Experimental conditions

Debris-flow discharge q_m (cm^2/s)	Flux sediment concentration (c_f)	Dam height H (cm)	Run	Supplied duration T (Sec)	Dam height H (cm)	Run	Supplied duration T (Sec)
50.0	0.2	10.0	1	5.0	15.0	18	10.0
50.0			2	15.0		19	36.0
50.0			3	26.0		20	49.0
50.0			4	34.6		21	83.0
50.0			5	51.0		22	96.0
50.0			6	54.6			
50.0			7	77.0			
50.0			8	104.6			
50.0			9	154.6			
50.0			10	184.6			
50.0	0.1		11	11.84		23	26.0
50.0			12	19.84		24	44.0
50.0			13	35.84		25	90.0
50.0			14	51.84		26	125.0
50.0			15	67.84			
50.0			16	83.84			
50.0			17	95.84			

The parameter of the horizontal axis, previously introduced in Figure 6(b), is modified as $V_{sin}/(c_*V_p)$ by adding the sediment concentration coefficient for the stationary layer, c_*, to the calculated potential storage volume (Eq. 34). This is because both V_{sout} and V_{sin} are measured in absolute volumes. Both flume data and computed results show that the relative outflow sediment volume increases with an increase of relative inflow sediment volume. Hence, the dam loses its control function gradually with an increase in the size of debris-flow. Focusing on the flume data, it is clearly seen that the relative outflow sediment volume with large value of V_{p*} is smaller under a wide range of $V_{sin}/(c_*V_p)$ in comparison to the data with small value of V_{p*}. Correspondingly, the computed results show the same tendency as flume data in the range of $V_{sin}/(c_*V_p) < 1.5$. In addition, the data obtained from flume tests using the 15-cm-high check dam agree with the previously proposed trend. This indicates that the proposed relationships are also suitable for different conditions. However, in two groups of the computations with different values of V_{p*}, there are no significant differences in the range of $V_{sin}/(c_*V_p) > 1.5$.

Judging from the results presented in Figure 7, the check dams function effectively when $V_{sin}/(c_*V_p)$ is <1.0. In this range, both the flume data and computed results are in accord. Such correspondence suggests that the relationships, which are formulated by three non-dimensional parameters for evaluating the sediment control function of check dams, are useful for assessing the sizes, locations, and numbers of check dams for sediment hazard reduction and mitigation, because V_{sin} and V_p can be determined using field surveys and numerical simulations, and in addition the size of a check dam is included in V_{p*}. These are representative results and are valid for other combinations i.e. debris-flow, initial conditions and size of check dam. The relationships formulated by the three proposed parameters can be used to evaluate the control function of a closed type check dam fairly well and can be useful for planning and designing check dams.

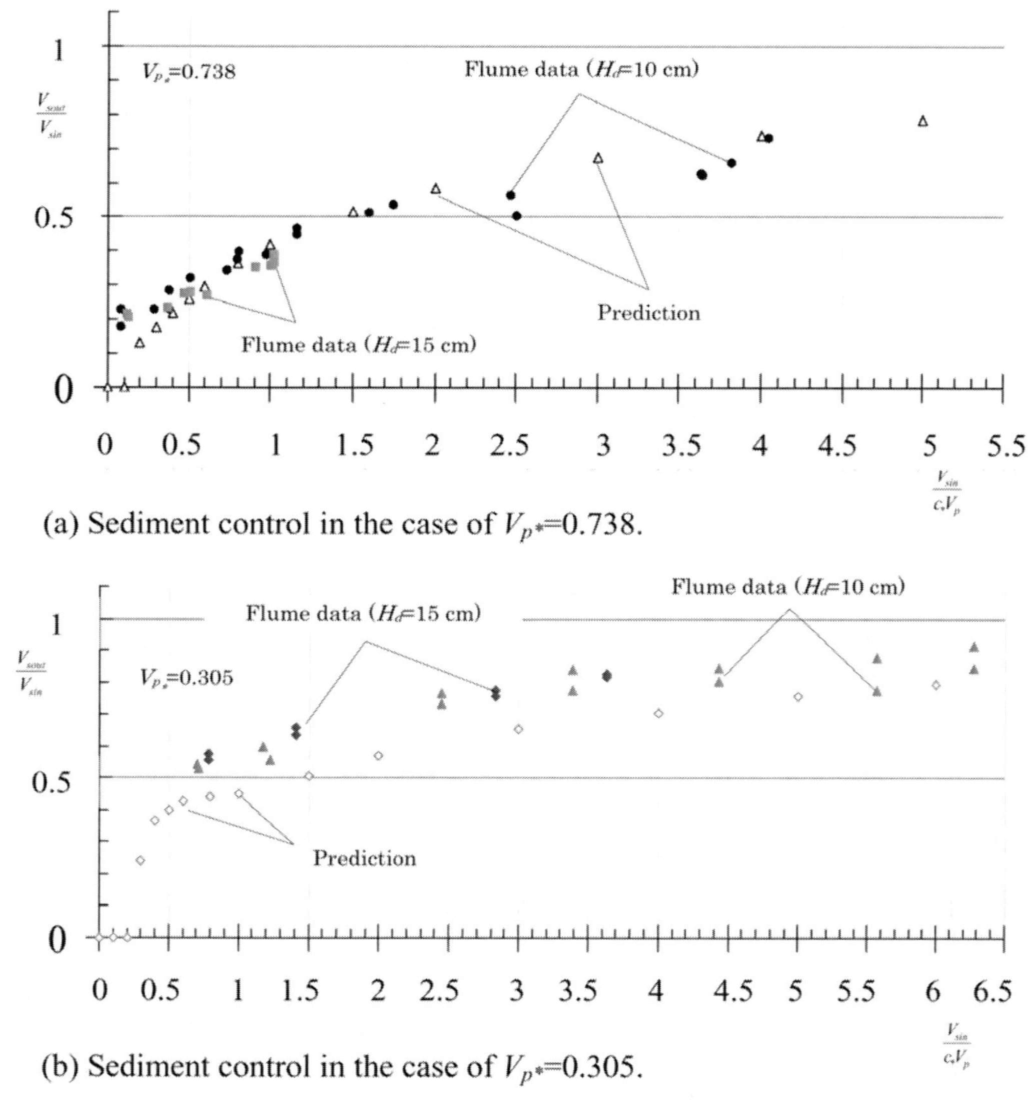

Figure 7. Generalized relationships between inflow and outflow sediment volumes for each value of non-dimensional potential storage volume.

4. Application of the Method to a Drainage Basin

4.1. Conditions of Computation

It might not be feasible to completely trap the sediment volume produced by a relatively large-scale debris-flow due to its amount. It is important, however, to reduce the amount so that the impact of the debris-flow hazard on an alluvial fan can be alleviated to a considerable degree. Overall planning of check dams involves several other factors and the method may also vary from place to place. In this chapter, we will introduce fundamental steps to analyze the problem, taking an example of the San Julian river basin in Venezuela.

The San Julian River is located in Vargas State in the northern coastal region of Venezuela. It drains an area of 23.6 km^2 to the sea and is composed of numerous tributaries. On December 15 and 16, 1999, large-scale floods and debris flows occurred in the river due to very unusual and localized rainfall. The San Julian fan was heavily damaged due to the event.

Three major tributaries are chosen for debris-flow simulations. The longitudinal bed profiles of tributaries denoted by T_1, T_2 and T_3, which were surveyed prior to the 1999 debris-flows, are illustrated in Figure 8. The upstream boundaries are set at 8210m, 7560m and 6140m from the river mouth for T_1, T_2 and T_3 respectively, and the corresponding drainage areas are 2.63km^2, 1.42km^2 and 1.43km^2. The common downstream boundary is set at 3000m from the river mouth. Important parameters, such as flow width, physical properties of sediment, and potential erosion depths, are determined by field surveys, laboratory experiments and, in some instances, expert knowledge or data from secondary sources. The physical properties of the bed material are specified as follows: uniform particle size diameter, (d)=20cm; inter-particle friction angle of sediment particles, (ϕ_s)=34^0; mass density, (ρ)=1.33g/cm^3; bed material concentration, (c_*)=0.52 (Egashira *et al.*, 2003). The potential erosion depth (D_p) throughout the reaches is estimated, based on expert knowledge, at 10m, which was cross-checked in some river sections in the field. The river-bed width (B) from all three upstream boundaries down to the confluence point at 5100m from the river mouth is specified as 20m, and that from the confluence point to the river mouth is specified as 40m. For simplicity, the input water discharge at each upper boundary is estimated using the following rational formula:

$$Q = \frac{1}{3.6} f_p r A \tag{49}$$

where f_p is the rainfall run-off coefficient, r is the rainfall intensity, and A is the drainage area. Based on the recorded hydro-meteorological data of December 1999, a rainfall intensity of 50mm/hr is assumed uniformly over the catchments. The estimated boundary discharges are 21.9m^3/s for T_1, 11.8m^3/s for T_2 and 11.9 m^3/s for T_3.

Firstly, debris-flow simulation in the absence of check dams is conducted for each tributary and the anticipated sediment volumes to be transported to the fan by the tributaries are calculated. Secondly, debris-flow simulation in the presence of check dams is conducted for each tributary, providing: a) the initial bed for the simulation is same as that left by the event of 1999, which was also computed; b) the rainfall runoff event is similar to the 1999 event and; c) the potential erodible bed depth is same as that left or developed by the first event, i.e. the debris-flow in 1999.

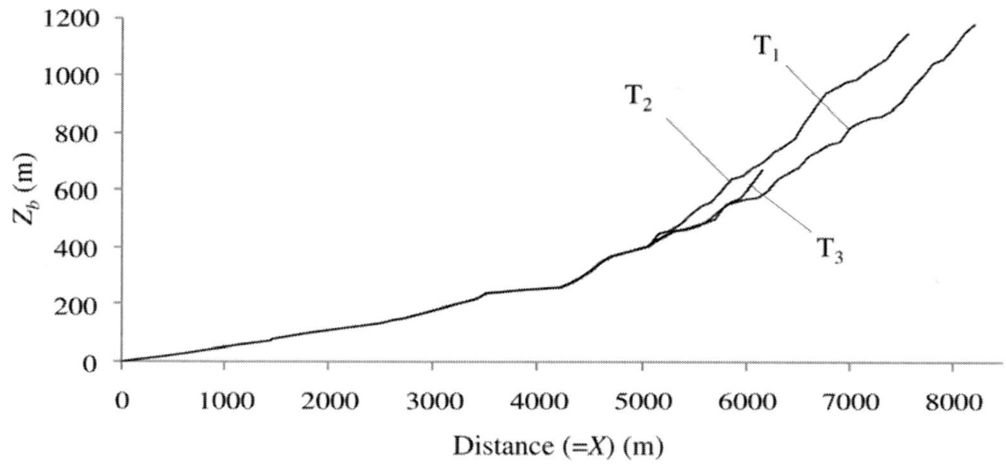

Figure 8. Bed profiles of T1, T2 and T3.

Tributary T_1 is especially considered to evaluate what could have happened if there had been already check dams constructed prior to the event of December 1999. Check dams' locations and sizes are proposed according to the check-list described in the following section. Since the evaluation of the control function of check dams is based on the assumption that check dams are initially filled up with sediment up to the crest level, the estimation of the sediment volume trapped by each check dam is carried out based on the potential storage volume (V_p) of each dam, which can be defined by equilibrium and the initial bed slopes of deposition. V_p can be calculated using Eq. 34. The average sediment concentration of a debris-flow ($\bar{c}$) is determined calculating the temporal change of the sediment concentration at each check dam section. Therefore, the equilibrium bed slope (θ_e) corresponding to a specific sediment concentration can be determined for each check dam (Eq. 35). While calculating V_p of each check dam, local topographical conditions, especially non-uniformities in bed slopes, are additionally considered (Figure 9). In Figure 9, x and y are the distances over which bed-slopes θ_1 and θ_2 expand respectively, c and d are the spans covered by θ_0 and θ_e respectively, and h is the check-dam height. By definition (Eq. 34), V_p of each check dam can be occupied fully only when a uniform supply of debris-flow exists for an infinitely long time, which seldom happens in nature. Therefore, only a part of V_p can be considered to be occupied during a debris-flow. In order to be on the safe side, 20-30% of V_p is reasonable to presume as the estimated trapped volume, which was estimated from the experimental results of different supply conditions (Osti *et al.*, 2005). Although V_p can be said as a virtual capacity of a check dam even after it is filled with sediment up to its crest level, the initial storage volume that sometimes refers to the dead volume of each check dam can also be considered as a trapped volume if a check dam is not fully occupied prior to the debris-flow.

For an example, the estimated sediment volume that can be trapped by each check dam is considered to be 20% of the potential storage volume.

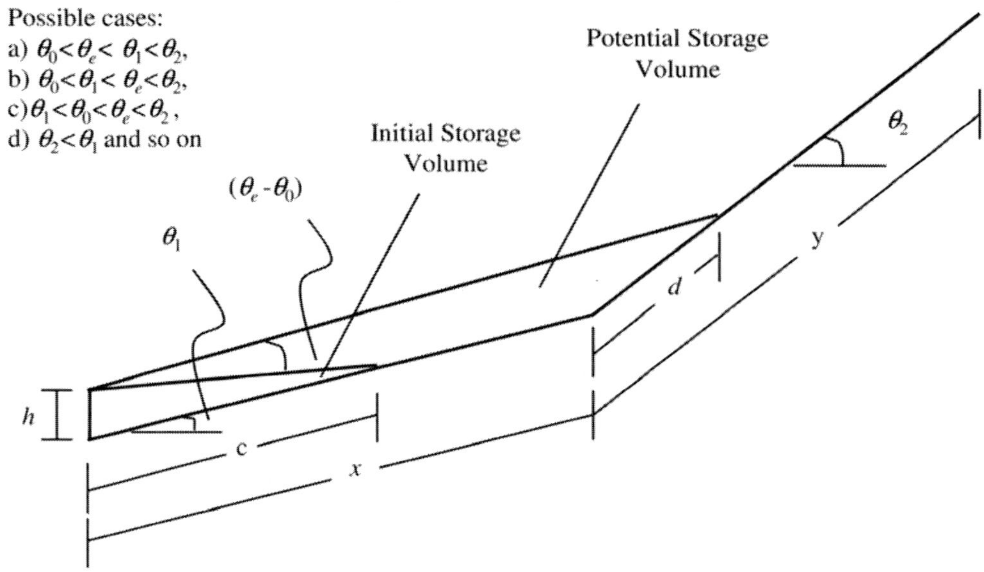

Figure 9. Potential storage volume in different topographical conditions.

After selecting possible locations, the number and/or height of the check dams are gradually increased until the expected degree of control is achieved. In practices, only the height of selected check dams can be increased considerably, while the other check dams remain same or can be completely eliminated. For simplicity, the required numbers of check dams are placed serially from the upstream end, and the dam height is considered to be uniform throughout the torrent. After deciding the location, number and size of the check dams, numerical simulations were performed to verify the effectiveness of the proposed system.

4.2. Results

The total sediment volume (V_{inflow}) produced from tributary T_1 during the event of December 1999 without the presence of any check dams is estimated at 109,000m^3. The main aim of this calculation is to find out what could have happened if there had been already a few check dams constructed prior to December 1999. All check dams numbered from 1 to 9 indicated are supposed to be 6m high (Table 2). The potential storage volume of each check dam is estimated using Eq. 34. The estimated cumulative sediment volumes trapped by a series of check dams (No. 1-9) are presented in Table 2. The total sediment volume that can be trapped by check dams 1-9 in T_1 is estimated at 68,000m^3, which is the cumulative figure of 20% of the potential storage volume of each check dam. Similarly, the total sediment volume (V_{inflow}) to be produced by each tributary during anticipated debris flows in all three major tributaries without check dams and the estimated cumulative sediment volumes trapped by a series of check dams in each tributary are also shown in Table 2. In order to simplify the analysis, a performance index, i.e. relative stored sediment volume (V_{stored}/V_{inflow}), is introduced and calculated for all series of check dams (Table 2).

Table 2. Estimated sediment volume trapped by check dams

Tributary/(Case)/	H_d (m)	No	20% of V_p (m ^{3}) Cumulative	V_{inflow} (m^3)	$\frac{V_{stored}}{V_{inflow}}$
T_1 (1999)	6	1	2,000	109,000	0.018
		2	5,000		0.046
		3	10,000		0.092
		4	18,000		0.165
		5	19,000		0.174
		6	25,000		0.229
		7	38,000		0.348
		8	52,000		0.477
		9	68,000		0.624
T_1 (anticipated)	3	1	3,500	68,000	0.051
		2	4,000		0.059
		3	6,000		0.088
		4	7,000		0.103
		5	11,000		0.162
		6	18,000		0.265
		7	27,000		0.397
		8	28,000		0.412
		9	42,000		0.618
		10	55,000		0.809
T_2 (anticipated)	3	1	3,000	47,000	0.064
		2	5,000		0.106
		3	11,000		0.234
		4	14,000		0.298
		6	21,000		0.447
		9	29,000		0.617
		5	40,000		0.851
		7	51,000		> 1
		8	53,000		> 1
T_3 (anticipated)	3	4	6,000	35,000	0.171
		5	12,000		0.343
		6	16,000		0.457
		7	20,000		0.571
		1	24,000		0.686
		2	26,000		0.743
		3	28,000		0.800

This suggests a different degree of control by the check dams in each tributary, e.g. check dams 1-9 in T_1 are estimated to control 62% of debris flow volume. This approach of analysis is fully based on the concept of potential storage volume and is yet to be verified by the numerical results.

Several runs of numerical simulation for each tributary in the presence of check dams were conducted to predict outflow sediment hydrographs and determine the total outflow sediment volume for each single check-dam and each series of check dams. Different combinations of check dams including a single check dam, as described in the previous sections, are proposed for each torrent. In this case, a non-dimensional parameter V_{stored}/V_{inflow} for each combination of check dams is calculated by subtracting relative outflow sediment volume ($V_{outflow}/V_{inflow}$) from unity (Table 3).

Table 3. Calculated outflow sediment volumes

Tributary	V_{inflow} (m^3) (No check dams)	Combinations/Height of Check Dams		$V_{outflow}$ m^3 (check dams)	$\frac{V_{stored}}{V_{inflow}}$
		No	H_d (m)		
T_1 (before 1999)	109,000	1-9	6	48,000	0.560
T_1 (probable event)	68,000	1-5	3	55,000	0.191
		1-10	3	16,000	0.765
		7	15	26,000	0.559
T_2 (probable event)	47,000	1-3	3	37,000	0.213
		1-4,6,9	3	19,000	0.596
		1-9	3	0	1.000
		9	15	18,000	0.383
T_3 (probable event)	35,000	4-7	3	23,000	0.343
		1-7	3	11,000	0.686
		7	12	22,000	0.371

In the case of the 1999 debris flow event, if there had been already nine check dams in tributary T_1 at 350, 850, 1150, 1550, 1850, 2150, 2350, 2650 and 2850m, each 6m high, the debris flow could have been mitigated significantly, reducing both the transported sediment volume (Table 2) and peak discharge, as well as increasing the lead time (Figure 10). In fact, the total sediment volume supplied from tributary T_1 into the fan could have been about 48,000m^3, instead of 109,000m^3, with the nine check dams installed. Similarly, the peak discharge could have been reduced to a fifth. This clearly indicates how check-dam systems of this kind could have protected the communities if implemented prior to December 1999. In Table 2, the estimated V_{stored}/V_{inflow} ratio in the presence of nine check dams is 0.62, whereas the calculated V_{stored}/V_{inflow} ratio in Table 3 is 0.56 (i.e. $V_{outflow}/V_{inflow}$=0.44), which indicates that the estimation was good because the value is close to the simulated result.

Selection of check dam locations and sizes against potential debris flow events in all three major tributaries are done in a similar manner. In tributary T_1, different combinations of check dams are tested: a) 5 check dams, 3m high each; b) 10 check dams, 3m high each; c) a single 15m-high check dam at 2300m from the river head. The results show that both the peak discharge (Figure 11) and total supplied sediment volume (Table 3) are significantly reduced by any combination of check dams, but at different scales. Figure 11(a) and Table 3 show that the single 15m-high check dam located at 2300m from the river head is also found to be relatively effective in debris flow control. In order to optimize the control systems, further examination is required to determine the most effective check dam combinations. In this example, numbers are written serially from upstream to downstream and for simplicity, combinations are considered accordingly, however selection of location among potential places can be done in a random basis and the height of dams could be differ from one location to another. For an easy understanding, a simple random combination (1-4, 6 and 9) is proposed in T_2. In addition, selection of check dams for each combination can be based on the maximum potential storage volume of each check dam calculated in Table 2.

Similarly, the effects of check dams in T_2 are evaluated. Calculated sediment volume without considering check dams at the fan head is 47,000 m^3, which is less than the estimated capacity of 9 check dams, i.e. 53,000m^3 (Table 2); therefore a debris flow can be fully controlled (Figure 11(b)). Similar results are obtained for tributary T_3 (Figure 11(c)).

In most cases, calculated V_{stored}/V_{inflow} (Table 3) fairly agrees with the estimated value of V_{stored}/V_{inflow} (Table 2). There are a few disagreements especially in the combinations, in which check dam locations are found in relatively steeper slopes. Check dams 1-3, 6, and 7 in T_3 are located in more than 12 degree bed slopes. It suggests that the chosen portion of potential storage volume (20%) of each check dam for estimation is somewhat higher for steeper slopes. In addition, differences between estimated and calculated values are due to the use of averaged sediment concentration of inflowing debris flows as well as due to the optimization of total computational time of debris flows in the study. This indicates that the analyses based on potential storage volumes without sophisticated numerical simulations are possible to evaluate the control function of check dam systems in applied engineering fields.

In order to evaluate whether an anticipated debris flow can be controlled by the existing check dam system, it is necessary to know the potential storage volume of the system which is determined by existing initial bed slope of the deposited sediment in each check dam and the sediment concentration of anticipated debris flow. The bed slopes can be surveyed to specify the potential storage volume of the system, whereas the sediment concentration of anticipated debris flow can be estimated or calculated in terms of numerical simulation, supposing potential erosion depth or thickness of unstable sediment along the reach. However, a regular monitoring of the system is required to evaluate the sediment control function and if necessary excavation should be done to ensure that the anticipated large scale debris flows can be controlled by the existing system.

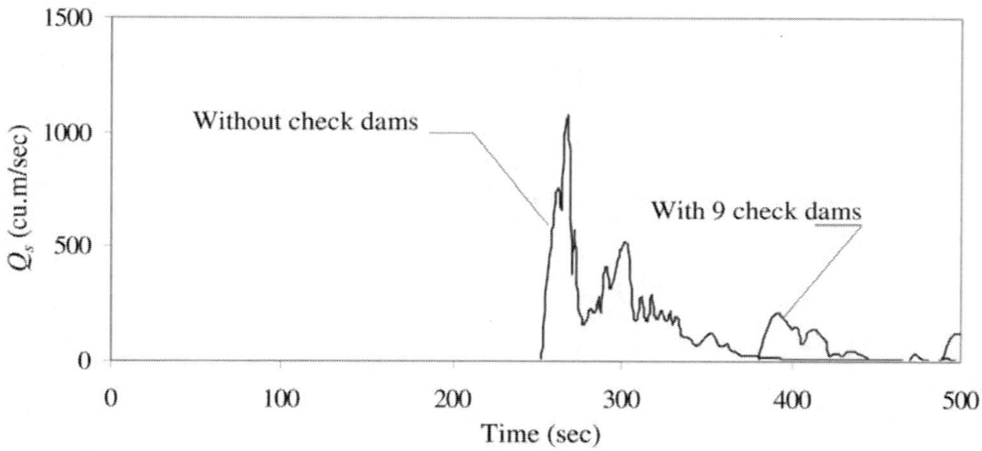

Figure 10. Reproduced hydrographs at 5200m in T_1 for debris-flow event 1999 with and without check dam provisions.

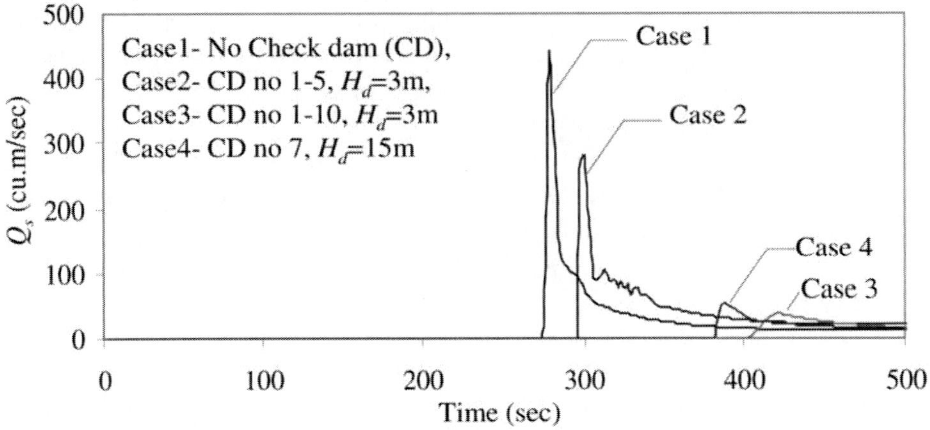

(a) Predicted hydrographs at 5200m in T_1

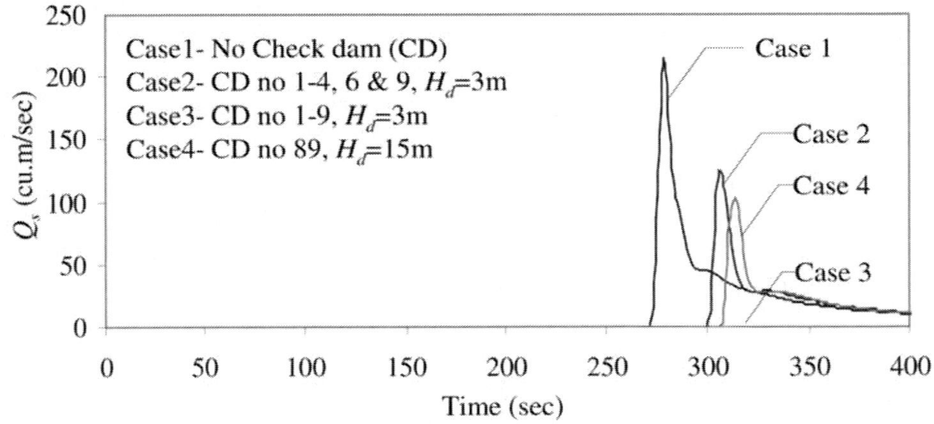

(b) Predicted hydrographs at 4550m in T_2

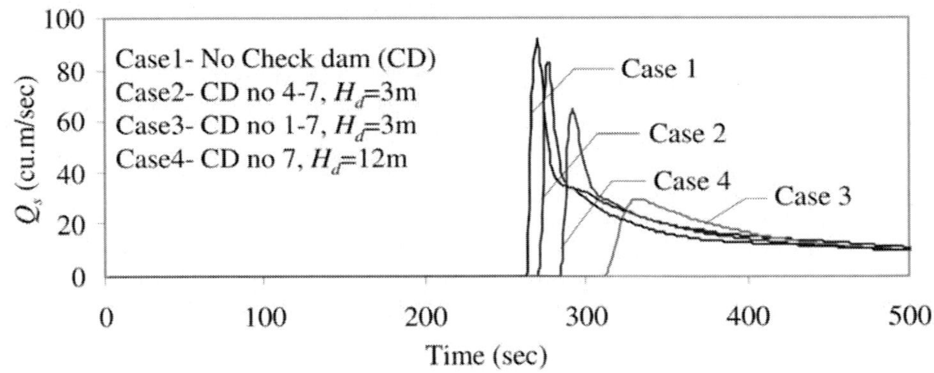

(c) Predicted hydrographs at 3130m in T_3

Figure 11. Control of sediment hydrograph by different combinations of proposed check dams in each tributary.

5. Optimization of Check Dam Locations

A series of check dams are preferable in a debris-flow transportation zone, because they can not only trap the sediment volume but additionally can stabilize the channel slope and therefore reduce the debris-flow size. However, construction of numerous dams is costly and may take considerable time. Therefore, a few or a single dam with an appropriate storage capacity would sometimes be a reasonable solution. Moreover, the effectiveness of an adopted system depends not only on the number and size of check dams but also on their locations. There are certainly many other criteria for the selection of check-dam locations; however the following general criteria are considered and verified in the context of the San Julian river basin in Venezuela.

(a) It is better to place a check dam downstream of a heavily-deposited or erosion-prone area so that the size of an upcoming debris-flow caused by bed erosion can be reduced significantly.
(b) A place where a torrent changes its gradient from steeper to milder is naturally potential for deposition, and the placement of a dam in the vicinity of such a location might not be much beneficial. However, if such a milder slope extends over a relatively long distance, considerable potential storage volume can be achieved.
(c) In connection with the criterion mentioned in (b), the amount of sediment to be deposited in the pocket of a check dam is greatly influenced by an existing torrent bed slope.
(d) Usually, if a check dam is placed in the vicinity of another check dam, the storage volume of those dams can be overlapping. The minimum spacing (*L*) required between two consecutive check dams can be estimated by Eq. 42, proposed by Osti and Egashira (2008):

$$L > \frac{H_d}{\tan\left(\bar{\theta} - \theta_e\right)} \tag{50}$$

where H_d is the height of the downstream check dam, $\bar{\theta}$ is the average bed-slope between the two check dams, and θ_e is the equilibrium bed slope.

(e) If large numbers of small size check dams are not affordable, a few or even a single check dam of considerable size can be recommended. The location of such check dams should be somewhere near the fan head satisfying all the above mentioned criteria.
(f) Drainage basins are usually composed of several tributaries. If resources are not enough to construct many check dams in each tributary, at least it is worth selecting one tributary, which has a significant contribution to debris-flow production.
(g) Even if the location is satisfied based on the criteria described in numbers (a)-(f), it may not be that much helpful to put the check dam if the section is narrow and stable enough as such sections usually promote channel storage fuction (Osti and Egashira, 2009) like as the function of any check dam.

Other factors, such as accessibility to check dam stations for inspection and maintenance, geological conditions, stability of surrounding slopes, availability of construction materials, environmental concerns, cost-benefit issues, may also influence the selection of check dam locations. These factors are site-specific, and therefore should be examined precisely by field inspections.

CONCLUSIONS

In this chapter, the concepts of bed-load transport and debris-flows are described by reviewing available constitutive equations. Constitutive relations for debris flow originally proposed by the second author of this chapter and his work of extending the proposed debris-flow equation to discuss active bed-load rates are discussed. It has been emphasized that the flux sediment concentration which is defined as sediment discharge divided by flow discharge changes continuously in uniform flows from bed load to debris flow range. Therefore, same sets of equations can be used to predict both debris-flow and bed-load. Utilizing this knowledge, a simple, reliable relationship for evaluating the control function of check dams was introduced. The chapter discusses the control function of a close type check dam with special attention focused on sediment volume transported by debris flows. A concept of potential storage volume is introduced to reduce the number of parameters needed for evaluating control of sediment runoff by a close type check dam. Three different non-dimensional parameters, namely relative outflow sediment volume, relative inflow sediment volume, and non-dimensional potential storage volume, are introduced based on dimensional analysis and results obtained from numerical simulations. The relationships formulated by the three proposed parameters can be used to evaluate the control function of a closed type check dam fairly well and may be useful for planning and designing check dams in any torrents. A planning tool for check dam systems against debris flows is also described. As an example, several possible combinations of check dams are proposed and tested in the San Julian river basin in Venezuela. Changes of debris flow characteristics, such as debris flow hydrograph, peak discharge, sediment volume, due to the presence of check dams are discussed. Method of evaluating the control function based on the potential storage volume of each check dam is introduced and successfully compared with the numerical results. The results show that the 1999 debris flows in the San Julian river basin would not have created that much severity if there had been a few check dams constructed before the event. It has been found that the effective control of debris flows can be achieved not only by increasing the number and size of check dams but also by selecting appropriate locations. General criteria for the selection of check dam sizes and locations are proposed.

REFERENCES

Arai M. and Takahashi T. (1988). Depositing process of mud-flow on gentle slope bed plunging from steep slope. *In Proc. of 6th Congress of IAHR*, Asia and pacific Division, Kyoto, 83-90.

Armamini A. and Larcher M. (2001). Rational criterion for designing opening of slit-check dam. *Hydraulic Engineering*, 127(2), 94-104.

Ashida K. and Egashira, S. (1989). The mechanism of sediment transport and control in mountain streams, In *Proc. of the Japan-China Joint Seminar on Natural Hazard Mitigation,* Kyoto, 383-397.

Ashida, K., Egashira, S., Kurita, M. and Aramaki, H. (1988). Mechanism of debris flow control with grid dam. In *Proceedings of the 32th Japanese Conference on Hydraulics*, JSCE, Japan, 723-728 (in Japanese).

Ashida, K. and Michiue, M. (1972). Study on hydraulic resistance and bed load transport rate in alluvial stream. In *Proc. Japan Society of Civil Engineers JSCE 206*, 59-69 (in Japanese).

Bagnold, R.A. (1966). An approach to the sediment transport problem from general physics, U.S. Geol. *Survey Prof. Paper*, 422-I.

Bovolin, V. and Mizuno H. (2000). Experimental study on the effect of a check dam against mudflow. In *Proc. of 2nd Int. Con. on Debris-flow Hazard Mitigation: Mechanics, Prediction, and Assessment*, Taipei, Taiwan, 573-577.

Busnelli, M. M., Stelling, G. S. and Larcher, M. (2001). Numerical morphological modeling of open-check dams. *Hydraulic Engineering*, 127(2), 94-104.

Cannon S. H. (2001). Debris-flow generation from recently burned watersheds, *Environmental and Engineering Geoscience*, 7(4), 321-341.

Cannon S. H., Bigio E. R., Mine E. (2001). A process for fire-related debris flow initiation, Cerro Grande fire, New Mexico. *Hydrological Processes*, 15(15),3011 – 3023.

Chen, C. (1988). General solutions for viscoplastic debris flow. *J. Hydraul. Eng.*, 114, 237-258.

Chen, R. and Ho, M. (1999). The Effects of Open Dam on Debris Flow. In *Proc. of 2nd Int. Conf. on Debris Flow hazard Mitigation: Mechanics, Prediction, and Assessment*, Taipei, Taiwan, 626 – 631.

Costa J. E. (1994). Multiple Flow Processes Accompanying a Dam-break Flood in a Small Upland Watershed, Centralia, Washington. U.S. Geological Survey Water-Resources Investigations Report, 94-4026: 24.

Egashira S. (1993). Mechanism of debris flow ceasing and sediment deposition. Lecture Series. *Journal of the Japan Society of Erosion Control*, 46 (1), 45-49 (in Japanese)

Egashira S. (2007). Review of research related to sediment disaster mitigation. *Journal of Disaster Research*, 2(1), 11-18.

Egashira S. and Ashida K. (1992). Unified view of the mechanics of debris flow and bed load. Shen H. H. et al. (eds.), *Advances in Micromechanics of Granular Materials*, Elsevier, 391-400.

Egashira S., Honda N. and Itoh T. (2001). Experimental study on the entrainment of bed material into debris flow. *Phys. Chem. Earth* (C), 26 (9), 645-650.

Egashira S., Miyamoto K., Ito T. (1997a). Constitutive equations of debris-flow and their applicability. In *Proceedings of 1st International conference on Debris-flow Hazards Mitigation,* Chen CL (eds), ASCE, NewYork, 340-349.

Egashira S. Miyamoto K. and Itoh T. (1997b). Bed-load rate in view of two phase flow dynamics, *Annual Journal of Hydraulic Engineering, JSCE*, Vol. 41, pp.789-794.(in Japanese)

Egashira, S., Itoh, T., Miyamoto, K. and Honda, N. (2002). Importance of correction factor associated with sediment concentration profile in debris flow simulation. In *Proceedings of 2nd International Symposium on Flood Defense,* Beijing, 1658-1666.

Egashira, S., Nakagawa, H., Itoh, T., and Osti R. (2003). Countermeasures and their Theoretical Background Against Sediment Hazard. In *Proc. of Int. Civil Engineering Conf. for Sustainbale Development in 21st Century*, Nairobi, Kenya, 425-436.

Egashira, S. and Itoh, T. (2005). Paradoxiacal discussions on sediment transport formulas. In *proceedings of River, Coastal and Estuarine Morphodynamics: RCEM.* Parker and Garcia (eds), Taylor and Francis Group, London, 33-38.

Einstein, A.H. (1950). The bed-load function for sediment transportation in open channel flows. US Department of Agriculture Technical Bulletin 1026, *Soil Conservation Service*, Washington, DC: p.71.

Engelund, F., and Hansen, E. (1967). A monograph on sediment transport in alluvial streams. *Technical Forlag*, Copenhagen, Denmark.

Graf, W. H. (1971). *Hydraulics of Sediment Transport*. McGraw-Hill Book Company. UK.

Graf, W.H. and Suszka, L. (1987). Sediment Transport in Steep Channels. *J. Hydroscience and Hydraulic Engrg.*, 5(1), 11-26.

Honda N. and Egashira S. (1997). Prediction of debris flow characteristics in mountain torrents. In *Proceedings of 1st International conference on Debris-flow Hazards Mitigation*, Chen CL (eds). ASCE: NewYork, 707–716.

Ito, T and Egashira, S. (2007). Influence of of debris flow discharge on equilibrium bed slope. In *Proceedings of 4th International Conference on Debris-flow Hazards Mitigation*, Chen CL and Major JJ (eds). Chengdu, China, 151-163.

Iverson, R.M. and Denlinger, R.P. (2001). Flow of variably fluidized granular masses across three-dimensional terrain: 1. Coulomb mixture theory. *Journal of Geophysical Research*, 106(B1), 37-552.

Johnson P. and McCuen R. H. (1989). Slit dam design for debris flow mitigation. *Journal of Hydraulic Engineering*, 15,293-1296.

Julien, P.Y. and Y.Q. Lan. (1991). Rheology of hyperconcentrations. *Journal of Hydraulic Engineering,* ASCE, 115(3),346-353.

Kershaw J.A., Clague J.J. and Evans S.G. (2005). *Geomorphic and sedimentological signature of a two-phase outburst flood from moraine-dammed Queen Bess Lake*, British Columbia, Canada. Earth Surface Process and Landforms, 30, 1-25.

Lin, P., Chang, W. and Liu, K. (1999). Retaining Function of Open Type Sabo Dams, In *Proc. of 2nd Int. Conf. on Debris Flow hazard Mitigation*, 636 – 645.

Marion A., Lenzi M. A. and Comiti F. (2004). Effect of sill spacing and sediment size grading on scouring at grade-control structures. *Earth Surface Processes and Landforms*, 29, 983-993.

Meyer-Peter, E., and Müller, R. (1948). Formulas for bed-load transport. In *Proc. of 2nd Meeting of International Association of Hydraulic Research*, Stockholm.

Mizuyama T., Oda A., Nishikawa S., Morita A. and Kasai S. (2000). Structures for controlling debris-flows in torrents where debris-flow does not occur frequently. In *Proceedings of 2nd International Conference on Debris-flow Hazards Mitigation*, Naeser W (eds). Balkema: Rotterdam, 579-582.

Mizuyama, T., Shimohigashi, H., Shimoda Y. and Ido K. (1984). Experimental study of a sabo dam with slit. *Proc. of 28th Japanese Conference on Hydraulics*, JSCE, 723-728 (in Japanese).

MLIT-Ministry of Land Infrastructure Transport and Tourism. (1984). Basics of planning the measures against debris flows and planning countermeasure facilities ageist debris flow. *Government of Japan*, Tokyo, Japan, p. 39.

Nakagawa, H. and Tsujimoto, T. (1980). Sand bed instability due to bed-load motion. J. Hydraul. Div., *Am. Soc. Civ. Eng*., 106(12), 2029–2051.

Nedelcu L. O. (1999). The usefulness of new model for gully-control structures effects prediction, In *Proc. of 10th Int. Soil Conservation Organisation Conference*, Lafayette, Indiana USA: 1000-1007.

Nsom, B. (2002). Horizontal Viscous Dam-Break Flow: Experiments and Theory, *J. Hydraulic Engineering*, 128, 543-546.

O'Brien, J.S. and Julien, P.Y. (1988). Laboratory analysis of mudflow properties. *Journal of Hydraulic Engineering*, 114, 877 – 887.

Osti R. and Egashira S. (2008). Method to improve the mitigative effectiveness of a series of check dams against debris flows. *Hydrological Processes*, 22, 4986-4996.

Osti R. and Egashira S. (2009). Hydrodynamic characteristics of the Tam Pokhari Glacial Lake outburst flood in the Mt. Everest region, Nepal. *Hydrological Processes*, 23, 2943-2955.

Osti R., Egashira S., Ito T. (2004). Prediction of 1999-San Julian debris flows based on dependent and independent occurrences. *Annual Journal of Hydraulic Engineering-Japan Society of Civil Engineers*, 48, 913-918.

Osti, R, Itoh, T. and Egashira, S. (2007). Control of sediment run-off volume through close type check dams. In *Proceedings of 4th International Conference on Debris-flow Hazards Mitigation*, Chen CL and Major JJ (eds). Chengdu, China, 616-625.

Osti, R., Egashira, S. and Itoh, T. (2005). Sediment control function of close type check dam. In *Proc. of 31st International Association of Hydraulic Engineering and Research (IAHR) Congress*, Seoul, Korea, 1, 3112-3121.

Parker, G., Kilingeman, P.C. and McLean, D.G. (1982). Bed load and size distribution in paved gravel-bed streams. *In J. of Hydraulic Division*, ASCE 108(4), 544-571.

Pierson, T.C. and Scott, K.M. (1985). Downstream Dilution of a Lahar: Transition from Debris Flow to Hyperconcentrated Streamflow. *Water Resources Research* 21(10),1511-1524.

Savage, S. B. and Hutter, K. (1991). The dynamics of avalanches of granular materials from initiation to runout. part I: Analysis, *Acta Mech*., 86, 201– 223.

Savage, S. B. and Hutter, K. (1989). The motion of a finite mass of granular material down a rough incline. *J. Fluid Mech*., 199, 177– 215.

Scott, K.M., Pringle, P.T. and Vallance, J.W. (1992). Sedimentology, Behavior, and Hazards of Debris Flows at Mount Ranier, Washington. USGS Open File Report, 90-385:106p.

Shinohara, K. and Tsubaki, T. (1959). On the characteristics of sand waves formed upon the beds of open channels and rivers. *Reports of Res. Institute of Applied Mechanics*, Kyushu University, 7(25), 15-45.

Takahashi, T. (1980). Debris flow on prismatic open channel. *J. of hydraulic Engineering, ASCE,* 106 (HY3), 381-396.

Takahashi T., Nakagawa H., Satofuka Y. (2001). Flood and sediment disasters triggered by 1999 rainfall in Venezuela; A river restoration plan for an alluvial fan. *Natural Disaster* Science, 23(2), 65-82.

Takahashi, T. (1991). *Debris flow, IAHR Monograph Series*, Rotterdam: Balkema.

Tsubaki, T., Hashimoto, H. and Suetsugi, T. (1982). Grain stresses and flow properties of debris flows. *Proceedings of the Japan Society of Civil Engineers*, 317, 79–91. (in Japanese).

VanDine, D.F. (1996). *Debris flow control structures for forest engineering*. Ministry of Forests Research Program, Working Paper 22/1996, Government of the Province of British Columbia, Vancouver.

Woolbiser A. D. and Arno T. L. (1965). Channel gradient above gully structure. Hydraulics *Engineering*, 91(3), 165-188.

Yang C.T., Sayre W.W. (1971). Stochastic model for sand dispersion. *Journal of the Hydraulics Division*, 97, 265–288.

In: Check Dams, Morphological Adjustments… ISBN: 978-1-62618-856-3
Editors: C. C. Garcia and M. A. Lenzi

Chapter 9

Filtering-Retention Check Dam Design in Mountain Torrents

Vincenzo D'Agostino*

TeSAF Department, University of Padova,
Viale dell'Università 16, 35020 Legnaro (PD), Italy

Abstract

During floods, mountain rivers are greatly affected by different forms of sediment transport, ranging from bedload to hyperconcentrated and debris flows. In the alpine region, there exists a long tradition (e.g., Thiery, 1891; Valentini, 1912) of protecting human settlements and activities against such flood events. The most effective strategy controls water-sediment flows by means of deposition basins associated with check dams located at their outlets. In the last fifty years, closed barriers have been replaced by open check dams, and thanks to well-focused research, a progressive improvement in their design and trapping efficiency has been observed. After introducing the historical evolution of retention check dams, the chapter will address current typologies of integrated works (deposition basin-check dam) that are capable of mitigating flood hazards over fan areas. Moreover, design parameters are also assessed in relation to the main characteristics of the flood: form of transport, expected sediment volume, impact force, and mode of deposition. The functioning hypotheses will be coupled with the corresponding types of constructive solutions in order to support correct design and construction. Some significant examples of open check dams operating in the Italian Alps will be presented and discussed.

Torrential Floods and Perspective of Their Control at the Basin Outlet

The hydrographic network of mountain basins usually presents steep slopes. Mountain streams are also named '*torrents*' in Europe. They are rivers with bed gradients ranging, in

* corresponding author: vincenzo.dagostino@.unipd.it

most cases, from 2-3% to 60-65% and with modal values from 10% to 30% in the Alps. Flood events in such rivers (torrential floods) are associated with the entrainment of large amounts of sediment: the adjective *torrential* indicates the impulsive response of the catchment to convective storm events. Many similarities exist between torrents and gullies of arid regions. Both are dominated by the mechanism of linear erosion that affects the bank-bed system, dominance of thalweg degradation in the upper reaches, retrogressive evolution of the heads, low bankfull width to flow depth ratios (<12 and, more frequently, <6), and intense sediment transport during floods. The main differences are the following: i) a gully network can be continuous or discontinuous, while torrents are always continuous and present a hydrographically connected network; ii) torrents may or may not have a base flow, while gullies are typically dry except during rain seasons or flood events; iii) gullies are always prone to erosion, while many torrents, even with high gradients, can be macro-armored. In fact, torrents can protect themselves when step pools and rapids form (D'Agostino and Lenzi, 1998).

In mountain torrents, the expected form of sediment transport is driven by the mean slope (S, m/m) of the bottom profile. As a first approximation, the sediment transport capacity for large floods can be expressed by omitting the threshold conditions of sediment motion. The simplified equation proposed by Smart and Jaeggi (1983) can be expressed as:

$$q_s = 2.5\ q\ S^{1.6} \tag{1}$$

where q is the unit discharge per unit width of the bed (m^3/s/m) and q_s is the unit bedload/hyperconcentrated flow rate (same units as q). Taking into account that C_v is the volumetric concentration of sediment:

$$C_v = \frac{q_s}{q + q_s} \tag{2}$$

and inserting Eq. (1) in Eq. (2), it follows that:

$$C_v = \frac{2.5\ S^{1.6}}{1 + 2.5\ S^{1.6}} \tag{3}$$

Eq. (3) describes the experimental values of S to a maximum of 0.2. An equation comparable to (3) can be obtained for S>0.2 using the experimental formula achieved by Ou and Mizuyama (1994) and assessing the mean C_v value of the surge and the tail of the debris flow. Substituting the static concentration of the debris flow with 0.65, Ou and Mizuyama's equation corresponds to Eq. (3) with a coefficient equal to 2.8 instead of 2.5 and an exponent equal to 1.5 instead of 1.6. Then, it is possible to combine the two relations. In Figure 1, the slope S is represented as a function of C_v and the associated form of sediment transport. For $C_v \leq 0.15$, the transport is assumed to be bedload transport; for $0.15 < C_v < 0.30$, it is assumed to be hyperconcentrated flow; and for $C_v > 0.25$, it is assumed to be debris flow. There is obviously a field of coexistence for hyperconcentrated flow and debris flow, where the occurrence of both is possible, depending on the type of sediment in motion: C_v=0.3 of a granular flow is expected to be hyperconcentrated flow, while a C_v of a silty-clay flow near 0.25-0.30 is a viscous muddy debris flow. It is worth observing in Figure 1 that the mean bed gradient is a powerful indicator of the transport form and, consequently, of the torrential phenomenon that must be managed in mountain areas.

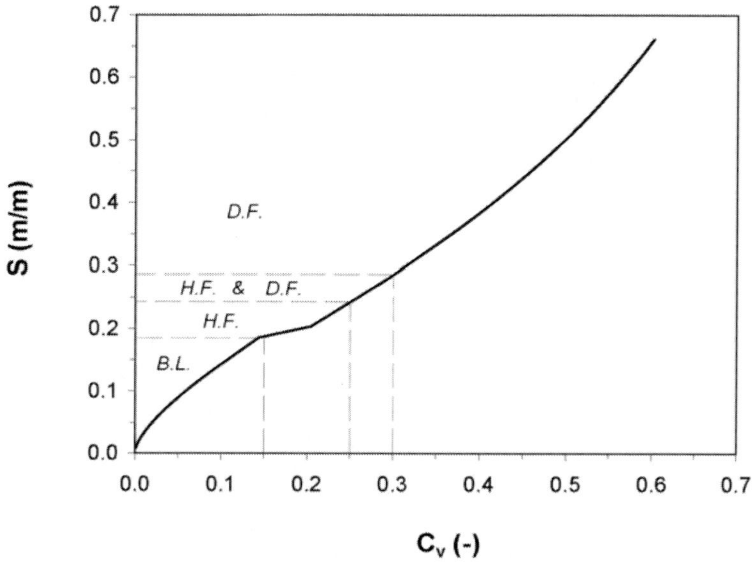

Figure 1. Expected form of sediment transport (B.L.= bedload transport; H.F.= hyperconcentrated flow; D.F.=debris flow) as a function of the mean channel slope *S*.

Figure 1 highlights four fields of competence in terms of mode of transport:
- bedload transport for $S<0.18$;
- hyperconcentrated flow for $0.18 \leq S < 0.24$;
- hyperconcentrated and debris flow (mixed zone) for $0.24 \leq S < 0.28$;
- debris flow for $0.28 \leq S < 0.65$.

An additional morphoenergetic factor reflecting the form of transport is represented by the Melton number *Me* ($Me = \Delta H / A^{0.5}$; ΔH is the difference in elevation between the highest point of the basin and the fan apex; *A* is the drained area in the fan apex) and the fan slope $tan(\alpha_f)$. Improving a previous finding of Marchi et al. (1993), D'Agostino (1996) used historic data concerning extreme transport events of the Eastern Italian Alps and obtained the threshold line (α_f in degrees):

$$\alpha_f = 7 - 14\ Me \tag{4}$$

For basin data above the threshold in Eq. (4), *mixed* transports (hyperconcentrated and debris flows) occurred, while for data below the threshold, bedload transport events were observed. When the fan is absent or coalescing, the value *Me*=0.5 may be used as the threshold criterion; that is, the condition $Me \geq 0.5$ suggests mixed torrents.

Recognition of the dominant sediment transport in a specific catchment must be carried out by means of field surveys of the river bed and deposition area, but the gravitational attitude to a mountain catchment that generates a certain form of transport (Figure 1 and Eq. 4) is of great importance to support field surveys and to design works for torrent control.

Protection actions against flash floods associated with sediment transport require a sustainable planning. The protection measures are grouped into two main categories collaborating at the catchment scale: *extensive, not structural, actions* dealing with forestry, soil bioengineering and shallow landslide management and *intensive hydraulic works*, which concern transversal and/or lengthwise structures along the river networks. Mountain hydraulic works are shared in two functional types: *consolidation* and *retention* structures.

The former are intended to stabilize the stream bed and the banks in the upper and medium part of the catchment, to slow degradation of the bed profile, and to reduce the sediment supply to the fan areas. Consolidation works consist of a sequence of check-dams, bed sills, grade control structures or man-made artificial step pool systems (Lenzi et al., 2000).

Retention structures are intended to control the transit of impulsive water-sediment flows, which otherwise would not be digested by the terminal reach of the torrent or by the man-made canalization flowing on the alluvial fan, or would not be discharged at the junction with the receiving river.

Retention structures consist of check dams of proper height (2.5-12 m) and trap and partly strain the estimated amount of sediment volume (V_s) in the expected form of transport (Figure 1). Moreover, they release slurries with lower C_v than the entering sediment flow rates. Standard localization of these check dams is around (just upstream and/or downstream of) the fan apex, with few exceptions due to particular topographic conditions. Functional failure of these structures may cause loss of human life and economic damage to houses and infrastructures because several phenomena can occur:

- direct impact of high density flux on buildings and vehicles;
- depositional areas or fans buried with vast amounts of sediment (Figure 2-a);
- deep erosion of the channel during the recessional phase of the event (Figure 2-b);
- direct clogging of the channel and/or culverts and the following avulsion, flooding and erosion of unprotected areas;
- clogging of the junction tributary-receiving river in the receiving river, causing lake formation in the river, backwater profile and inundation upstream of the clog, as well as a dam break hazard that affects the downstream river reach.

Traditional retention check dams were constructed as closed barriers (Thiery, 1891; Valentini, 1912), but for 20-30 years they have been replaced by straining check dams. Because the dam barriers present openings in their vertical face, such works are called *open check dams*.

a b

Figure 2. Quebrada el Cojo (Macuto, Venezuela) two months after the December 1999 flood event: a) part of a village is buried at the fan apex; b) main channel deepening occurred during the recession phase of the event.

FROM CLOSED TO OPEN CHECK DAMS

The first organized actions at the basin scale to control sediment transport in mountain rivers date back to the sixteenth century (Benini, 1990). In this period, the construction material consisted of wood and stone. Consolidation check dams were placed in the upper and intermediate reach of the basin, where degradation processes were active, while retention check dams (RCD) were built in the fan apex or in an appropriate reach of the torrent along the fan. The ancient Italian word to name a retention check-dam is "serra," which means a work that closes the torrent. The optimum locations for a RCD were a rocky gorge and the downstream end of a wide reach. The rocky gorge warrants valid static conditions, and it allows the original check dam height to be raised. The upstream widening maximizes the sediment storage capacity for a given dam height. A famous check dam, the Ponte Alto's serra, was built in 1537 on the Fersina torrent to protect the town of Trento (Italy) from floods transporting large amounts of gravel. After successive elevations, this dam has reached the notable height of 38 m. The original aim of RCDs was to give temporary protection (Thiery, 1891; Valentini, 1912) while the settlement of sediment source areas via reforestation, soil bioengineering works and river network consolidations became effective. In the beginning, RCDs were considered unsinkable by sediment flows, and only Gras (in Valentini, 1912) suggested construction of the fan apex of partly sinkable structures with lower heights and, subsequently, sinkable check dams downstream when the previous ones were filled up. From 1935-1970, material for constructing check dams changed according to the improving technology: boulders or stones joined with mortar transitioned to concrete and reinforced concrete. RCDs achieved an increased ability to bear dynamic thrusts associated with slurry flows. Along with such innovation, the functional goals of RCDs changed, and they were directed to look for a more selective sediment entrapment. In fact, the behavior of closed RCDs has shown several negative outcomes:

(1) The efficiency is temporarily limited to a few flood events if the storage capacity is not reestablished by cleaning up the deposits.

(2) Unimportant, ordinary flood events may reduce the storage capacity and decrease the expected trapping efficiency for subsequent severe events.
(3) It is not necessary to trap all sediment sizes because only the coarsest components cause damage and the inability to convey flows.
(4) The continuous retention of sediments, even for ordinary floods, can reduce the sediment supply by causing bed degradation and lack of sediment for lowland rivers and shores.
(5) The adoption of RCDs does not allow the trapping efficiency to be a function of the prevalent form of transport (bedload, debris flow with or without logs).
(6) The RCDs interrupt the river continuum in mountain streams and damage the river ecosystem, especially fish migration.

Taking into account points 1 and 2, French designers (Genet, 1953; Monet, 1966) proposed the first filtering-retention check dams (FRCDs) by inserting rectangular openings crossed by horizontal beams of steel in the vertical wall (Figure 3). The adoption of FRCDs spread in France and in the Alpine region (Austria, Switzerland, Italy) (Puglisi, 1967; Kronfellner-Kraus, 1970; Fattorelli, 1971) and also in Japan (Watanabe et al., 1980). These works benefited from progress in the mechanization of construction sites and the quality of building materials, better knowledge of sediment transport processes and their physical modeling and, finally, feedback regarding their successes and failures over years of application under different conditions.

Figure 3. Example of an open retention Genet check dam: tributary torrent in Susa's valley (Torino, Piemonte Region, Italy).

An FRCD classification results from the main design goals:

- *Dosing check dams*, when the function of reduction of sediment discharges around the flow peak is predominant (Üblagger, 1972);
- *Trapping check dams*, with the aim of stopping most sediment volumes and particularly the coarsest components.

Dosing Open Check Dams

During floods, the bedload transport persists longer than the debris flow does. The amount of total sediment introduced into the tributary may exceed the transport capacity and/or the storage capacity of the receiving river, particularly when the river is not in flood conditions. In other cases, the torrent presents many critical cross-sections (culverts, contractions, sudden reductions of slope) along its final stretch, and such cross-sections are prone to be clogged during the peak of the hydrograph when equal mobility of the bed is more probable and the coarsest sediments are entrained.

Dosing check dams are designed to partially reduce and modulate sediment discharge and the coarsest sediments passing downstream of a given channel cross-section. The objective is achievable through the hydraulic backwater induced by openings inserted into the wall of the check dam. The *reference discharge* (Q_d) to design an open check dam must be greater than the critical discharge for sediment motion (Q_c). The openings have to become a section to control the water surface profile. The resulting backwater profile reduces the mean flow velocity upstream of the barrier and then induces:

- sedimentation and deposition behind the dam;
- C_v reduction in the flow downstream of the structure.

During the falling phase of the flood, when the backwater is no longer active, part of the deposit tends to be slowly eroded and facilitates a certain self-cleaning of the storage area and the downstream reach sediment supply.

The geometry of the openings often consists of a unique rectangular slit (Figure 4) or single/multiple windows (rectangular, circular, etc.) associated with an overhanging spillway. The spillway can be designed to collaborate with the openings to bring out the backwater profile, but it is more frequently designed to discharge the maximum flow rate under design (from Q_{100} to Q_{200}; subscript indicates the return period in years) as a spare spillway in case of complete obstruction of the openings. The opening design can be carried out by forcing the flow to approach the narrowing to recover the specific energy (H, head above the bottom) requested by the control section ($H=H_c$). For a slit dam, H_c corresponds to the flow energy at critical condition; for flow under a gate (e.g., a window check dam), H_c depends on the head-discharge relationship of the opening geometry.

The openings are designed according to the energetic approach based on the following equation:

$$H_c = k\, H_1 \qquad (5)$$

Figure 4. Open slit dam built in the terminal reach of the Prampèr torrent (Belluno, Italy). The barrier was designed to reduce the bedload rate during floods at the junction between the Prampèr and the Maè torrents (Zoldana valley). Two subsidiary check dams were also constructed to control local erosion at the outlet of the slit.

where k is a coefficient of amplification - set between 2.0 and 2.5 - that expresses the necessary energy increase with respect to that available (H_1) upstream of the depositional reach.

The energy H_c in Eq. (5) can be assessed for water discharge such that the bed load transport is quite active. D'Agostino and Lenzi (1996) and Mao and Lenzi (2007) analyzed the passage from a selective transport to a condition of quasi-equal mobility in an alpine torrent of the Dolomites (Rio Cordon experimental basin, Italy). They found that sediment motion achieved quasi-equal mobility when the water discharge is close to 3-4 times the critical discharge of motion of the mean diameter. According to these findings, the reference water discharge (Q_d) to insert in Eq. (5) should be a value just below the quasi-equal mobility condition: Q_d =(2-2.5) Q_c. The goal is to not consume the storage volume before it is necessary and to only use entrapment when the coarsest sediments are entrained during the flood.

A practical example has been developed while designing a slit check dam in a river reach with the following characteristics: slope gradient upstream of the slit dam, S=0.10 m/m; upstream cross-section (storage reach) similar to a triangular shape; top width W from 0.0 to 24 m for flow depth h from 0.0 to 0.6 m and constant width W_{max}=24 m for $h \geq 0.6$ m. It is worth noting that, hydraulically, the schematized cross section represents a likely real situation, because at low flow conditions, all gravel-bed torrents occupy a small part of the bankfull width with single or multiple channels. Manning's roughness of the torrent reach is n=0.05; grain sizes of the bed surface for which 40%, 50%, 90% are finer are $D40$ =0.07 m; $D50$ =0.08 m; and $D90$ =0.32 m; peak discharge under design is Q_{200}=20 m^3/s. A short reach of the torrent, just upstream (20 m) and downstream (10 m) of the check dam, has a mean width W_m of 15 m and is almost rectangular. If the Schoklitsch's (1962) equation ($Q_c \cong 0.6\ W\ D40^{1.5}\ S^{7/6}$) is applied to the triangular cross section, the threshold discharge results in Q_c=1.5

m^3/s (with h=0.225 m, and W=9 m). Using the symbols of Figure 5, Eq. (5) can be rewritten in the form:

$$\left(\frac{Q_d^2}{B^2 g}\right)^{1/3} + z = k\left(h_1 + \frac{V_1^2}{2g}\right) \tag{6}$$

where g is the acceleration due to gravity (9.81 m/s^2), z is the height of the slit bottom above the foundation, B is the slit width, h_1 and V_1 are water depth and mean flow velocity in section 1, respectively. After assuming a value of k=2.5 and three tentative values for z (0.0 m; 0.2 m; 0.4 m), Eq. (6) can be solved to determine the unknown width of the slit B. The variables h_1 and V_1 also depend on Q_d through Manning's equation applied to the schematized triangular cross section. Figure 6 represents an interval of the solutions of Eq. (6) for several Q_d/Q_c ratios (h_1 ranges from 0.1 to 0.6 m) versus the narrowing ratio B/W_{max}. In the final assessment of B and z values, the designer can select the proper solution. For Q_d/Q_c=2, the possibilities under analysis are (Figure 6): B=1.50 m and z=0.0 m; B=1.95 m and z=0.2 m; or B=2.84 m and z=0.4 m. The first solution is preferable where the preservation of the river continuum is mandatory. After this choice (z=0.0 m), the slit height can be designed.

The height of the slit is equated to the specific energy in section 3 of Figure 5 ($H_3 \cong h_{3'}$; $h_{3'}$ =water depth before the deposit arrival). It is assumed that the dissipation due to the contraction between section 3 and section c is equal to 20% of the energy in c; it follows that:

$$H_3 = 1.2\left[1.5\left(\frac{Q^2}{B^2 g}\right)^{1/3}\right] + z \tag{7}$$

In the case study, the design discharge Q=Q_{100}=20 m^3/s (z=0.0 m; B=1.50 m) gives H_3=4.73 m (Eq. 7) (mean velocity V_3=0.28 m/s; $h_{3'}$ =4.72 m), and after the addition of a freeboard, the slight height can be set to 5 m. In the beginning of the flood and of the slit dam functioning, for Q=Q_d=2 Q_c=3 m^3/s, Eq.(7) gives H_3 =1.33 m (V_3=1.05 m/s; $h_{3'}$ =1.27 m). When the water depth in section 3 (before the arrival of the deposit) is close to $h_{3'}$=4.7 m, the wetted area A_w offered by the reach topography just behind the dam (section 3, Figure 5) is quasi-rectangular with a mean width (W_m) equal to 15 m (A_w= W_m $h_{3'}$ =70.5 m^2) and a flow velocity close to the bottom (V_b) around 0.21 m/s (V_b=0.75×Q/A_w=0.75×V_3). Comparing this velocity with the threshold velocity (V_t) given by Sternberg's relationship: V_t=4 $D^{0.5}$ (diameter D in meters; the coefficient 4 was suggested by Graf, 1971), it follows that, for an extraordinary flood, the barrier has a great retention efficiency because diameters passing downstream are finer than 3 mm. The same computation for Q=2 Q_c=3 m^3/s gives V_b=0.79 m/s, a passing diameter equal to 39 mm and less grain-size selectivity when the sediment transport rate is lower. During the flood, the deposit increases progressively until an equilibrium bed slope (S_c) is achieved. An empirical evaluation of S_c, according to field research carried out by Heede and Mufich (1973) and Della Lucia et al. (1980) for consolidation check dams, ranges between 0.6 and 0.7 of the original bed slope S_o (thus, $S_c \cong$ 2/3 S_o).

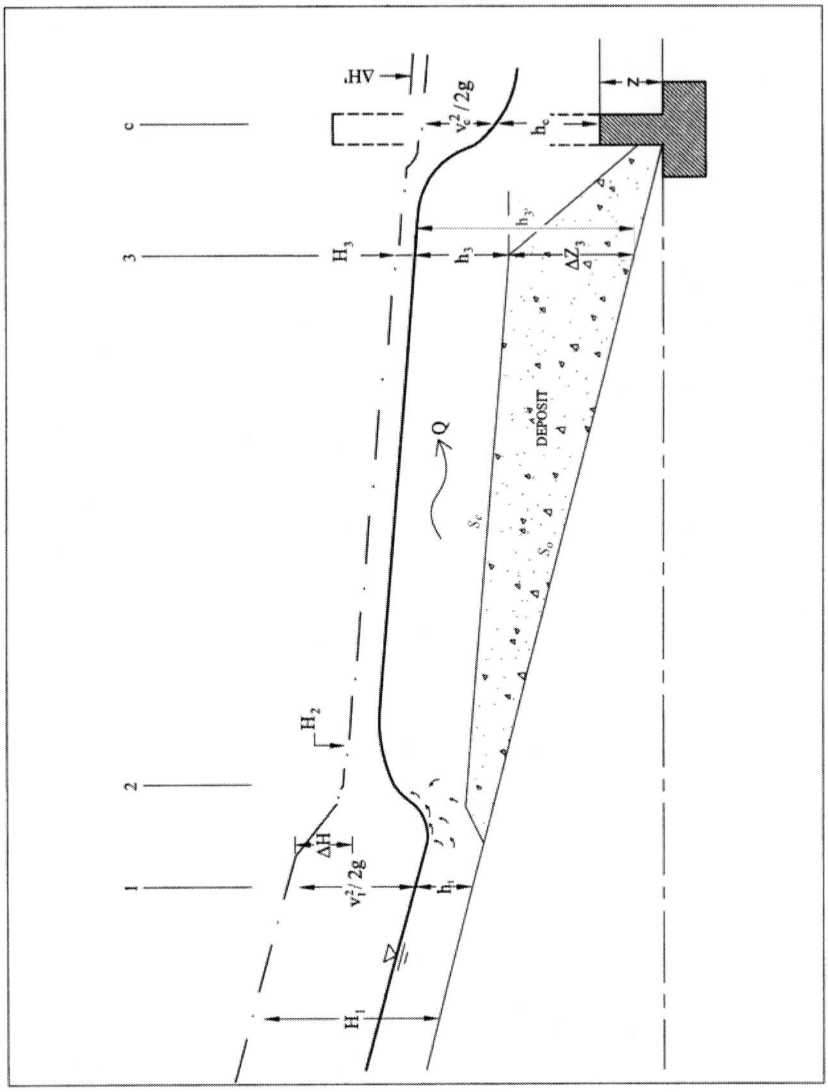

Figure 5. Sketch of a functioning open-slit check dam (energy line before the deposit; $\Delta Z_3 + h_3$ is approximated to $h_{3'}$).

After the computation of S_c, it is possible to evaluate the maximum height of the front deposit (ΔZ_3), which causes a reduction in the water depth from $h_{3'}$ to h_3 (Figure 5); the energy balance between sections *3* and *c* can be written:

$$\Delta Z_3 = \Delta H' + 1.5\left(\frac{Q^2}{B^2 g}\right)^{1/3} - \left(h_3 + \frac{V_3^2}{2g}\right) \quad (8)$$

where $\Delta H'$ is the total energy dissipation due to contraction and the velocity V_3 and depth h_3 are given by Manning's equation applied for an energy slope equal to S_c. With the example data: Q=20 m^3/s; S_c=2/3×S_o=0.067; n=0.05; and $\Delta H'$=*1.2* times the critical energy in the section c; it follows that h_3=0.46 m and V_3=2.96 m/s from Manning's equation applied to the rectangular cross section just upstream of the check dam (W_m=15 m). As a result (Eq. 8), ΔZ_3=3.82 m (ΔZ_3+h_3=4.3 m).

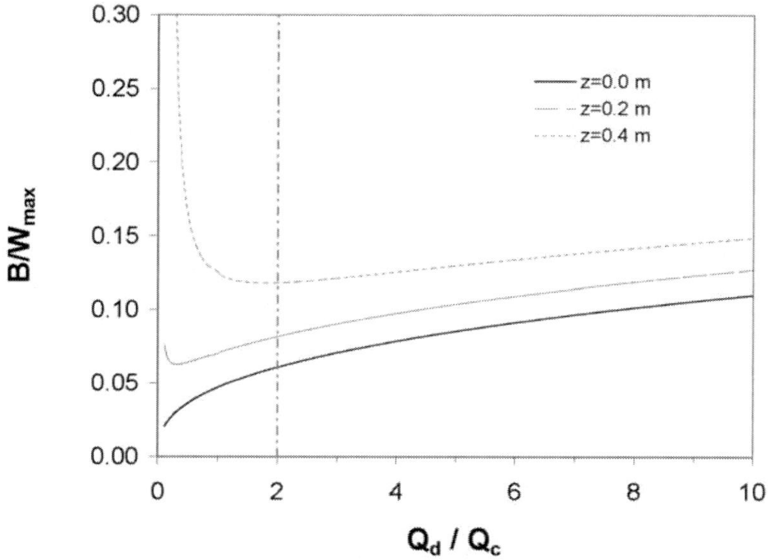

Figure 6. Solution of Eq. (6) for a case study: relation between the reference discharge Q_d over the threshold discharge Q_c and the narrowing ratio (B/W_{max}) for three different positions (z) of the bottom of the slit.

An additional conservative approach to evaluating ΔZ_3 may be obtained by applying momentum conservation in the direction of motion between section *3* (here assumed to be rectangular with width W) and a section c' at the toe of the deposit that also corresponds to the first cross section of the slit (width B). According to the scheme in Figure 7 and assuming that the sum of the external forces (F_e) acting on the control volume is zero, the following equation results:

$$\sum F_e = \rho\, Q \left[\frac{\zeta_{c'}\, Q}{B\,(h_3 + \Delta Z_3)} - \frac{\zeta_3\, Q}{W\, h_3} \right] = 0 \qquad (9)$$

where ρ is the water density and ζ is the Boussinesq coefficient for the section indicated in the subscript. Rearranging Eq. (9) for Boussinesq coefficients equal to unity, it follows that:

$$\Delta Z_3 \cong h_3 \left(\frac{W - B}{B} \right) \qquad (10)$$

Eq. (10) was also tested by Armanini and Benedetti (1996) by means of a physical model: satisfactory estimations were obtained for $B<0.4\,W$. In any case, the general relation (9) could be used by setting $\zeta_{c'}$ =1.33 and ζ_3 =1.0 (Chow, 1959). The application of Eq. (10) to the example data (W=W_m=15 m, B=1.5 m, h_3=0.46 m) gives ΔZ_3=4.14 m (ΔZ_3+h_3=4.6 m), which is close to that obtained using Eq. (8). Both results also indicate that the selected slit height (5 m) is appropriate. Finally, the maximum length of the deposits (L_{max}) is estimated as:

$$L_{max} = \frac{\Delta Z_3}{(S_0 - S_c)} \tag{11}$$

and the corresponding maximum storage volume for an equivalent rectangular cross section (width W_r) is the following:

$$V_{max} \cong \frac{L_{max}\ \Delta Z_3\ W_r}{2} = \frac{\Delta Z_3^2\ W_r}{2\,(S_0 - S_c)} \tag{12}$$

The data under examination with W_r=W_{max}=24 m, and $\Delta Z_3 \cong$4 m provide the values L_{max}= 121 m and $V_{max} \cong 5800$ m^3 (Eqs. 11-12).

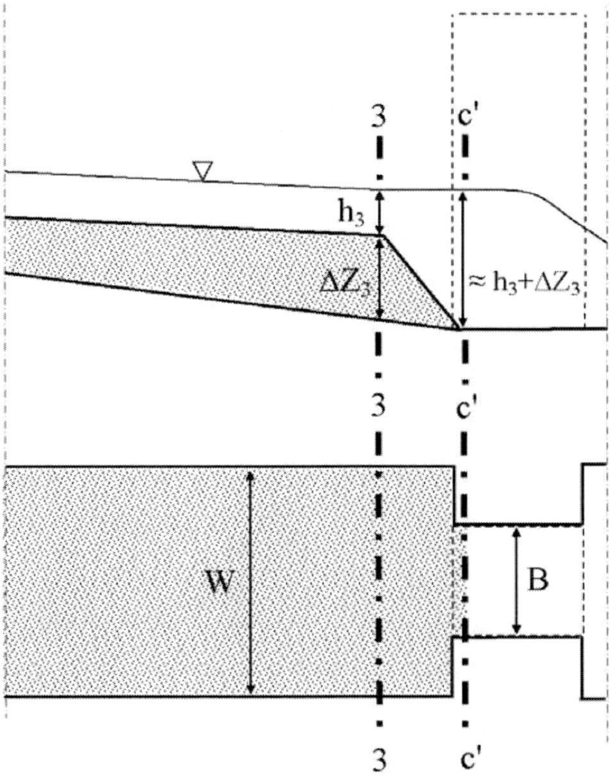

Figure 7. Simplified scheme to compute the height ΔZ_3 of the deposit.

This case study proves that dosing check dams achieve temporary sediment storage during the flood by exploiting the length of the reach upstream of the dam. This reach has to maintain constant slope and maximum width for a long stretch to maximize the dosing function. When the accumulated feeding volume exceeds the maximum storage capacity, which is controlled by the dynamic equilibrium slope, the retention function of the open dam halts, and only the consolidating function is maintained, because of the geometry of the control section *c* (Figure 5). The increment of the dosing function when transporting sediment volume $V > V_{max}$ demands the adoption of a series of cascaded open dams.

The complete hydraulic design of a dosing check dam requires the protection of the bank toes inside the storage basin (Figure 3), particularly in the case of slitted dams. In fact, when part of the deposit is eroded during the recessional phase of the flood, a central bar often remains, and secondary currents produce skimming forces that stress the river bed close to the banks.

Further erosive phenomena occur downstream from the dam, and they may make it necessary to use a reinforced concrete tooth or a line of piles as the underpinning of the foundation. A subsidiary check dam is also advisable to stop the downstream migration of the local erosion (Figure 4). The cautionary maximum scour (s_{max}) caused by the critical flow coming out from the slit can be assessed by means of the equation (D’Agostino, 1994):

$$s_{max} = \left(0.7\frac{B}{W} + 0.58\right)\left[0.94\left(\frac{Q}{B}\right)^{2/3} - 1.6\, D_{90}\right] \tag{13}$$

Eq. (13) is valid regardless of the fall height and the difference in height between the upstream water level and the tailwater level above the erodible bed. For the example (Q=20 m^3/s; $D90$=0.32 m; B=1.5 m; W=W_m=15 m), s_{max}=3.1 m below the upper surface of the foundation.

TRAPPING CHECK DAMS

Mountain streams are particularly hazardous during the transit of impulsive sediment fluxes. The protection against torrential floods associated with hyperconcentrated and debris flow can be obtained by constructing a deposition area closed by FRCD, which works by favoring the trapping function of the expected sediment volumes (Figures 8-9). These works are mainly located at the fan apex and upstream of the village and infrastructures at risk. Trapping FRCDs protect the junction stream-river but are also very helpful when the channel conveyance of the stream is not able to warrant the passage of peak discharges at high sediment concentrations. Thus, only a sediment flux turned into a bedload can pass through the channel with a freeboard. If there is a lack of a sufficient storage volume (V_{max}), the deposition area can be excavated, and its contours can be raised with properly protected and reinforced levees. In some circumstances, the required storage capacity is shared out into a series of detention basins, and each of these is closed by FRCDs that have an increasing trapping capacity from upstream to downstream. A debris-flow breaker can be adopted (Figure 10) as the initial retention structure, if it is necessary to reduce the flow energy and its potential force of impact on buildings and roads. The structure is designed using single or multiple buttresses (reinforced concrete), with the upstream face protected by steel plates. Flow-breaker systems also assist in addressing the flux and spreading the debris flow inside the deposition area. When check dams are designed to control debris flow, five main factors have to be estimated with regard to the characteristics and the magnitude of the flood.

1 - Debris flow composition and rheology. It is important to know if mudflow, turbulent granular flow or mixed mud-debris flow is expected and if these phenomena will come with a certain amount of large woody debris. The catchment geology and morphology of the lateral levee (ridges) and deposits support the analysis. Igneous and massive metamorphic rocks primarily generate granular debris flows, while schistose metamorphic and sedimentary silicic-clastic rocks are associated with muddy flows (Moscariello et al., 2002). Debris flows composed of dolomites and calcareous often behave as macroviscous flows. A levee with sharp crests is indicative of frictional-collisional flows, while rounded crests reflect viscous-plastic mudflows (Bardou et al., 2003). It is also compulsory for the filter design to measure the grain size distribution of previous deposits and assess the coarsest sediment size that makes up the debris-flow surges.

2 - Frequency of occurrence. Information on past events and field surveys that investigated the sediment sources areas feed the analysis. In addition, the degree of connection between landslides, bare soil zones and the hydrographic network supports the classification of the catchment as either *sediment-supply limited* or *sediment-supply unlimited* (Bovis and Jacob, 1999). Infrequent occurrences are typical in the first case (from 1/30 to 1/200 year^{-1}), while high-medium frequencies (1/2-1/10 year^{-1}) are common in the second case.

3 - Peak discharge and overall volume of debris flow. Current research has improved knowledge of this difficult topic. A geomorphologic approach is preferable for supply-limited basins, which take into account the expected recharge rate of the torrent for a certain period of time (30-100 years). Literature data on the European Alps (Spreafico et al., 1999; Marchi and D'Agostino, 2004) and British Columbia (Hungr et al., 1984) report channel debris yield rates ranging from 0.5-1 m^3/m for networks that generally do not experience erosion to 30-50 m^3/m for a deep gully in the valley (>20 m) and widespread bank instabilities. Values of 3-10 m^3/m are quite common for active debris-flow catchments, where the slope of the stream banks is close to their stability limit and the channel incision in the range of 4-8 m. For supply unlimited catchment, conceptual transformation of a flash flood hydrograph into a debris-graph demonstrates whether the hydraulic capacity of the transport matches the available geomorphic volumes. D'Agostino and Marchi (2003) proposed the use of the volumetric method (Hashimoto et al., 1978) integrated over time (t) through the relationship:

$$Q_d(t) = \left[\frac{c}{c - c_e(t)}\right] Q(t) \tag{14}$$

where Q_d is the debris-flow discharge (m^3 s^{-1}) associated with the water discharge Q (V_d and V are the corresponding volumes of the whole event); c and c_e are the static concentration of bed sediments before the triggering (c=0.60-0.70) and the debris-flow sediment concentration in motion, respectively. According to laboratory experiments, the ratio c_e/c reaches a maximum close to 0.9 (Hashimoto et al., 1978; Tubino and Lanzoni, 1993) for channel step slopes (slope angle β>20°). The assumption of a constant ratio c_e/c=0.9 for the whole duration of the flood is excessive in relation to real debris-flow surges, and a trend of $c(t)$ could be defined. The function $c_e(t)$ is simulated by setting a maximum value of c_e (Ou and Mizuyama, 1994):

$$c_{e\,max} = \frac{4.3\ c\ (tan\ \beta)^{1.5}}{1 + 4.3\ c\ (tan\ \beta)^{1.5}} \tag{15}$$

and a minimum, as an upper limit of the bedload transport of $c_{e\ min}$=0.20. Then, a linear variation of c_e can be assumed from $c_{e\ min}$ to $c_{e\ max}$ by starting from the time when the Q value greatly exceeds Q_c (Q=2-3 Q_c) to the peak time of $Q(t)$. Similarly, c_e decreases linearly from the maximum to the minimum in the falling portion of the hydrograph. The first term in Eq. (14) contains the triggering water discharge Q and an additive water discharge (Q') generated by the collapse of the hillslope-channel that is fully saturated with water: $Q_d=Q_s+Q+Q'$; with Q_s=sediment discharge. The value of Q' can be expressed as:

$$Q' = (1 - c)(Q_d - Q) \tag{16}$$

which, integrated over the time (Q' produces the volume V'), allows the real sediment volume ($V_s=V_d - V - V'$) to be determined. As an alternative, the hypothetical solidification of Q' gives a cautionary overestimation of V_s as the difference between the total volume, V_d produced by $Q_d(t)$, and the triggering water volume V associated with $Q(t)$.

Empirical equations, when applied to the pertinent geographical region, are complementary and strengthen the estimation of volumes and peak discharges, which have been assessed via geomorphic and hydrologic approaches. Marchi and D'Agostino (2004), after filtering the data and multiple regression analysis of 125 debris-flow events of the Eastern Italian Alps (V_s range of data: 10^3-10^6 m^3), obtained the equation:

$$V_s = 65000\ A^{1.35}\ S^{1.70} \tag{17}$$

where V_s is in m^3, A is the basin area drained at the fan apex (km^2) and S is the mean gradient of the main stream (m/m). The authors attempt to disconnect Eq. (17) from the region of calibration. After rearranging the equation and minimizing the sum of the quadratic losses between the computed and observed volumes, they obtained:

$$V_s = 2.9\ V\ S^2 \tag{18}$$

where V is the probable runoff volume (m^3) associated with debris flow. D'Agostino (2006) remarked that, though it has an empirical origin, Eq. (18) is similar to Eq. (15). In fact, at the time scale of the whole event, a mean volumetric concentration $c_{e\ mean}=V_s/V_d$, (V_d =total volume of water plus sediment) may be introduced. Thus Eq. (18) becomes:

$$c_{e\,mean} \cong \frac{2.9\,(tan\,\beta)^2}{1 + 2.9\,(tan\,\beta)^2} \tag{19}$$

which gives values of $c_{e\,mean}$ not far from the C_v concentrations of Eq. (3).

A notable data collection of peak discharge values $Q_{d\,max}$ (m^3/s) versus V_s (m^3) was carried out by Rickenmann (1999). The author proposed a semi-theoretical line that satisfies the Froude scaling and is very close to the trend of granular debris-flow data:

$$Q_{d\,max} = 0.1\,V_s^{5/6} \tag{20}$$

It is also possible to match an empirical equation, which was proposed by Mizuyama et al. (1992) for muddy debris flow, with an equation having the exponent of Eq. (20) and thus fitting the Froude similarity. In this case, a coefficient equal to 0.011, instead of 0.1, results. This comparison proves that the data samples of different parts of the world clearly converge to different durations that are associated with granular or muddy flows. For example, the association of Eq. (20) with a triangular debris-graph including 50% of V_s and 50% of water volume provides surge transit durations equal to 2 and 7 minutes for V_s=10^3 m^3 and 10^6 m^3, respectively. Under the same hypotheses, the duration of a mudflow passage ranges from 19 to 61 minutes, which is one order of magnitude longer than that of granular flows.

4 - Probable angle of deposition. The angle of deposition (γ_d) is a crucial variable that drives the cost-benefit analysis of the designed FRCD. In fact, for a given dam height (Z), the storage volume V_{max} depends quadratically on Z and linearly on the reciprocal of the difference between the bed slope (S_o) and the slope of deposit (see Eq. 12 with Z instead of ΔZ_3 and $tan\ \gamma_d$ instead of S_c). From a physical point of view, γ_d is almost coincident with the dynamic equilibrium angle of a slurry flow, which is the minimal bottom profile angle that does not reduce the flow velocity. When the debris front comes up against the check dam filter, it loses water and increases both C_v and friction slope. Deposition angles take values between 3° and 10° (12°); the lowest values concern mudflows rich in clay and silt, while the highest values are associated with granular flow that contains many boulders larger than 0.3 m (intermediate diameter). Mean values from 5° to 8° are associated with well-sorted mud-debris flows. Rheologic tests carried out in the laboratory can drive the choice of γ_d, but field observations of occasionally stopped surges are a more reliable method to refine the γ_d estimation.

5 - Dynamic forces. The correct structural design requires a certain degree of safety with regard to the external forces in play. In fact the dam must be stable with respect to sliding, overturning and collapse of the foundation terrain. Furthermore, normal and shear stresses have to be borne by the reinforced concrete and by the steel beams (Figure 8). During debris flow, most severe conditions occur, and the impact force (F_d) exerted against a closed barrier must be computed carefully. Several researchers have proposed the computation of the maximum dynamic pressure (p_{max} in N m^{-2}) associated with F_d using the relation:

$$p_{max} = \lambda\,\rho_b\,u^2 \tag{21}$$

where u is the mean debris-flow velocity in m s^{-1}, λ is a dimensionless impact factor depending on the rheological properties of the debris flow and ρ_b is the bulked flow density in kg m^{-3}.

A first theoretical deduction of Eq. (21) is derived from the momentum conservation of a fluid jet that has normal impact on a large extension plate at the steady state condition. In this case, λ is equal to 0.5. Taking into account the unsteady condition of an impacting debris flow, Armanini (1997) envisaged two different behaviors: the surge impact on the wall is mainly diverted in the vertical direction or is reflected in the opposite, generating a wave propagating upstream.

a) b)

Figure 8. Filtering check dam (reinforced concrete and H steel beams) to retain debris flows in alpine torrents (Trento, Italy): a) Fersina river valley ; b) Avisio river valley.

Figure 9. Filtering net check dam in the Kamikamihori valley (Japan) (courtesy of Lorenzo Marchi).

After some simplifications of the momentum balance equation, in both cases the mean pressure is proportional to the mixture density and to the square of the front celerity. The approximation in Eq. (21) for unsteady impact is partly counterbalanced by the value assigned to λ. Impact factor values between 1 and 2 are admitted for surges with low-sediment concentration, and values equal to 2 are suggested for bentonite (Watanabe and Ikeya, 1981). Armanini and Scotton (1992) presented impact tests carried out in a flume with a rough bed and 47% slope. They obtained the highest value (λ=2.2) for a mixture of water and a cylindrically shaped PVC material and the lowest (λ=0.45) for a granular flux (100% of anionic resin, density equal to 1080 kg m^{-3}).

Figure 10. Debris-flow breaker (Dugal mountain stream, Trento, Italy).

From model tests with a calcareous-dolomitic mixture, Holzinger and Hübl (2004) found maximum λ values around 2.5-3.5 for Froude numbers between 1 and 2.

After using Eq. (21) to compute the impact force F_d, the following results:

$$F_d = \lambda \rho_b u^2 A_w = \lambda \rho_b Q_{d\,max} u \tag{22}$$

where $Q_{d\,max}$ is the front debris-flow discharge, A_w is the cross-sectional area of the flux at the impact and u is the front velocity associated to $Q_{d\,max}$. The coefficient λ of Eq. (22) is usually set to 1, considering that the average value (p_m) of the dynamic pressure against the wall is less than p_{max}. Hungr et al. (1984) supposed the surge as a prism which travels with a uniform mean velocity, and proposed a similar relation to Eq. (22), assuming λ=1. The authors advised distributing the load "over an area as wide as the expected surge, but about 1.5 times greater in height to account the formation of a stagnant wedge in front of the toe of the barrier". The last terms of Eq. (22) also imply that, for λ= 1, the dynamic force is computed by accounting for the momentum of the flux before the impact. When the wall is downstream inclined at an angle ε (<90°) with respect to the main direction of u, the horizontal component of F_d is derived by multiplying Eq. (22) by $sin(\varepsilon)$.

Lichtenhahn (1971) and the Service Federal des Routes et des Digues (1973) reported an estimate of F_d deduced from the structural collapse of check dams that were not filled with debris before the front arrival. By assuming a triangular distribution of the pressure, the back-calculated dynamic force falls in the following magnification range of the hydrostatic thrust:

$$F_d = (7 - 10)\ \frac{1}{2}\rho\, g\ Z^2 \tag{23}$$

where ρ is the density of the water and Z is the height of the structure or the surge depth when the dam height exceeds the front height. Eq. (23) envisages maximum pressure near the base of the dam:

$$p_{max} = 100\ \rho\ Z \approx 50\rho_b\ Z \tag{24}$$

where $\rho_b \cong 2\ \rho$. D'Agostino and Casonato (2000) stressed that laboratory experiences carried out with granular material (coal slack) by Scotton and Deganutti (1997) corroborate Eq. (24); the latter authors found the following maximum pressures:

$$p_{max} = (24.5 - 73.5)\ \rho_b\ Z \tag{25}$$

with a mean value of p_{max}, for tests concerning an inter-granular solution of carboxy-methil-cellulose in water, equal to 52 ρ_b Z, which is similar to the result obtained from Eq. (24). The same study observed that the mean value of the maximum pressures at different locations along the walls equals 0.76 p_{max}.

Supplementary field observations relative to the measure of the impact force by means of an indentation meter are reported by Suwa et al. (1973) and Okuda et al. (1977). The indentation meter consists of a 0.15 m square steel panel attached to a steel cone with a pointed apex. When a force is applied to the external panel, the steel cone is pushed into a measuring panel. The external force is computed with a pre-calibrated relation between the indentation mark and the force acting on the panel. Okuda et al. (1973) measured highly variable F_d values ranging from 3 kN to 60 kN. The Sabo Department also reports mud flow impact forces on net check dams in the Kamikamihory valley (Figure 9). In this case, the indentation meter gave values ranging from 0.8 to 48.4 kN. Values up to two orders of magnitude greater are quoted in the literature for the contact thrust of a boulder against a structure (Van Dine, 1996).

After the assessment of the five described basic features, the design focuses on the check-dam filter, its height and typology and the dimensions of its openings.

The height Z of 10 m is often a static limit for the terrain foundation to bear the load of the dynamic force, which depends on the same Z (Eqs. 23-25). On the other side, the efficiency of the dam is more satisfactory if the maximum storage volume V_{max} is close to the estimated volume V_s (Eqs. 14, 17, 18).

In general, the construction of FRCD systems with an overall storage $V_{max}< 0.35\ V_s$ must be avoided. In fact, the arrangement of a deposition area with a retention capacity that is too low can lull the inhabitants into false security, because the mitigation rate is largely exceeded by the residue hazard. In such circumstances, a greater public perception of the potential hazard is preferable to a weak remedy.

Many filter types are in use. They have the shape of gratings that normally face the flow or, more often, are inclined with ε angles close to 40°-50° (Figure 8). Thus, the horizontal component of F_d is reduced and the total clogging of the filter after the debris surge is avoided. The filter geometry is composed of a series of openings (slits, windows) that may or may not have the same area. The correct filter function during the flood event has to be such that the barrier is permeable for bedload transports and selective when the bulked discharge reaches sediment concentrations of hyperconcentrated flows and debris flows. According to the event back-analyses carried out in Japan (Ikeya, 1989) and in Italy (D'Agostino et al., 2004), the main parameter governing the behavior of the filter is its *relative spacing*, which is expressed by the ratio between the minimum dimension B_{min} (regardless of the vertical or cross-horizontal position) and a reference diameter of the transported sediments. The permeability condition for bedload transport can be expressed as:

$$\frac{B_{min}}{D_{bl}} > 3 \tag{26}$$

where D_{bl} indicates the maximum transported material size for a medium scale flood (i.e., D_{bl} = from $D75$ to $D84$ of the armored bed surface), while the condition for a quasi-total trapping of a debris flow occurs if:

$$\frac{B_{min}}{D_{df}} < (1.5 - 2.0) \tag{27}$$

where D_{df} is the maximum boulder diameter of the debris flow.

Further investigations are necessary to better predict the maximum bulked discharge passing downstream of the filter (Q_{out}) as a function of the relative spacing B_{min}/D_{df}. By combining the well-known findings published by Watanabe et al. (1980) (three data) with the research of Ishikawa et al. (2004) dealing with the catchment rate of a FRCD physical model (eight experiments), the following equation (Figure 11) can be inferred for granular debris flows:

$$\frac{Q_{out}}{Q_{in}} = \left(0.47\,\frac{B_{min}}{D95} - 0.60\right) \tag{28}$$

where Q_{in} is the maximum debris-flow discharge approaching the filter. Eq. (28) has a determination coefficient equal to 0.71. Fitted data have some scatter and require confirmation at a real spatial scale. Nevertheless, the relation can result in a practical first tool to assess, to a first approximation, the efficiency of check dams while straining granular flows. The application of Eq. (28) to mud flows is unadvisable.

Figures 12 and 13 show an example of a trapping FRCD designed by D'Agostino for the Rio Chiesa stream (Belluno, Italy). For the high dynamic force of impact (F_d=810 kN; Eq.22: Q_d=90 m^3/s; u=5 m/s; ρ_b = 1800 kg/m^3), the foundation terrain was reinforced by means of micropiles (Figure 12). The central rectangular slit is 3.7 m high and 3.0 m wide and presents an overhanging trapezoidal spillway that is 2 m deep. The slit is equipped with five square steel beams (side of 0.2 m, thickness 16 mm), which can be placed in the slit because of two C beams inserted along the slit contour (Figure 13). The solution offers progressive assessment of the filter geometry in case the barrier is too permeable or not permeable enough, and it allows the excavator to enter the deposition area through the filter by restoring the storage capacity after each event.

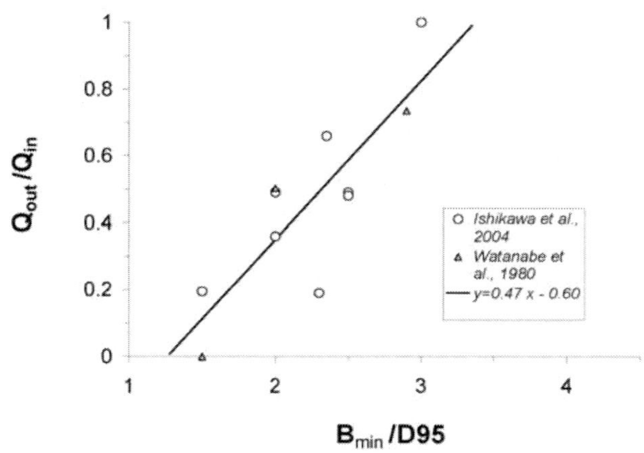

Figure 11. Rate of reduction of the debris-flow peak discharge (Q_{out}/Q_{in}) versus relative spacing ($B_{min}/D95$) of the filter openings.

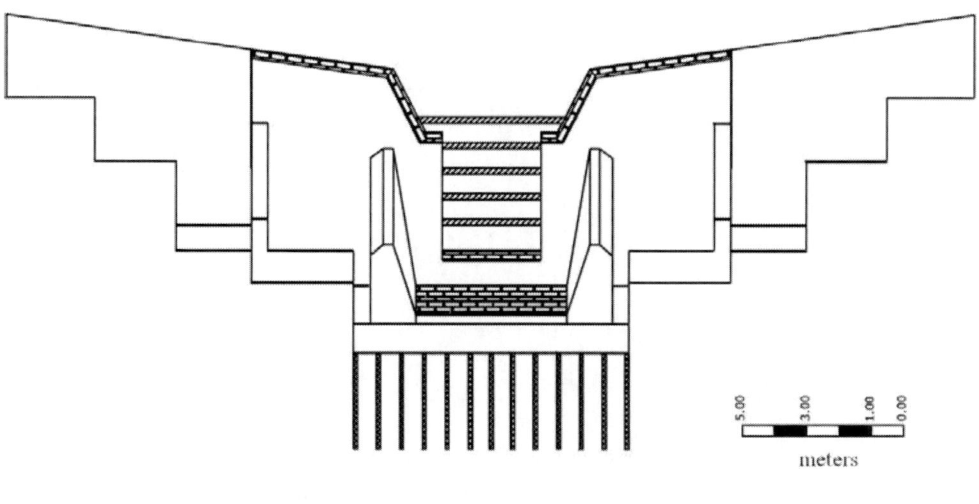

Figure 12. Frontal view of the filtering-retention check dam in the Rio Chiesa (Belluno, Italy).

Figure 13. Filtering-retention check dam in Rio Chiesa stream after a debris-flow event (Belluno, Italy).

A grain size distribution of the bed surface upstream of the work under design was carried out using the pebble count sampling technique (Wolman, 1954) and measuring a previous debris-flow deposit. The following sizes were obtained: $D50$=0.05 m; $D75$=0.13 m; $D84$=0.18 m; $D90$=0.31 m; $D95$=0.50 m. The filter was designed by setting out a clear opening at the base equal to 0.8 m in height and with upper openings with a spacing of 0.6 m (the expected height of the front deposit is Z=4.2 m). Assuming that $D_{bl}=D84$ and $D_{df}=D95$, the corresponding relative spacing results agree with conditions (26) and (27): B_{min}/D_{tr}=0.8/0.18=4.4 for bedload events, and B_{min}/D_{df} =0.6/0.5=1.2 (1.6 for the lowest opening) for debris-flow events. Since its construction, the Rio Chiesa FRCD has been in operation and has experienced debris flow and muddy-debris flow; in both cases, the main surge has always been stopped by the filter (Figure 13), and the storage capacity (5000 m^3) has been almost completely utilized. Field surveys were carried out after the debris-flow event; the mean diameter (intermediate axis) of the five coarsest boulders behind the filter is 0.65 m; the angle of deposit (γ_d) was 10.9° along the centerline of the deposit and 8.3° close to the banks (14.6° is the bed bottom sloping upstream of the check dam). The surface $D50$ in the deposition reach was much finer than that of the natural debris-flow deposit before construction (2.6 cm instead of 5.0 cm), proving that such trapping check dams, according to companion post-event surveys (D'Agostino et al., 2004), are scarcely grain-size selective.

CONCLUSION

Retention check dams (RCDs) have a long tradition in the in the Alps of mitigating the effects of torrential floods. At the beginning of the twentieth century, these dams were deemed temporary because they are passive works that did not prevent the occurrence of high sediment-transport rates during the flood events The main actions in managing a mountain catchment were stabilization, restoration and reforestation of sediment source areas and consolidation of those mountain network reaches where degradation processes were active (Demontzey, 1891). These basic concepts still remain fully valid, but the increasing anthropogenic pressures on mountain areas have encouraged many improvements and the functional specialization of RCDs. The structures were transformed from closed to 'open' filter-retention check dams (FRCDs), and their use spread because they do not require waiting for an integral restoration of the basin and because they offer a quick mitigation of the torrential hazard.

Chapter 9 reported the main design features of the FRCDs and highlights how their functional specialization can be shared between the bedload and the debris-flow control. In several cases, both types of events coexist in the same basin, and a unique FRCD must cope with them. In such circumstances, the defense against debris flow prevails, but the designer must design a filter that is capable of managing all forms of sediment transport, which even may include large woody debris. The RCDs in Figures 8 and 9 represent the solution adopted in northern Italy to control the mixed forms of transport and to warrant, at the same time, low-medium selectivity for the bed load transport, retention of floating woody debris and a strong trapping capacity for debris flows. Even though the current FRCDs appear less temporary if compared to the first closed barriers, in the long run they still need to be aided by systematic maintenance of the stabilization measures operating in the basin (forest management, soil bioengineering and hydraulic works) and by sustainable soil use that looks well into the future.

REFERENCES

Armanini, A. (1997). On the dynamic impact of debris flows. In A. Armanini, and A., M. Michiue (Eds.), *Recent Developments on Debris Flows* (208-225). Berlin, German: Springer-Verlag.

Armanini, A., and Benedetti, G. (1996). Sulla larghezza di aperture di briglie a fessura. *Atti del XXV Convegno Nazionale di Idraulica e Costruzioni Idrauliche: Vol. 3* (13-24).Torino, Italy: MAF Servizi.

Armanini, A., and Scotton, P. (1992). Experimental analysis on dynamic impact of a debris flow on structure. *Intern. Symp. Interpraevent: Bern, Vol. 6.*

Bardou, E., Ancey, C., Bonnard, C., and Vulliet, L. (2003). Classification of debris flow deposits for hazard assessment in alpine areas. In D. Rickenmann, and C. L. Chen (Eds.), *Third Int. Conference on Debris-Flows Hazards Mitigation: Mechanics, Prediction and Assessment*: *Vol. 2* (799–808). Rotterdam, Netherlands: Millpress.

Benini, G. (1990). *Sistemazioni idraulico-forestali.* Torino, Italy: UTET.

Bovis, M. J., and Jacob, M. (1999). The role of dubris supply conditions in predicting debris flow activity. *Earth Surface Processes and Landforms, 24* (11), 1039-1054.

Chow, V. T. (1959). *Open-channel hydraulics* (Int. Ed. 1973). Singapore: McGraw-Hill.

Demontzey, M. P. (1891). *Le reboisement des montagnes et l'extinction des torrents: Conférence faite à l'Association francaise pour l'avancement des sciences, le 14 mars 1891. Extrait del la revue scientifique*. Paris, France: Administration des Deux Revues.

D'Agostino, V. (1994). Indagine sullo scavo a valle di opere trasversali mediante modello fisico a fondo mobile. *L'Energia Elettrica, 71*(2), 36-50.

D'Agostino, V. (1996). Analisi quantitativa e qualitativa del trasporto solido torrentizio nei bacini montani del Trentino Orientale. *Scritti dedicati a Giovanni Tournon* (111-123), Italia: Associazione Italiana di Ingegneria Agraria and Associazione Idrotecnica Italiana.

D'Agostino V., and Lenzi, M. A. (1996). La valutazione del trasporto solido di fondo nel bacino attrezzato del Rio Cordon. *L'Acqua, 4*, 23-40.

D'Agostino, V., and Lenzi, M. A. (1998). La massimizzazione della resistenza al flusso nei torrenti con morfologia a step pool. *Atti del XXVI Convegno Nazionale di Idraulica e Costruzioni Idrauliche: Vol. 1* (281-293). Catania, Italy: CUECM.

D'Agostino, V., and Casonato, A. (2000). Dynamic impact hypotheses for designing check dams involved in debris flow phenomena. *Proc. Int. Workshop Los alludes torrencial de Diciembre 1999 en Venezuela*. Caracas, Venezuela: Facultad de Ingegneria, Instituto de Mecanica de Fluidos.

D'Agostino, V., and Marchi, L. (2003). Geomorphological estimation of debris-flow volumes in alpine basins. In D. Rickenmann, and C. L. Chen (Eds.), *Third Int. Conference on Debris-Flows Hazards Mitigation: Mechanics, Prediction and Assessment*: *Vol. 2* (1097–1106). Rotterdam, Netherlands: Millpress.

D'Agostino, V., Dalla Fontana, G., Ferro, V., Milano, V., and Pagliara, S. (2004). Briglie aperte. In V. Ferro, G. Dalla Fontana, S. Pagliara, S. Puglisi, S., and P. Scotton (Eds.), *Opere di sistemazione idraulico-forestale a basso impatto ambientale,*(283-384). Milano, Italia, McGraw-Hill.

Della Lucia, D., Fattorelli, S., Nardin, D., Provasi, C., and Tomasetti, R. (1980). *Ricerca sulla determinazione della pendenza di compensazione nei torrenti del Trentino.* Trento: Provincia Autonoma di Trento- Azienda Speciale di Sistemazione Montana.

Fattorelli, S. (1971). Opere per la correzione dei torrenti nelle Alpi Francesi. *Monti e Boschi, 22* (5), 33-44.

Genet, E. (1953). Un noveau type de barrage de correction: le barrage-peigne, *Compte rendu général du voyage d'etudes dans les Alpes Francaise du Groupe de Travail FAO/EFC/TORR: I session, Nancy, 1952,* France: FAO.

Graf, W. H. (1971). *Hydraulics of Sediment Transport* (1st ed.). New York: Mc-Graw Hill.

Hashimoto, H., Tsubaki, T., and Nakayama, H. (1978). Experimental consideration on debris flow. *Proc. 33rd National Meeting of Japan Society of Erosion Control Engineering* (574-575).

Heede, B. H., and Mufich, J. G. (1973). Functional relationships and a computer program for structural gully control. *Journal of Environmental Management, 1*, 321-324.

Holzinger, G. , and Hübl, J. (2004). Impact force on a debris flow breaker: derived from laboratory experiments. *Interpraevent 2004: Riva del Garda/Trient, Vol. 3* (131-139). Villach, Austria: Kreiner Druck.

Hungr, O., Morgan, G. C., and Kellerhals, R. (1984). Quantitative analysis of debris flow torrent hazards for design of remedial measures. *Canadian Geotechnical Journal, 21*(4), 663-677.

Ikeya, H. (1989). Debris flow and its countermeasures in Japan. *Bull. of the International Association of Engineering Geology, 40*, 15-33.

Ishikawa, N., Fukawa, G., Katsuki, S., and Yamada, T. (2004). Catch effect of debris flow for the open type steel check dam by physical and numerical modeling. *Interpraevent 2004: Riva del Garda/Trient, Vol. 3* (153-163). Villach, Austria: Kreiner Druck.

Kronfellner-Kraus, G. (1970). Über offene Wildbachsperren. *Mitteilungen der forstlichen Bundesversuchsanstalt- Wien, 88*, 7-77.

Lenzi, M. A., D'Agostino, V., and Sonda, D. (2000). *Ricostruzione morfologica e recupero ambientale dei torrenti: Criteri metodologici ed esecutivi.* Cosenza, Italy: Bios.

Lichtenhahn, C. (1971). Calcolo delle briglie in calcestruzzo basato su nuove ricerche eseguite in Svizzera negli ultimi anni. *Annali Accademia Italiana di Scienze Forestali, 20*, 89-113.

Mao, L., and Lenzi, M. A. (2007) Sediment mobility and bedload transport conditions in an alpine stream. *Hydrological Processes, 21*, 1882-1891.

Marchi, L., and D'Agostino, V. (2004). Estimation of the debris-flow magnitude in the Eastern Italian Alps. *Earth Surface Processes and Landforms 29*, 207-220.

Marchi L., Pasuto, A., and Tecca, P. R. (1993). Flow processes on alluvial fans in the Eastern Italian Alps. *Z. Geomorph. N. F., 37*, 447-458.

Mizuyama, T., Kobashi, S., Ou, G. (1992). Prediction of debris flow peak discharge. *Intern. Symp. Interpraevent: Bern, Vol. 4* (99-108).

Monet, P. (1966). Une experience de correction torrentielle donnerait-elle la clé du phénomemène des transports solides par les eaux courantes?. *Revue de Géographie Alpine, 43*, 455-460.

Moscariello, A., Marchi, L., Maraga, F., and Mortara, G. (2002). Alluvial fans in the Italian Alps: sedimentary facies and processes. In: P. Martini, V.R. Baker, and G. Garzon (Eds.), *Flood and Megaflood Processes and Deposits - Recent and Ancient Examples* (141-166). Oxford, UK: Blackwell Science.

Okuda, S., Suwa, H., Okunishi, K., Nakano, M., and Yokoyama, K. (1977). Synthetic observation on debris flow, Part 3, observation at Valley Kamikamihorizawa of Mt. Yakedake in 1976. *Annuals of Disaster Prevention Research Institute, Kyoto University, 20B-1*, 237-263.

Ou, G., and Mizuyama, T. (1994) Predicting the average sediment concentration of debris flow. *Journal of Japan Society of Erosion Control Engineering, 47*(4), 9-13.

Puglisi, S. (1967). L'impiego dei dispositivi filtranti nella correzione dei torrenti. *L'Italia Forestale e Montana, 22* (1), 12-24.

Rickenmann, D. (1999). Empirical relationships for debris flow. *Natural Hazard, 19*, 47-77.

Schoklitsch, A. (1962). *Handbuch des Wasserbaues* (3rd ed.). Wien, Austria: Springer.

Scotton, P., and Deganutti, A. M. (1997). Phreatic line and dynamic impact in laboratory debris flow experiments. In Cheng-lung Chen (Eds.), *Proceedings of the first international Conference on Debris-Flow Hazards Mitigation: Mechanics, Prediction, and Assessment*, New York: ASCE.

Service Federal des Routes et des Digues (1973). *Dimensionnement des barrages en torrentes en béton et beton armé.* Berne, Switzerland: Office Centrale des imprimes et du materiel.

Smart, G. M., and Jaeggi M. N. R. (1983). *Sediment Transport on Steep Slopes: Vol. 64. Mitt. Versuchsanstalt für Wasserbau Hydrologie und Glaziologie*, Zürich: ETH.

Spreafico, M., Lehman, C., Naef, O. (1999). *Recomandations concernant l'estimation de la charge sédimentaire dan les torrents.* Groupe de travail pour l'hydrologie opérationelle, Berne, Switzerland.

Suwa, H., Okuda, S., and Yokohama, K. (1973). Observation system on rocky mudflow. *Bulletin of the Disaster Prevention Research Institute, Kyoto University*, 23, 59-73, 1973.

Thiery, E. (1891). *Restauration des montagnes, correction des torrents, reboisements*. Paris, France: Baudry et C. Libraires.

Tubino, M., and Lanzoni, R. (1993). Rheology of debris flows: experimental observations and modelling problems. *Excerpta, 7*, 201-236.

Üblagger, G. (1972). Retendieren, Dosieren und Sortieren: Kolloquium über Wildbachsperren, *Mitteilungen der forstlichen Bundesversuchsanstalt- Wien*, 102, 335-372.

Valentini, C. (1912). *Sistemazione dei torrenti e dei bacini montani*. Milano, Italy: Hoepli.

VanDine, D. F. (1996). *Debris flow control structures for forest engineering: Vol. 22, Working Paper*. Victoria, Canada: Res. Br., B. C. Ministry of Forests.

Watanabe, M., and Ikeya, H. (1981). Investigation and analysis of volcanic mud flows on Mt Sakurajima, Japan. *Proceedings of the Florence Symposium Erosion and sediment transport measurement: IAHS-AISH Pubbl. No.133* (245-256).

Watanabe, M., Mizuyama, T., Uehara, S., and Suzuky, H. (1980). Experiments on sabo structures for debris flow. *Civil Engineering Journal, 22-2*, 64-70.

Wolman, M. G. (1954). A method for sampling coarse river-bed material. *Transactions American Geophysical Union, 35*, 951-956.

In: Check Dams, Morphological Adjustments… ISBN: 978-1-62618-856-3
Editors: C. C. Garcia and M. A. Lenzi

Chapter 10

THE EFFECTIVENESS OF TORRENT CHECK DAMS TO CONTROL CHANNEL INSTABILITY: EXAMPLE OF DEBRIS-FLOW EVENTS IN CLAY SHALES

A. Remaître and J.-P. Malet
School and Observatory of Earth Sciences,
Institute of Earth Physics, UMR 7516,
University of Strasbourg & CNRS.
5 rue René Descartes, 67084 Strasbourg Cedex, France

ABSTRACT

Debris flows consist of a fully saturated mixture of water, sediment and debris that can travel several kilometers. They are very dangerous phenomena claiming thousands of lives and millions of Euros each year over the world. One of the most effective techniques to manage torrential flood and debris-flow hazards is to construct series of check dams. While a huge amount of work has been performed on debris-flow triggering mechanisms, there are still few studies discussing the quantification of erosion by debris flows and the influence of the series of check dams on debris-flows scouring and deposition mechanisms. As well, theoretical and numerical work has been performed on the size, the shape and the structure of torrent check dams, allowing the definition of general design criteria; it is worth noting that less research has focused on the optimal location of these dams along the debris-flow track.

The first part of this chapter reviews the historical aspects of torrential mitigation in the French Alps; the second part presents some geomorphologic and kinematical characteristics of debris-flows events with a specific focus on the influence of check dams on scouring and deposition intensity. The third part proposes a methodological framework to evaluate the impacts of the number and the location of the check dams on the reduction of debris-flow intensity (in terms of flow thickness, flow velocity and volume). This framework is based on simple numerical calculations with a 1D debris-flow runout model. It is demonstrated that a few check dams located nearby the source area may decrease substantially the debris-flow intensity on the alluvial fans.

1. INTRODUCTION

Debris flows are found worldwide in various mountainous environments; they are a common type of mass movement and consist of fully saturated mixtures of water, sediment and debris that can travel, in a series of surges, several kilometers (Iverson, 1997). Observations of past events have revealed debris-flow velocities varying from $0.5m.s^{-1}$ to over $20m.s^{-1}$ (Goudie 1995; Hungr et al., 2001). Debris flows are widely recognized as one of the dominant geomorphic processes in steep mountainous terrain (Pierson, 1980; Costa, 1984; Johnson and Rodine, 1984; Takahashi, 1991; Scott et al., 1995; Corominas et al., 1996; VanDine and Bovis, 2002; Godt and Coe, 2007). The generic term “debris flow” can be broadly divided into “open slope debris flow” and “channellized debris flow” (Evans, 1982; Pierson and Costa, 1987).

This chapter addresses channelized debris flows only. Unlike other types of mass movements, the evaluation of the hazard requires, besides the prediction of the time occurrence and of the source material location, the assessment of the run-out susceptibility (e.g. O’Brien et al., 1993; Ayotte and Hungr, 2000; Laigle and Marchi, 2000; D’Ambrosio et al., 2007).

Debris flows most commonly mobilize from landslides, which occur when rainfall increases pore water pressure causing failure. However streamflow falling onto colluvium can also trigger debris flows (Johnson and Rodine, 1984). Various factors, including natural or human factors, influence debris flow activity (Figure 1).

The effects of debris flows are often catastrophic to the inhabitants. They can cause serious casualties and damages to properties. Impacts of debris-flow activity could potentially be reduced with the development of debris-flow mitigation actions (Figure 1). Debris-flow mitigation can be divided in (1) debris-flow forecast and (2) debris-flow protection. Classically, protection methods (Figure 2) can be broadly divided into passive and active control measures.

Passive measures are mainly land-use planning keeping the endangered areas free of settlements and infrastructure to prevent damages or public notification and education. Active methods of protection require some engineering once the hazard has been identified and assessed (VanDine, 1996); these methods start at the debris-flow source area and are continued through the debris-flow path down to the apex and the debris cone. They can involve remediation to reduce or eliminate the potential of a debris flow from occurring like reforestation (bio-engineering), or the design and construction of defence works to reduce the effects of the event (Figure 1, 2).

Considering this last point, one of the most important active protection works in order to reduce the debris-flow impact consists to modify the topography of the track by building series of check dams. Check dams are usually built in series in the transportation zone of a debris flow and/or on the debris-fan. In this chapter, we will discuss only check dams built near the source area or in the transportation zone; in such locations, check dams are used to: (1) minimize entrainment (essentially banks undermining, channel degradation and slope instabilities) along the stream (Figure 3A) and (2) reduce steep channel gradients locally in order to decrease the velocity and the kinetic energy (Figure 3B).

Figure 1. Concept of risk in a torrential watershed (modified from van Effenterre, 1982). A) Hydrological and geomorphological processes controlling the triggering/run-out/spreading of torrential event: (A1) active landslide (volume > 1000 m^3), (A2) runoff, soil erosion and gullying, (A3) area affected by several shallow landslides, (A4) Natural dam, (A5) bed incision, banks undermining and associated slope instabilities, (A6) overflowing and creation of a debris fan. B) Human impact on torrential hazard: (B1) deforestation, (B2) urbanization, (B3) defective drainage, water leakage and infiltration, (B4) abandonment of cultivated land, pasture and land degradation. C) Torrential mitigation actions: (C1) warning system, (C2) series of check dams, (C3) storage basin and straining structure, (C4) reforestation, (C5) periodic cleaning out of the storage systems, (C6) deflection walls, (C7) land use planning.

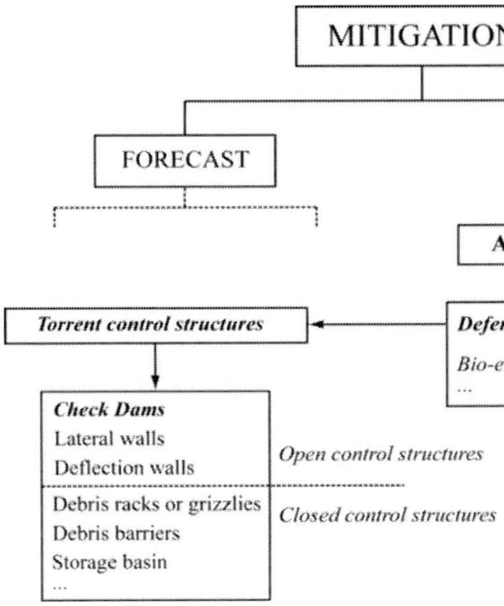

Figure 2. The mitigation concept (modified from Flageollet, 1988 and VanDine, 1996).

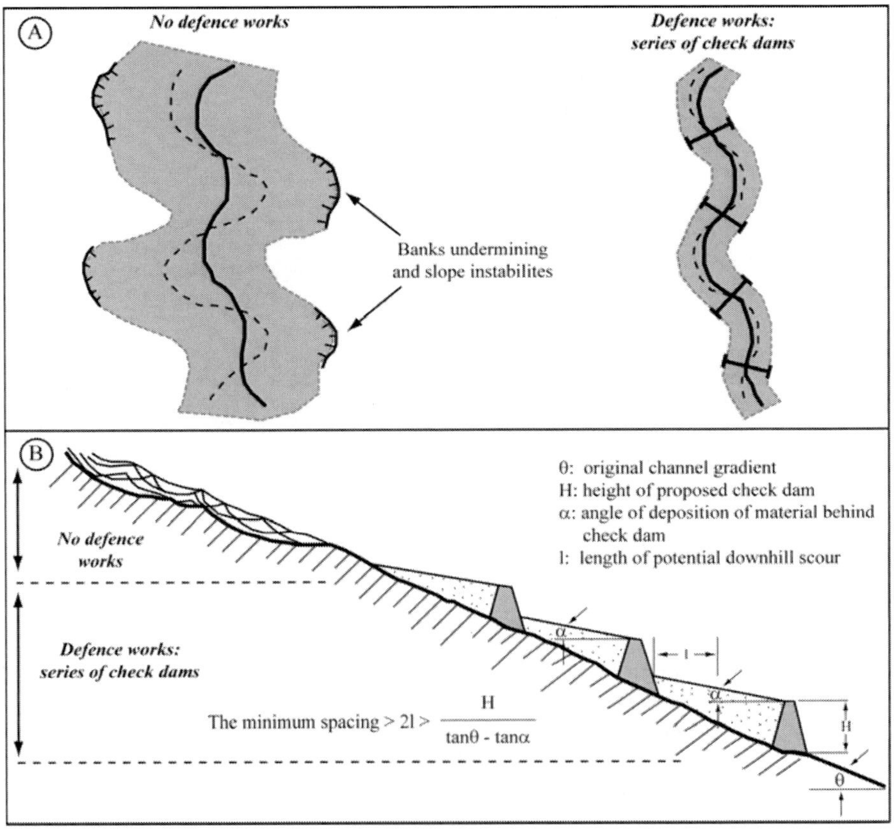

Figure 3. Usefulness of check dams to (A) minimize scouring phenomenon and (B) reduce channel gradients (modified from Deymier et al., 1994 and VanDine, 1996).

If considerable theoretical and numerical works have been performed on the size, shape and structure of torrential check dams, allowing the definition of general design criteria (Leys and Hagen, 1971; Eisbacher and Clague, 1984; Government of Japan, 1984; Heierli and Merk, 1985; Whittaker et al., 1985; Johnson and McCuen, 1989; Couvert et al., 1991; Armanini and Scotton, 1992; Chatwin et al., 1994; Deymier et al., 1994; Miyazawa et al., 2003), it is worth noting that less research has focused on the optimal number and location of these dams along the debris-flow track. For instance, some previous works have been focused on the spacing between check dams which depends on stream gradient, dam height, angle of deposition of material behind the dam, and downstream extent of potential scour. Chatwin et al. (1994) provide a formula for the spacing of check dams (Figure 3B). Nevertheless, there is still a lack for the optimal location and number of check dams to increase the efficiency of defence works.

The present chapter discusses some aspects of the effectiveness of torrent check dams to control channel instability through examples of debris-flow events in clay shales basins in the South French Alps. First an historical review of torrent check dams construction in the French Alps is presented. Then, some geomorphological and kinematical characteristics of two debris-flows events are discussed, with a specific focus on the influence of check dams on scouring intensity. Finally, a methodological framework, based on simple numerical calculations with a 1D debris-flow run-out model, to evaluate the impacts of the number and the location of the check dams on the reduction of debris-flow intensity is presented.

2. An Historical Overwiew of Torrential Mitigation in the French Alps

During the 17th century in France, the forests were considered as an economic resource, especially for the construction of warships. Moreover, deforestation permits the extension of arable lands which were rare in mountainous region. These deforestations led to dramatic consequences with an important increase of land degradation and mass movement occurrence.

In France, the French Ministry of Agriculture and Forest is in charge to manage natural hazards and risks in mountainous areas. This dates back to 1860, when a special public service (RTM) was created by an act of the government as result of a flood disaster four years before. Indeed, a major hydrogeomorphological crisis occurs in 1856; intense thunderstorms have threatened the French Alps causing numerous severe floods. Two of the biggest French rivers, the Rhône River and the Saône River, caused such heavy life and property damage in the French Alps that neither the inhabitants nor the communities were able to deal with the aftermath.

From 1860 until today control measures are planned and performed by the RTM, because the interactions between runoff, erosion and torrential events on the one hand and vegetation, especially forest cover, on the other hand were already known (Figure 1). From the beginning the French RTM service tries to prevent or reduce mountainous hazards or their effects by a combination of different kind of watershed and land use management measures and technical defence works. Ten RTM offices were created for the all country, six in the Alps and four in the Pyrenees.

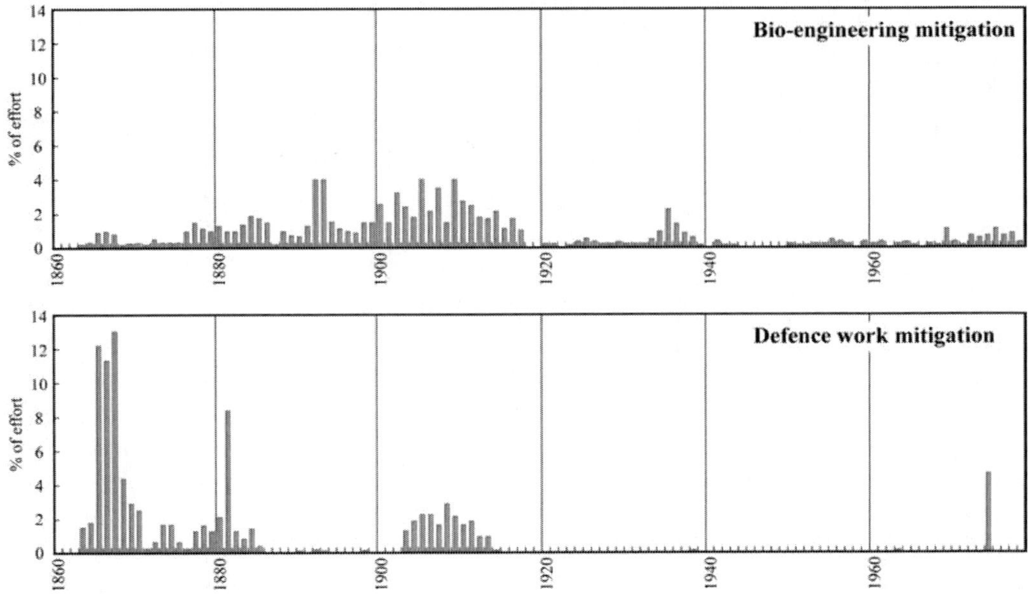

Figure 4. RTM effort from 1860 to 1940 in the South French Alps for (A) bio-engineering mitigation and (B) hydraulic engineering mitigation (in Liebault and Taillefumier, 2000).

The action of the RTM services has been conducted in three steps (Brugnot and Cassayre, 2002), (Figure 4):

- 1860 – 1882: a first law is adopted in 1860. This law creates "the RTM areas" which correspond to deforested regions that should be reforested. The main objective was to reforest 13,000 km^2 of bare soils. Additionally, the RTM service offers its technical support to local communities for the construction of technical countermeasures;
- 1882 – 1914: a second law is voted in 1882, it reinforced the RTM services through a considerable increasing of governmental funding. Gradually, the deforested areas are bought by the state. The construction of technical countermeasures (especially series of check dams) is increasing. Approximately one hundred torrents in the French Alps are equipped with defence works in 1905 (Figure 5);
- 1914 – 1940: the two world wars led to decrease governmental funding. The "RTM" services stop to build new series of check dams, but they still performed maintenance of the defence works and bio-engineering.

Most of the series of check dam built during these periods were masonry check dam made of dry stones (Figure 6A). This type of check dams have been progressively replaced by concrete check dams (Figure 6B, 7). Nevertheless, masonry check dams are still built nowadays in the French Alps and Pyrenees but mostly in the upper part of the torrential streams.

After the Second World War, the RTM policy enters into a long period (30 years) of inactivity. In fact, the national funding decreased substantially causing a lack of maintenance of all the defence works. Moreover, the RTM had severe difficulties to find workmen; most of them did not agree to work like hard labors.

Figure 5. Construction of series of check dams in the Barcelonnette basin, South French Alps (A) and (B) Local inhabitants building a check dam in 1898 in the Faucon torrent, (C) A view on a series of check dams in the upper part of the Faucon torrent. (Courtesy of the RTM Services of Alpes-de-Haute-Provence).

Figure 6. The two main types of check dams in the South French Alps: (A) photograph of a masonry check dam (Faucon torrent); (A) photograph of a concrete check dam (Champerousse torrent).

In parallel, during the past four decades (1960–2000), the French mountainous areas experienced major changes in population size, development patterns, economic conditions, and social characteristics. This is mostly related to the intensification of the tourism activity. These social, economic, and built-environment changes altered the French mountainous hazardscape in profound ways, with more people living in high-hazard areas than ever before. The strong development of specific equipments and accommodation spaces (ski resorts, hostels, holiday villages) increased the vulnerability of such mountainous areas. In the early 1970's two major events occurred:

- On February 10, 1970; an important snow avalanche killed 39 people in the ski resort of Val d'Isère;
- On April 16, 1970; a complex flow-type landslide threats a youth hostel and killed 72 people (most of them were children).

These two catastrophic events had a huge impact on the French population. In 1971, the French government decided to redefine the tasks of the RTM services. Three new technical tasks were added to RTM: (1) participation in the elaboration of natural risks maps, (2) collaboration with the local authorities for urban planning and (3) providing information and knowledge to inhabitants concerning the prevention and rehabilitation in case of danger and disaster. Additionally, the RTM services get new national funding dedicated to the rehabilitation and maintenance of the defence works.

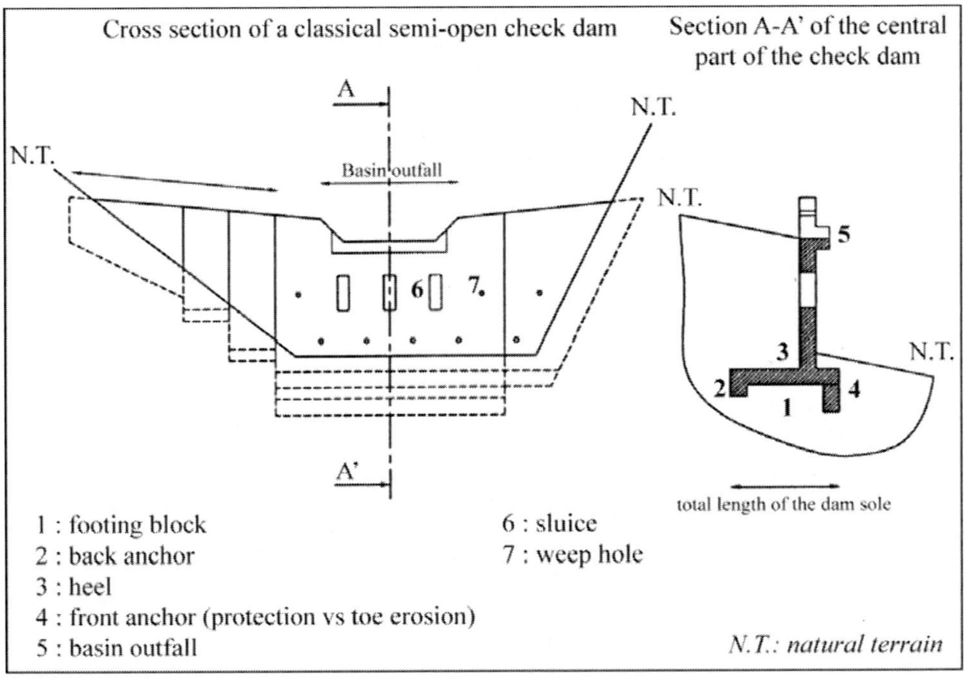

Figure 7. Sketch of a typical semi-open check dam (in Deymier et al., 1994).

3. Influence of Check Dams on the Dynamics of Debris Flows: Example of the Faucon Torrent

The triggering of a debris flow has commonly described as occurring in one of two ways. The first considers the triggering of a debris flow as the result of a landslide that steadily transforms itself into a debris flow by liquefaction during its movement (Johnson and Rahn 1970). The second describes the triggering of a debris flow as a sudden failure of coarse debris in a high altitude channel or gully bed (Takahashi 1981). The input, or trigger, for this later mechanism stems from either snowmelt or intense rainstorms. The two ways of triggering mechanisms have been observed during two events which have occurred in the

Faucon torrent (South French Alps). During these two events, check dams conditions have played an important role on the debris-flow dynamics.

3.1. Study Site

The study was conducted within the Faucon torrent in the Barcelonnette basin, a steep, mountainous region. The Barcelonnette basin extends from 1100 to 3000 m a.s.l. (Figure 8), it is representative of lithological, geomorphological, climatic and landuse conditions observed in the South French Alps.

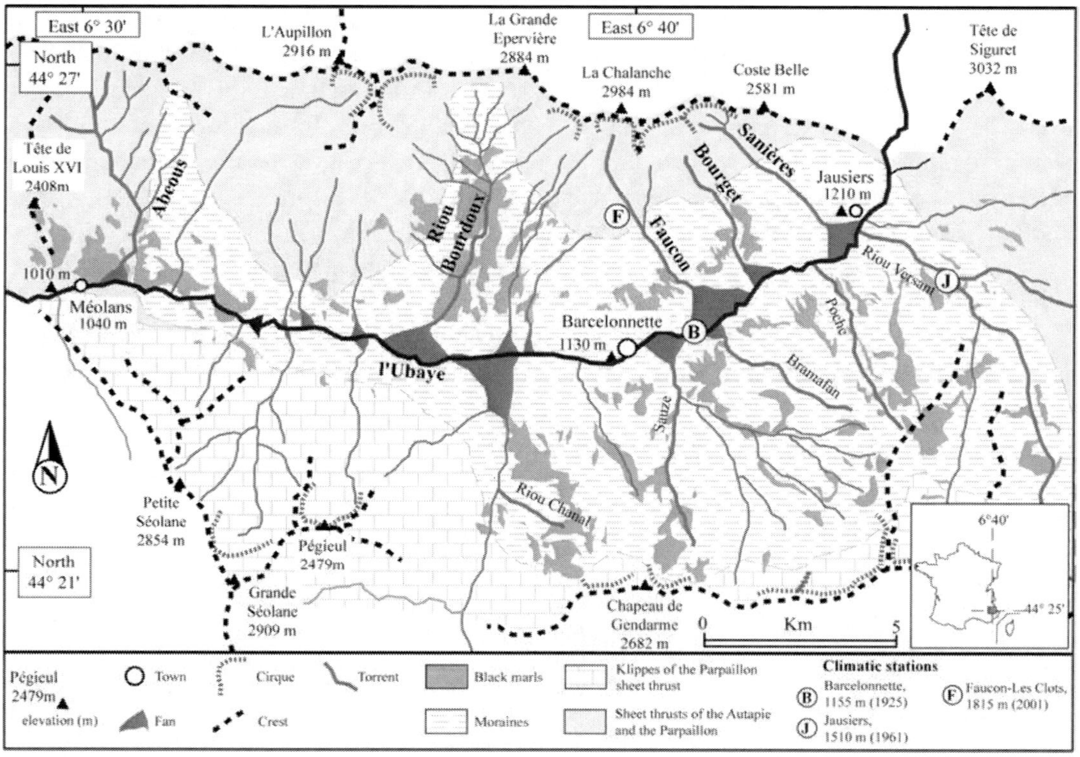

Figure 8. Morphological sketch of the Barcelonnette basin.

The basin is located in the dry intra-Alpine zone, characterized by (1) a mountain climate with a marked interannual rainfall variability (735 ± 400 mm over the period 1928-2005) and 130 days of freezing per year, (2) a continental influence with significant daily thermal amplitudes (> 20°) and (3) a Mediterranean influence with summer rainstorms yielding more than 50 mm.h-1 on occasion. The basin is drained by the Ubaye River. Local slopes are characterized by a specific morphology:

(a) At the upper part (1900 to 3000 m a.s.l.), slopes are steeper than 45° and consist of sheet thrusts of faulted calcareous sandstones. These slopes are often covered by non-consolidated debris varying in thickness between 0.5 to 5.0 m. Several debris tracks are affecting these slopes.

(b) At the lower part (1100 to 1900 m a.s.l.), gentle slopes (10-30°) consist of Callovo-Oxfordian black marls, mainly composed of fragile plates and flakes packed in a clayey matrix. Slopes are covered by various quaternary deposits: thick taluses of poorly sorted debris, moraines deposits and landslide debris.

These factors have given rise to the development of several mass movements (Flageollet et al., 1999), active torrential streams and debris tracks (Remaître et al., 2005a, 2008). Historical data concerning the torrential activity were collected in order to improve the current information about hazard. In the period between 1850 and 2005, 561 historical torrential events were found through the review of catalogues, newspapers, monographs, technical reports, bulletins and scientific papers (Figure 9, 10). The type and quality of information collected, and the methodologies used to analyze the data can be found in Flageollet et al. (1999) and Remaître (2006). The analysis has allowed verifying that flash floods (461 events) have been the most common types of torrential activity. 100 debris-flows have been recorded. The spatial distribution of historical debris-flows shows that they have occurred mainly in the torrents located on the south-facing slope of the Barcelonnette basin. Indeed, 90% of the debris-flow events were recorded in five torrents (Figure 9): Riou-Bourdoux, Sanières, Faucon, Bourget and Abeous.

The Faucon torrent is a tributary of the river Ubaye and it is one of the most active streams in the Barcelonnette basin (Figure 11). The channel is 5500 m in length and its mean slope is more or less 12° ranging from 50° (headwater basin) to 3° (debris fan). Altitudes range between 2984 and 1130 m a.s.l.

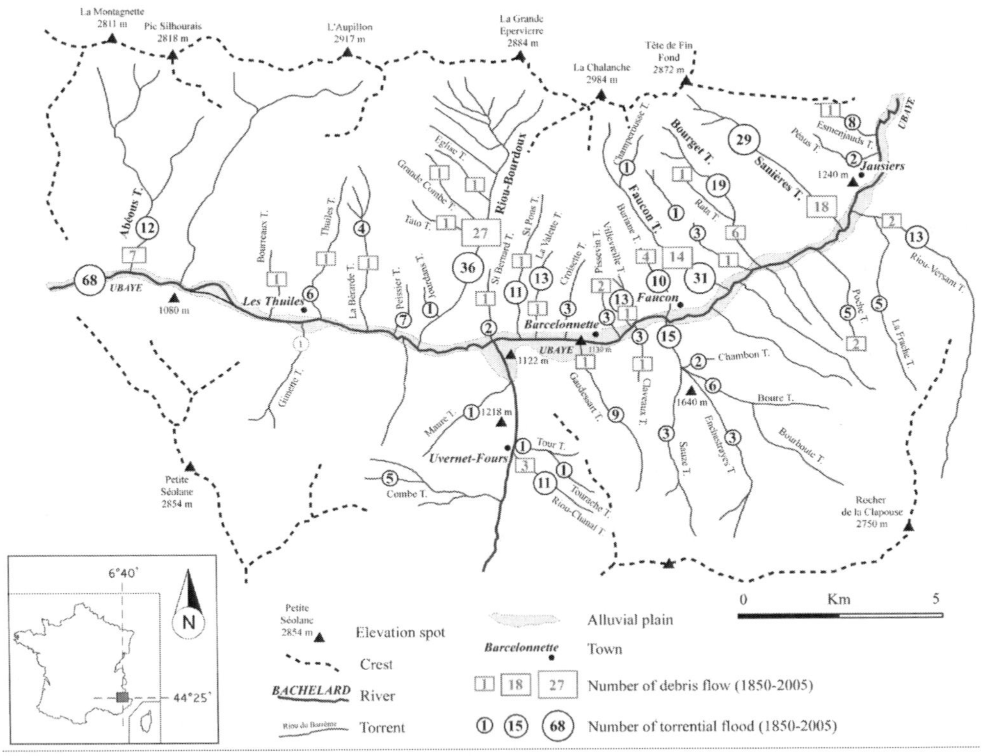

Figure 9. Map of the torrential activity in the Barcelonnette basin.

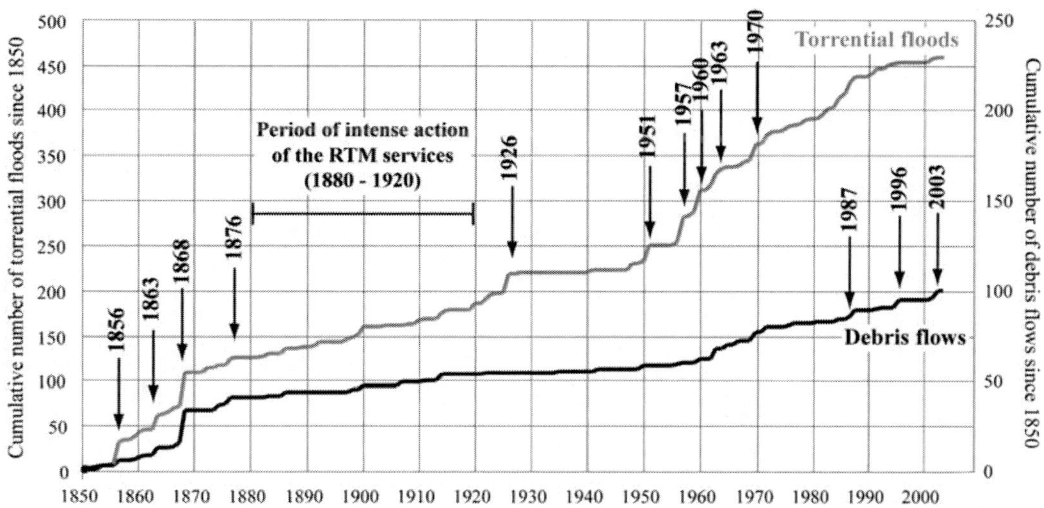

Figure 10. Torrential activity in the Barcelonnette basin (1850-2004). Torrential floods and debris flows have been considered.

The Faucon torrent exhibits a steep gradient (ranging from step–pools to waterfalls), a bed-rock channel with gorge and cascading sections and a basal debris fan. Channel morphology is characterized by two main types of cross-section, a V-shaped profile and a flat-floored cross profile. It contains abundant sediment (including multiple streamflow and debris-flows deposits), and has significant slope–channel coupling (screes and shallow stream-side landslides).

Since 1850, fourteen major debris flows have occurred in the torrent. In order to prevent flooding, authorities have built a hundred check dams on the torrent since the 1890's; but only a half of them are still efficient. Two main types of check dams can be observed: concrete check dams and masonry check dams. Three major events have occurred in the two last decades; two debris flows in 1996 and 2003 and one torrential flood in 2002 (Remaître, 2006; Remaître et al., 2005a, 2008, 2009). The check dams have been inventoried in a GIS database. For each dam, several characteristics have been recorded like the type, the size, and the year of construction. A sample of the database is available in Remaître et al. (2008); the complete database is available on request.

3.2. Methodology

Two detailed case studies of debris-flow events are presented here. In each case, the debris flow took place in the Faucon stream. Each of the events reveals strong relations between the debris-flow kinematic and the conditions of the series of check dam before the event that allow a better understanding of the processes involved in debris-flow generation (initiation and scouring). The erosional and depositional processes have been analyzed through a methodology associating detailed morphological mapping and sedimentological analysis of both the source material and the debris-flow deposits.

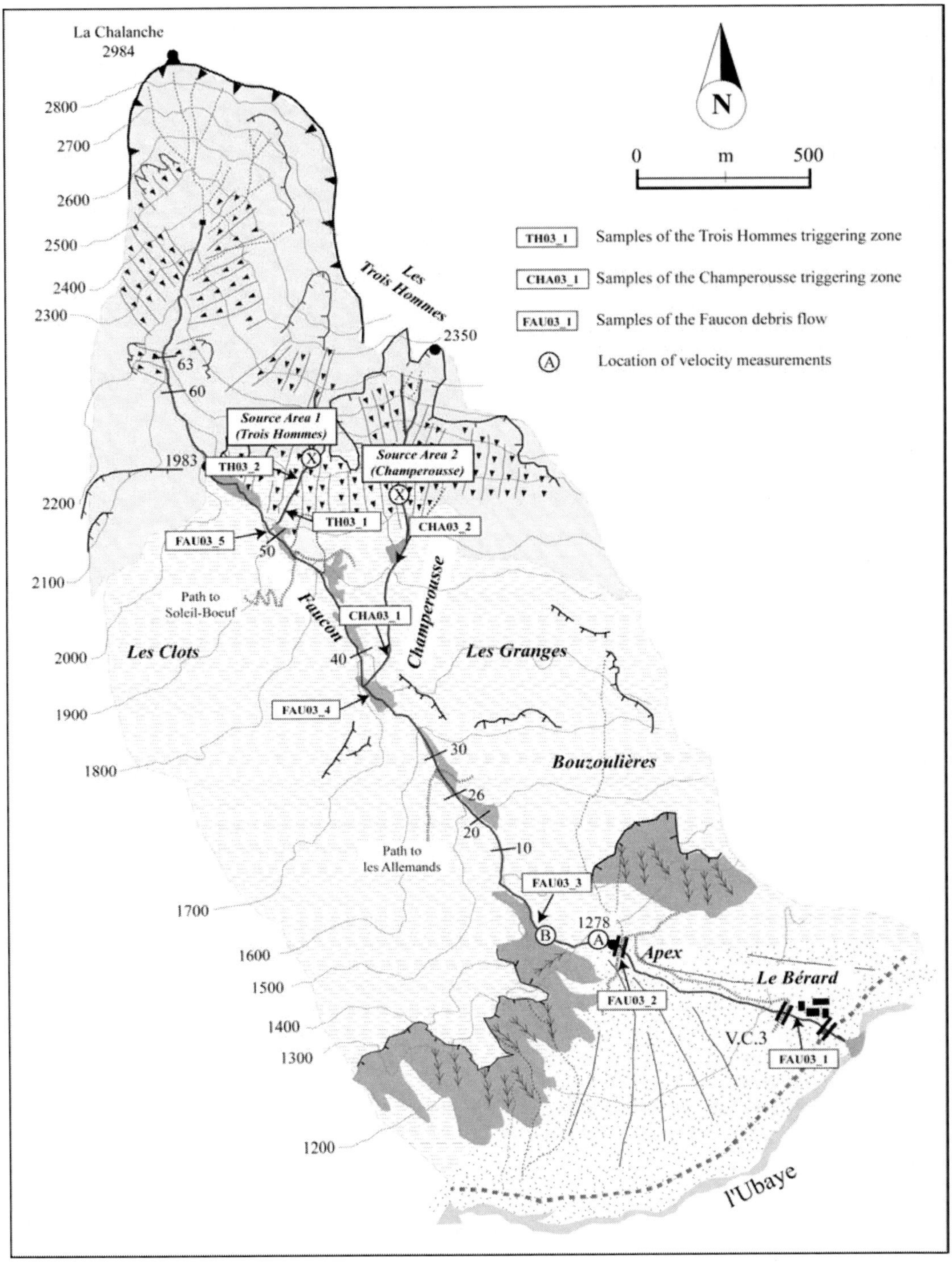

Figure 11. Morphological map of the Faucon watershed.

3.2.1. Geomorphological Observations

Herein, the debris-flow volume variations along the flow track have been carefully estimated. Indeed, knowledge of the volume is fundamental for the hazard assessment on the alluvial fan. The estimation of the volume has been realized on the basis of a field survey before and after the debris-flow event. A two-step methodology has been adopted:

1) A general appraisal between the volume of the source area and the volume of the deposits in the channel and the debris fan/cone;
2) A more detailed study of the debris-flow volume variations during the run-out based on the volume estimation for each torrential stretch. In order to simplify the volume calculations, morphology of the debris-flow (Figure 12A) and check-dams location and characteristics (Figure 12B) have been implemented in a GIS.

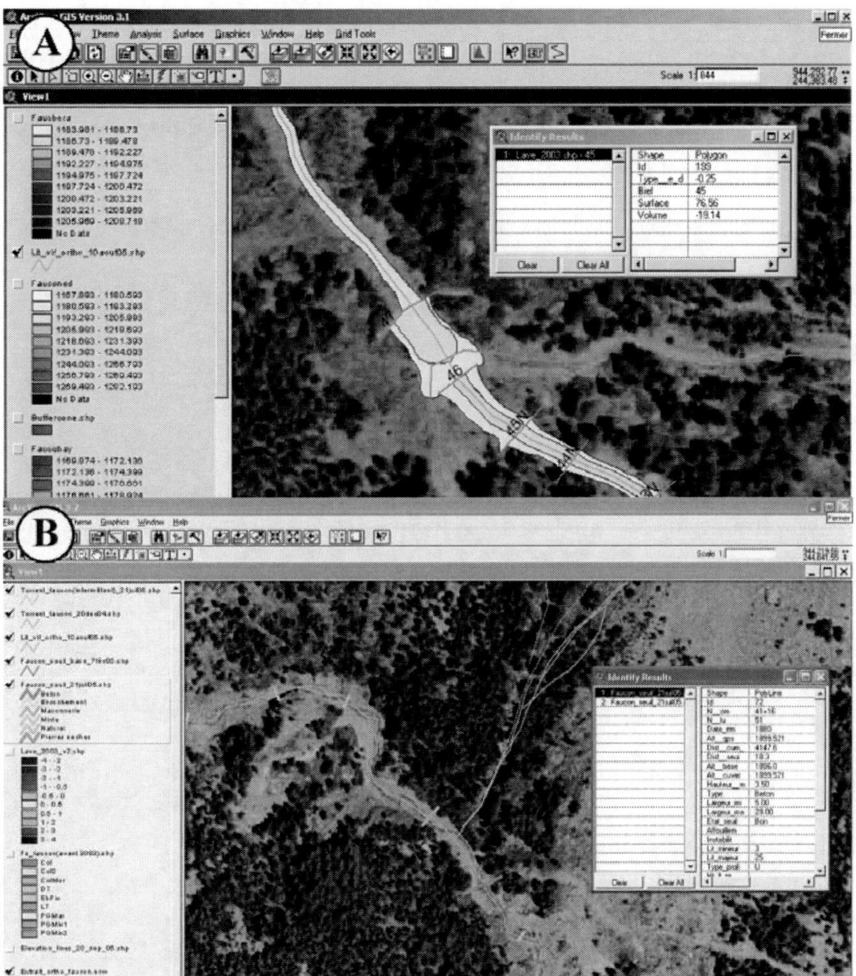

Figure 12. GIS database of the Faucon torrent. (A) Morphology of the 2003 debris flow (height of deposits and scouring depth), (B) Check dam location and characteristics.

3.2.2. Material Characterization

The sedimentological properties of the debris-flow deposits have been investigated in order to:

(1) Define the evolution of the debris-flow characteristics during the run-out in terms of grain-size distribution;
(2) Identify the main outcropping surficial formations involved in these events.

For the 1996 debris flow, five samples of debris-flow matrix have been collected (Remaître et al., 2005a). For the 2003 event, nine samples of debris-flow matrix were gathered at widely spaced locations along the main axis of the debris-flow event (Figure 8). Debris-flow deposits were sampled at 20 and 50 cm depth; the average weight of the samples was about 50 to 150 kg. All samples were oven dried and sieved from less than 20 mm to 0.050 mm, the fraction > 20 mm has been measured and characterized (petrography) in the field. The proportion of fines (<0.050 mm) was analyzed by laser diffractometry. Some additional samples were collected using the sand replacement method in order to define the bulk density (dry and wet). A complete description of the methodology and results can be found in Remaître et al. (2005b) and Remaître (2006).

3.3. RESULTS

3.3.1. The 1996 Debris-Flow Event: Influence of Check Dams on the Triggering

On Monday August 19, 1996, between 4:00 and 6:30 p.m. in the middle Ubaye valley, a debris flow was triggered by an intense and local thunderstorm. Indeed no rainfall was recorded by the pluviograph located at the Faucon alluvial fan. Several inhabitants provided eye-witness descriptions of debris flow.

They describe flowing masses of muddy-bouldery debris, first moving with pulsating waves “slower than a running man”, and then rushing downward at high speed. According to eye-witnesses and the French Forest Office, the total duration of the event was about 2 ½ hours. The debris flow caused moderate damage and the main road across the alluvial fan was cut for several hours.

Concerning the interpretation of the 1996 debris-flow event, the reader must keep in mind that all the field observations were collected two years after the event. Thus, all the interpretations have to be carefully considered.

This warning is also valid for the analysis of run-out and deposition. Nevertheless, the sampled debris-flow site showed good preservation of flow track/levee curvature and shape. Additional observations and data concerning the characteristics of the 1996 event (rheology, sediment budget, back-calculated velocities and discharges, run-out modeling) can be found in Remaître et al. (2005a, b).

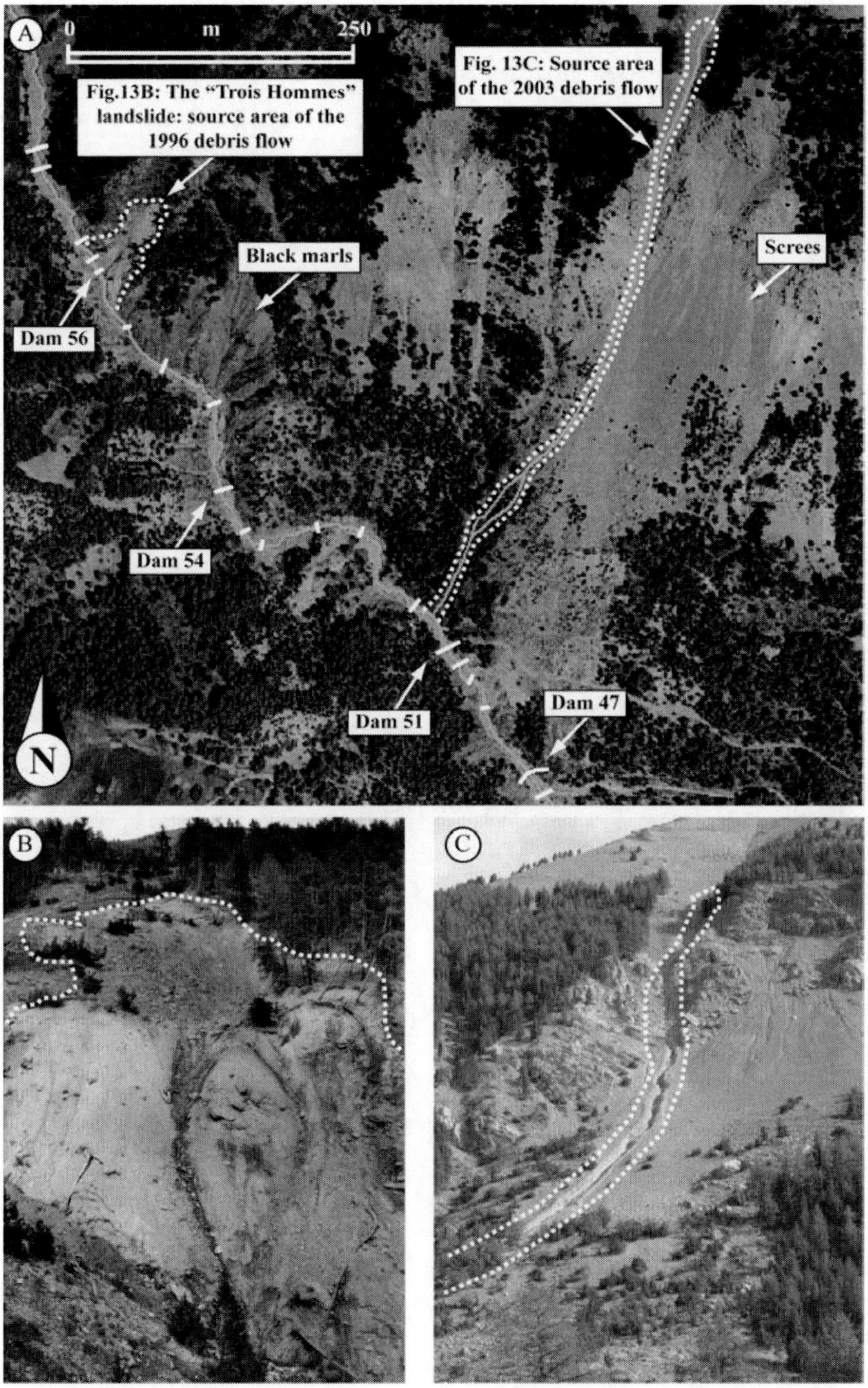

Figure 13. The source areas of the 1996 and the 2003 debris flows at the Faucon torrent. (A) Orthophotograph of the upper part, (B) photograph of the landslide which is consider to be the source area of the 1996 debris flow, (C) photograph of the 2003 debris- flow source area in a scree talus.

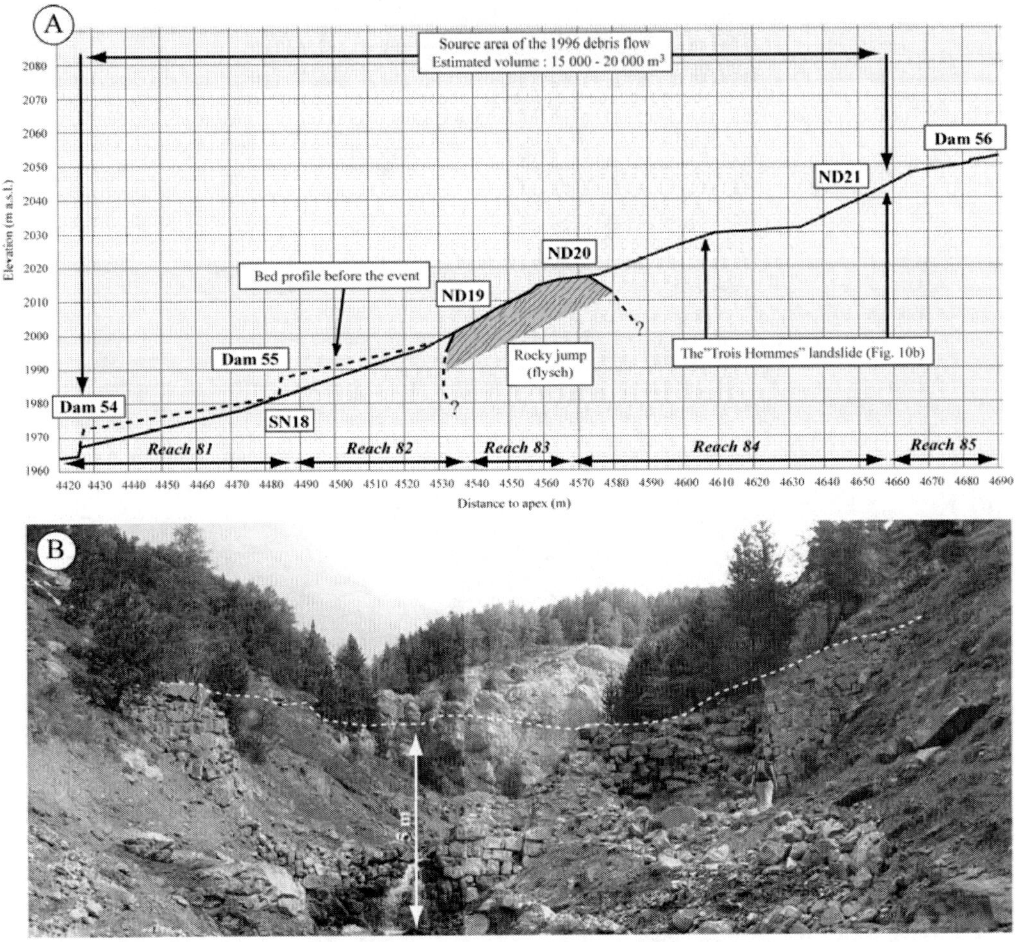

Figure 14. Morphology of the source area of the 1996 debris flow. (A) Bed profile of the Faucon torrent between check dam 54 and check dam 56. This area is considered as the main source area of the 1996 debris flow. (B) Photograph of check dam 54. Field observations suggest that his failure caused the triggering of the 1996 event.

Field evidences and aerial photographs study suggest that the debris-flow source area was located in the upper part of the torrent (above 2100 m a.s.l.), between check dam 54 and check dam 56 (Figure 13A):

- Check dam undermining and channel scouring in this area is particularly strong and reached nearly 5 m depth at check dam 54 (Figure 14B). Check dams 54 (Figure 14A), 55, 56 and 57 are completely destroyed. These check dams correspond to dry stone masonry type and they were built in 1880. Some vestiges of the wings are still visible, but the central part almost completely disappeared;
- An important quantity of debris fulfilled the channel. These debris mostly consist of coarse materials (some blocks exhibit a volume > 1 m^3) and various vegetable debris (mostly tree trunks). Such type of debris and morphology is characteristic of a natural dam collapse. The thickness of these debris decreases upstream;

- Mud splatters were observed on the tree trunks located on both sides of the channel. These mud splatters testify that the thickness of debris trapped by the temporary dam was at least 5 m.

These observations are in agreement with a dam breach which could have triggered the debris flow. The grain-size analysis of the source material and the debris-flow deposits suggest that the material involved in the debris-flow triggering had two sources:

1) A strong channel scouring was observed. Upstream of check dam 54, the channel was incised and widened. There is evidence of multiple temporary stream blockages, especially between the check dam 55 and the natural dam 21;
2) A hillslope sediment source which consist of a shallow landslide located near the check-dam 56 and affecting the “Trois Hommes” slope (Figure 13B). The analysis of an aerial photograph taken in May 1996 shows that the “Trois Hommes” shallow landslide occurred between May 1996 and August 1996.

The volume of the landslide and the volume of scoured sediments in the reaches located between check dams 54 and 56 range between 15’000 and 20’000 m^3 (Figure 14A). The main causes for the triggering of the 1996 event are associated to (a) the occurrence of a severe thunderstorm, (b) the occurrence of a shallow landslide, (c) the failure of the check dam 54. In this case, the general bad condition of the check dams 54, 55 and 56 is perhaps responsible for the debris flow triggering.

3.3.2. The 2003 Debris-Flow Event: Influence of Check Dams on Scouring Intensity

In this section, only the run-out stage of the 2003 debris-flow event will be considered. Nevertheless, some information about the triggering and the spreading stage are provided in order to discuss the sediment budget of the debris flow. Additional observations and data of the characteristics of the 2003 debris flow (rheology, sediment budget, back-calculated velocities and discharges, run-out modeling …) are detailed in Remaître et al. (2008, 2009).

On Monday August 5, 2003, between 6:00 and 8:00 p.m., a debris flow occurred. This event has been triggered on two specific spots on the east flank of the Faucon catchment: the Trois Hommes area (Figure 13C), and the upper part of the Champerousse torrent. For both cases, the morphology of the source area corresponds to a strong incision in scree slopes. The volume of the Trois Hommes debris-flow ranged approximately from 4000 to 5000 m^3, while the volume of the material in the source area of Champerousse ranged from 6000 to 7500 m^3. Unlike the Trois Hommes event, all the material did not reach the Faucon torrent; in fact 3000 m^3 of material has been trapped by the check-dams network. The observations at the Trois Hommes slope and the Champerousse torrent indicate that the source volume ranges from 7500 to 9500 m^3, a value of 8500 m^3 can be considered.

On the upper part of the debris fan/cone, most of the deposited debris has filled the channel; the debris flow has started to spread below the VC3 Bridge (Figure 11). Mudlines and splatters of fine-grained liquid sediments along both sides of the path revealed a nearly constant cross-sectional area (20 to 30 m^2). Field measurements and evidences of the residents indicate that the event evolved into 5 surges, for a time interval ranging between 2 and 5 minutes. These debris-flow surges filled progressively the channel; the last surge overflowed and caused the occlusion of the VC3 Bridge. Eyewitnesses indicate that the debris-flow height of the last surge reached 5 to 6 m. Most of the debris spread over the left bank, causing some substantial damages on five houses. Some residents were in their houses as the overflowing occurred but remained uninjured. The thickness of the debris deposits ranged from 1.0 to 2.0 m. Moreover, the debris flow also breached the main road of the valley; the traffic was stopped during several hours and remained difficult for three weeks while authorities cleaned the channel and the fan. We have measured the area and the depth of the deposits both in the channel and on the fan. Length, width and depth were recorded and mapped. Total volume of debris-flow deposits was estimated to be 45,000 m^3 on the debris fan and 15,000 m^3 in the upper channel.

Channel scour is responsible for the great difference between the 8500 m^3 of the two source areas and the 60,000 m^3 of the total volume. The length of the debris-flow track is about 3500 m, with a slope gradient of about 15°. Several areas of scouring were observed, they are mainly located in the 1996 debris-flow deposits area. The channel scour rate amounts to 15 $m^3.m^{-1}$. This value is in agreement with values observed by other authors in various geological environments (Hungr et al., 1984; Jibson, 1989; Jakob et al., 2000). Observations of the channel after the event indicate that the scour depth ranges between 0.5 and 4 m.

The most intense scour depth was observed at the toe of some check dams (Figure 15), as a result of regressive channel erosion. Field evidences show that 26 check dams were scoured or partially destroyed by the 2003 event.

Figure 15. Example of two scoured check dams after the 2003 debris flow event.

The 2003 event started as a granular flow, bulked increased in fine elements by incorporating marly sediments along the debris-flow track and transformed into a muddy debris flow (Remaître et al., 2008). Such process has been observed during the 1996 event (Remaître et al., 2005a). In order to analyze more precisely relations between the 2003 debris-flow kinematics and the morphological factors of the torrential stretches, the spatial evolution of the sediment budget and conditions of the check dams before the 2003 event have been compared (Figure 16).

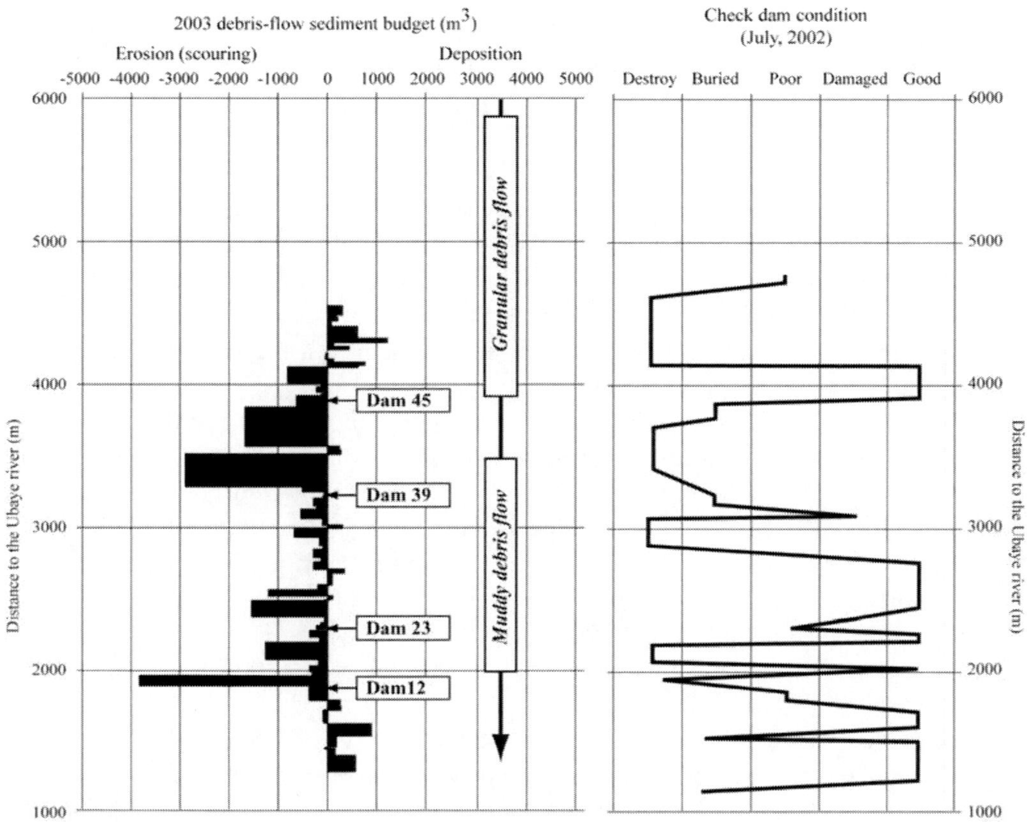

Figure 16. The 2003 debris-flow event at Faucon torrent: sediment budget and check dam conditions.

Results show a correlation between the volume of scoured sediments and the condition of the check-dams. Indeed the two main areas where the bulking was particularly intense (between check dams 45 and 39 and between check dams 23 and 12), are characterized by a general bad conditions of the check dams (buried or destroyed). These stretches do not exhibit a steep longitudinal profile, the average slope of the torrent in these areas is about 7-9°.

It could be expected that the steepest stretches would be characterized by an intense scouring during the event, but this was not always observed in the field. In fact, the stretches characterized by an intense scouring have been fully or partially filled by the 1996 debris-flow deposits (Remaître et al., 2005) and the 2002 streamflow deposits (Remaître, 2006) producing a large source of 'available' sediments for the 2003 debris flow. Relations between the 1996, the 2002 and the 2003 events seem to be very close. The spreading and the overflowing on the fan of the 2003 event were much more important than in 1996, although the volume of both events was very close. This must be put in relation with the occurrence of the streamflow in 2002 which filled the lower stretches located on the debris/fan producing an increasing of the bed profile level and a decreasing of the flow sections.

The example of the 2003 event shows the fundamental role of the series of check dams on the scouring and the enlargement of the debris-flow volume during its run-out. The general bad condition of the check dams and the presence of large available material (deposits from the previous events) have both contributed to increase the intensity of the debris flow.

4. Influence of Check Dams on Debris-Flow Run-Out Intensity: A Modeling Approach

In this section, some modeling exercises have been carried out to evaluate numerically the influence of the number and the location of the check dams on the debris-flow run-out intensity (in term of flow thickness, flow velocity and volume). The JDFM-1D model (van Asch et al., 2004), is used to simulate the run-out of the debris flow. The constitutive equation used in the model is a simplified 2-parameters viscoplastic rheology. The model uses the Janbu force diagram to resolve the force equilibrium equations; a Bingham fluid rheology is introduced and represents the resistance term.

The JFDM-1D model can take into account the amount of material entrained by the flow along the path (scouring) in order to increase the final volume deposited. According to Rickenmann et al. (2003), the intensity of the scouring is assumed to be a function of the integrated mean shear stress of the debris-flow mixture which passed through sections of the torrent, and is controlled by the slope gradient, the volume and the density of the mixture which enters this section. Therefore, breaking the energy of the flow in the earlier stage of the debris-flow event kinematics would reduce the total amount of entrained material. A complete description of the model can be found in van Asch et al. (2004) and Remaître et al. (2008).

The model has been calibrated both on the debris-flow events that occurred in 1996 and 2003 at the Faucon stream. Influence of the check dams on the debris-flow intensity is quantified taking into account several check dams configurations (number and location) as input geometrical parameters.

4.1. Modeling Scenarios

For each modeling test, the same triggering scenario has been used based on the observations after the 2003 event. A volume of 5000 m^3 of material has been considered, which corresponds to one of the source area (Trois Hommes area). The rheological characteristics of the debris-flow material can not be changed during the run-out. Therefore, we considered that the flow exhibits viscoplastic behaviour for the entire simulation. The source area is located at the upper part of the profile (point A on Figure 17) while the check point (point B on Figure 17) location corresponds to the upper part of the fan where the flow-track shows a clear flattening of the slope gradient. The run-out distance is approximately 4000 m.

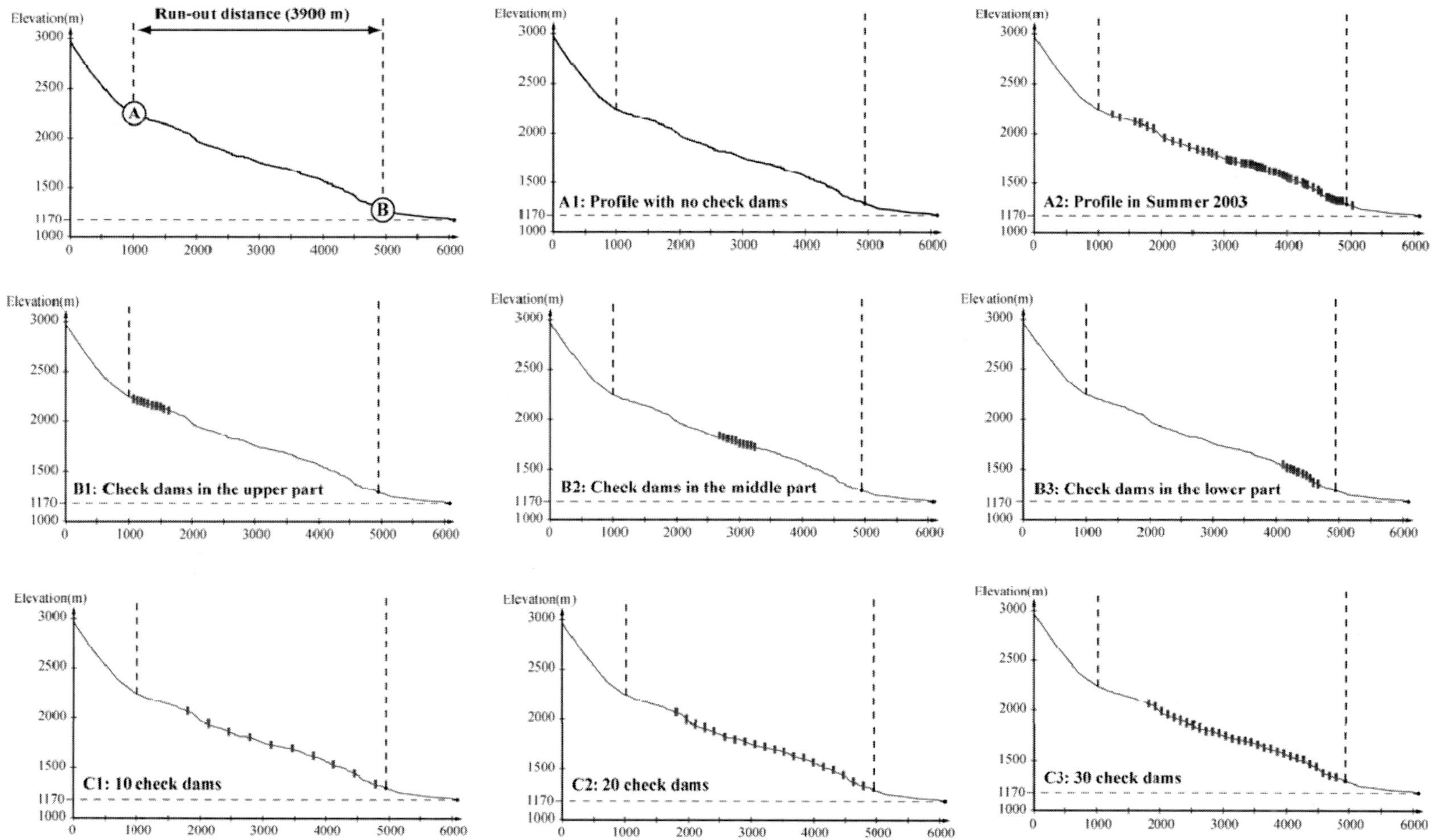

Figure 17. Settings of the three modeling scenarios. The bed profile corresponds to the one of Faucon torrent.

In the model, the series of check dams influence the intensity of the debris-flow through topographic variations of the flow track (slope angle). For the scenario A, the height of check dams corresponds to the height observed in the field in July 2003. For the scenarios B and C, a 5m height has been considered for all the check dams. Three main run-out scenarios have been tested (Figure 17):

(1) Scenario A: effect of the check dams on the intensity of the 2003 debris-flow. This scenario is a kind of "back analyses". We compare modeling results for two configurations of debris-flow pathway; a profile with no check dams (A1), and the profile with the check dams observed before the 2003 debris-flow event;
(2) Scenario B: effect of the location of check dams on the intensity of a debris flow. Three configurations of check dams location were tested: check dams located in the upper part of the torrential pathway (B1), in the middle part (B2) and in the lower part (B3);
(3) Scenario C: effect of the number of check dams on the intensity of a debris flow. Three configurations of check dams number were tested: a profile with 10 check dams (C1), a profile with 20 check dams (C2) and a profile with 30 check dams (C3).

4.2. Modeling Results

For each scenario, the maximal flow height, the maximal velocity and the total volume of debris where analyzed and compared (Figure 18).

For the scenario A, logically, the intensity of the debris flow is decreasing when the torrent is equipped with check dams. The maximum flow height is decreased from 5.95 m (A1: no check dams) to 2.21 (A2: 75 check dams), while the maximum velocity is decreased from 1.58 m.s^{-1} (A1) to 0.53 m.s^{-1} (A2). The total volume of the debris-flows is decreasing from 69,000 m^3 (A1) to 33,000 m^3 (scenario A2).

The run-out modeling results for the Scenario C show a decrease of the flow intensity. The decrease is particularly strong between the scenario A1 (no check dams) and the scenario C1 (10 check dams): decreasing of the maximal velocity, the maximal flow height and the volume are respectively 29% (1.58 to 1.12 m.s-1), 25% (5.95 to 4.41 m) and 26% (69,000 to 51,000 m^3); while the decreasing is gently moderate when the number of check dams is increasing (scenarios C1, C2 and C3). For instance, between the C1 and the C3 scenarios, decreasing of the maximal velocity, the maximal flow height and the volume are respectively 11% (1.12 to 1.00 m.s-1), 7% (4.41 to 4.12 m) and 10% (51,000 to 46,000 m^3).

The comparison of debris-flow intensity for the three cases (B1, B2 and B3) indicates that the location seems to have a strong influence on the debris-flow intensity. Indeed, the differences are significant between the B1 scenario (dams located on the upper part) and the B3 scenario (dams located on the lower part): decreasing of the maximal velocity, the maximal flow height and the volume are respectively 37% (1.19 to 0.74 m.s^{-1}), 36% (4.97 to 3.18 m) and 33% (62,000 to 41,000 m^3).

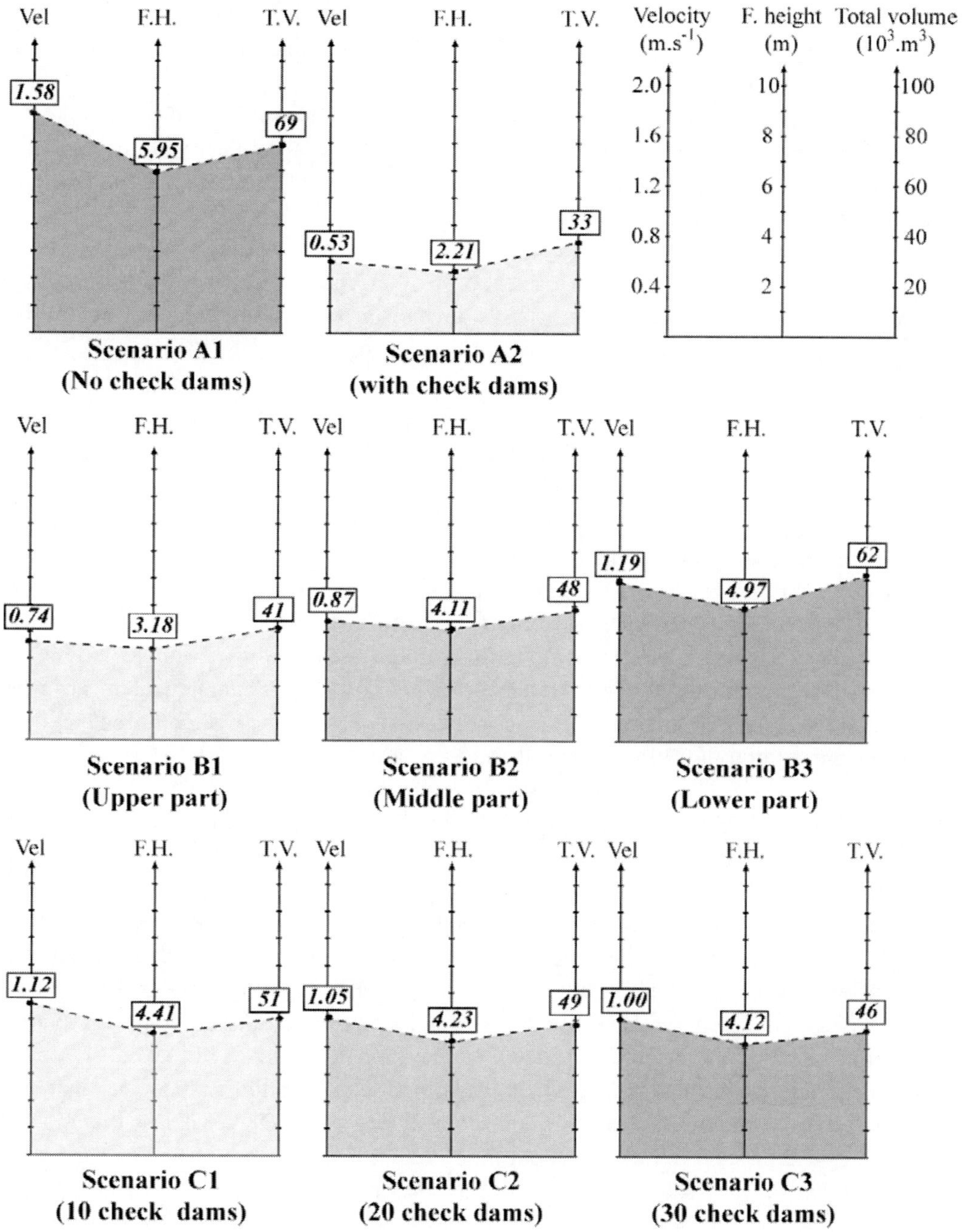

Figure 18. Results of the three modelling scenarios.

4.3. Discussion and Conclusions

The sensitivity analysis of the relative influence of, on one hand the number of check dams, and on the other hand the location of the check dams shows interesting results. The simulation which provides the lowest debris-flow intensity corresponds to the case where the check dams are located at the upper part of the flow track (Scenario B1), near the source area. These results suggest that a relative small number of check dams located near the potential source areas could be very efficient by breaking early the energy of the debris-flow. For this reason, additional river engineering measures could be proposed, such as construction of new check dams in the upper reaches of the Faucon torrent. These conclusions are valid for a debris flow that exhibited a muddy behavior; some additional modeling tests have to be provided for granular debris-flows.

This new approach opens a new direction for future research. Additionally work remains to be done in order to (i) analyze the influence of other check dam characteristics (height, width) on the debris-flow intensity and (ii) to develop a robust and efficient methodology which can be applied for routine debris-flow hazard assessment.

CONCLUSION

Studies of debris flows throughout the world indicate that many factors have been responsible for the triggering and the run-out of debris. Numerous studies have demonstrated that rainfall, lack of vegetation and other factors had a strong influence on debris-flow processes. It is ridiculous to relate a single parameter, such as intensity of rainfall or landcover to the occurrence of debris flows in a given area (Johnson and Rodine, 1984). During the two last centuries several strategies for debris flow risk management have been developed to fight against the consequence of such phenomena which threaten all the mountainous areas, although with different intensity and recurrence. The strategies include hazard and risk zoning methods, non-structural measures and defence works.

As it concerns the last point, the experience gained and the results obtained in several countries encourage the use of defence work to improve the safety of vulnerable elements and inhabitants. The role of series of check dams in regulating torrential and hillslope processes is fundamental. The results reported here indicate a strong relation between debris-flows triggering and run-out processes and the characteristics of series of check dams, particularly those located in the upper basin near the source areas. These series of check dams have to be specifically suited to the character of the debris flow, the resources, and equipment available for the maintenance of the structure. Indeed, if they are not regularly cleaned or repaired, series of check dams can increase significantly both (i) the probability of occurrence of an event and (ii) the volume of an event through scouring processes. Nevertheless there is still a great benefit to use series of check dams but their conditions have to be checked carefully by the authorities. This is particularly true in clay-shale basins where debris-flow scouring can be very intense.

ACKNOWLEDGMENTS

This paper has been supported by the project "*Mountain Risks*" funded by the European Commission. The authors gratefully thank Pr. Olivier Maquaire for his support during the field campains.

REFERENCES

Armanini, A. and Scotton, P. (1992). Experimental analysis on the dynamic impact of a debris flow on structures. In Proc. of the International Symposium *Interpreavent* (pp. 107-116). Bern, Switzerland.

Ayotte, D. and Hungr, O. (2000). Calibration of a runout prediction model for debris-flows and avalanches. In N. Wieczorek and N.D. Naeser (Eds.), Proceedings of the 2nd International Conference on *Debris-Flow Hazards Mitigation: Mechanics, Prediction, and Assessment*, Taipei, China (pp. 505-514). Rotterdam: Balkema.

Brugnot, G. and Cassayre, Y. (2002). De la politique française de Restauration des Terrains de Montagne à la prévention des risques naturels. MSH Alpes.

Chatwin, S.C., Howes, D.E., Schwab, J.W. and Swanston, D.N. (1994). A guide for management of landslide-prone terrain in the Pacific Northwest. (2nd Edition). *B.C. Min. For., Victoria B.C. Land Manage. Handb. 18.*

Corominas, J., Remondo, J., Farias, P., Estevao, M., Zezere, J., Diaz de Teran, J., Dikau, R., Schrott, L., Moya, J. and Gonzales, A. (1996). Debris Flow. In R. Dikau, D. Brundsen, L. Schrott and M.-L. Ibsen (Eds.), *Landslide Recognition: Identification, Movement and Causes* (pp. 161-180). Chichester: Wiley.

Costa, J.E. (1984). Physical geomorphology of debris flows. In J.E. Costa, and P.J.Fleisher (Eds.), *Developments and Applications in Geomorphology* (pp. 268-317). Heidelberg: Springer-Verlag.

Couvert, B., Lefebvre, B., Lefort, P. and Morin, E. (1991). Etude générale sur les seuils de correction torrentielle et les plages de dépôt. *La Houille Blanche 46(6)*, 449-456.

D'Ambrosio, D., Iovine, G., Sparato, W. and Miyamoto, H. (2007). A macroscopic collisional model for debris-flows simulation. *Environmental Modelling and Software 22*, 1417-1436.

Deymier, J.-M., Tacnet, J.-M. and Mathys, N. (1994). Conception et calcul de barrages de correction torrentielle. *Collection Etudes, Série Equipement pour l'Eau et l'Environnement*, Cemagref Editions.

Eisbacher, G.H. and Clague, J.J. (1984). Destructive mass movements in high mountains: hazard and management. *Geol. Surv. Can., Pap. 84-16.*

Evans, S.G. (1982). Landslides and surficial deposits in urban areas of British Columbia. *Canadian Geotechnical Journal 19*, 269-288.

Flageollet, J.-C. (1988). Les mouvements de terrain et leur prévention. Paris: Masson.

Flageollet, J.-C., Maquaire, O., Martin, B. and Weber, D. (1999). Landslides and climatic conditions in the Barcelonnette and Vars basins (Southern French Alps, France). *Geomorphology 30*, 65-78.

Godt, J.W. and Coe, J.A. (2007). Alpine debris flows triggered by a 28 July 1999 thunderstorm in the central Front Range, Colorado. *Geomorphology 84*, 80-97.

Goudie, A. (1995). *The changing Earth: rates of geomorphological processes*. Oxford: Blackwell.

Government of Japan (1984). Basics of planning the measures against debris flows and planning countermeasures facilities against debris flow. *Min. Contruction.*

Heierli, W. and Merk, A. (1985). Barriers for erosion control of mountain torrents. In Proc. International Symp. on *Erosion, debris flow and disaster prevention* (pp. 375-380), Tsukuba, Japan.

Hungr, O., Evans, S.G., Bovis, M.J. and Hutchinson, J.N. (2001). A review of the classification of landslides of the flow type. *Environmental and Engineering Geosciences 7*, 221-238.

Hungr, O., Morgan, G.C. and Kellerhals, R. (1984). Quantitative analysis of debris torrent hazards for design of remedial measures. *Canadian Geotechnical Journal 21*, 663-667.

Iverson, R.M. (1997). The physics of debris flows. *Review in Geophysics 35*, 245-296.

Jakob, M., Anderson, D., Fuller, T., Hungr, O. and Ayotte, D. (2000). An unusually large debris-flow at Hummingbird Creek, Mara Lake, British Columbia. *Canadian Geotechnical Journal 37*, 1109-1125.

Jibson, R.W. (1989). Debris flows in southern Puerto Rico. In A.P. Schultz and R.W. Jibson (Eds.), Proceedings of the International Conference on *Landslide processes in Eastern North America and Puerto Rico* (pp. 29-55). Geological Society of America Special Paper 236.

Johnson, A.M. and Rahn, P.H. (1970). Mobilization of debris flows. *Z. Geomorphol. Suppl. 9*, 168-186.

Johnson, A.M. and Rodine, J.R. (1984). Debris flows. In D. Brundsen and D.B. Prior (Eds.), *Slope Instability* (pp. 257-361). Chichester: Wiley.

Johnson, P.A. and McCuen, R.H. (1989). Silt dam design for debris flow mitigation. *Journal of Hydraulic Engineering 115 (9)*, 1293-1296.

Laigle, D. and Marchi, L. (2000). Example of mud/debris-flow hazard assessment using numerical model. In N. Wieczorek and N.D. Naeser (Eds.), Proceedings of the 2nd International Conference on *Debris-Flow Hazards Mitigation: Mechanics, Prediction, and Assessment*, Taipei, China (pp. 417-424). Rotterdam: Balkema.

Leys, E. and Hagen, G. (1971). Design of gravity dams against water pressure and earth pressure. *Imst Regional Construction Office, Torrent and Avalanche Control, Austria.* (in German).

Liébault, F. and Taillefumier, F. (2000). L'évolution contemporaine de la bande active des principaux affluents de la Drôme, de l'Eygues et du Roubion (Préalpes du Sud, France). *Revue de Géographie de Lyon, 75 (4)*, 327-336.

Miyazawa, N., Tanishima, T., Sunada, K., and Oishi, S. (2003). Debris-flow capturing effect of grid type steel-made sabo dam using 3D distinct element method. In D. Rickenmann and L.C. Chen (Eds.), Proceedings of the 3rd International Conference on *Debris-Flow Hazard Mitigation: Mechanics, Prediction and Assessment*, Davos, Switzerland (pp. 527-538). Rotterdam: Milpress.

O'Brien, J.S., Julien, P.Y. and Fullerton, W.T. (1993). Two dimensional water flood and mudflow simulation. *Journal of Hydraulic Engineering 119 (2)*, 244-259.

Pierson, T.C. (1980). Erosion and deposition of debris flows at Mount Thomas, North Canterbury, New Zealand. *Earth Surface Processes and Landforms 5*, 227–247.

Pierson, T.C. and Costa, J.E. (1987). A rheological classification of subaerial water-flows. In J. Costa and G.F. Wieczorek (Eds.), *Debris flows/ avalanches: process, recognition and mitigation* (pp. 1-12). Geological Society of America, Reviews in Engineering Geology 7.

Remaître, A. (2006). Morphologie et dynamique des laves torrentielles : applications aux torrents des Terres Noires du bassin de Barcelonnette. *PhD Dissertation, University of Caen Basse-Normandie.*

Remaître, A., Maquaire, O. and Malet, J.P. (2005a). Morphology and sedimentology of a complex debris flow in clay-shales basin. *Earth Surface Processes and Landforms 30*, 339-348.

Remaître, A., Malet, J.-P., Ancey, C., Locat, J. and Maquaire, O. (2005b). Flow behaviour and run-out modelling of a complex debris flow in clay-shales basin. *Earth Surface Processes and Landforms 30*, 479-488.

Remaître, A., van Asch, Th. W.J., Malet, J.-P. and Maquaire, O. (2008). Influence of check dams on debris-flow run-out intensity. *Natural Hazards and Earth System Sciences 8 (6)*, 1403-1416.

Remaître, A., Malet, J.-P. and Maquaire, O. (2009). Sediment budget and morphology of the 2003 Faucon debris flow (South French Alps): scouring and channel-shaping processes. In J.-P. Malet, A. Remaître and T. Bogaard (Eds.), Proceedings of the International Conference on *Landslide Processes: from geomorphologic mapping to dynamic modelling*, Strasbourg, France (pp. 75-80). Strasbourg: CERG Editions.

Rickenmann, D., Weber, D. and Stepanov, B. (2003). Erosion by debris flows in field and laboratory experiments, In D. Rickenmann and L.C. Chen (Eds.), In D. Rickenmann and L.C. Chen (Eds.), Proceedings of the 3rd International Conference on *Debris-Flow Hazard Mitigation: Mechanics, Prediction and Assessment*, Davos, Switzerland (pp. 883-894). Rotterdam: Milpress.

Scott, K.M., Vallance, J.W. and Pringle, P.T. (1995). Sedimentology, behaviour, and hazards of debris flows at Mount Rainier, Washington. *U.S.G.S. professional paper 1547.*

Takahashi, T. (1981). Debris flow. *Annu. Rev. Fluid Mech. 13*, 57-77.

Takahashi, T. (1991). *Debris Flows*. Rotterdam: Balkema.

Van Asch, Th. W.J., Remaître, A., Malet, J.-P. and Maquaire, O. (2004). Numerical modelling of the run-out of a muddy debris flow. The effect of rheology on velocity and deposit thickness along the run-out track. In W. Lacerda (Ed.), Proceedings of the 9th *International Symposium on Landslides*, Rio de Janeiro, Brazil (pp. 1433-1438). Rotterdam: Balkema.

VanDine, D.F. (1996). Debris Flow Control Structures for Forest Engineering. *B.C. Min. For., Research Paper 22.*

VanDine, D.F. and Bovis, M. (2002). History and goals of Canadian debris-flow research. *Natural Hazards 26 (1)*, 67-80.

Van Effenterre, C. (1982). Les barrages perméables de sédimentation. *Revue Forestière Française 5.*

Whittaker, J.G., Schaelchli, U. and Jaeggi, M.N.R. (1985). Design problems with torrent check dams in Switzerland. In Proc. International Symp. on *Erosion, debris flow and disaster prevention* (pp.387-394), Tsukuba, Japan.

In: Check Dams, Morphological Adjustments…
Editors: C. C. Garcia and M. A. Lenzi

ISBN: 978-1-62618-856-3

Chapter 11

INDUCED MORPHOLOGICAL CHANGES AFFECTING THE STABILITY OF CHECK DAM SYSTEMS

Xiang-Zhou Xu[1]*, Hong-Wu Zhang[2], Guang-Qian Wang[2], Yi Xu[1] and Ou-Yang Zhang[3]

[1]School of Civil and Hydraulic Engineering,
Dalian University of Technology, Dalian 116024, P. R. China
[2]Department of Hydraulics and Hydropower Engineering,
Tsinghua University, Beijing, 100084, P. R. China
[3]Bureau of Hydrology, Changjiang River Water Resources Commission,
Wuhan, 430010, P. R. China

ABSTRACT

The building of check-dams is one of the most effective measures for the conservation of soil and water on the Loess Plateau of China, and the hydro-sedimentologic balance is the most important factor influencing the relative stability of the check-dam systems. This means that soil and water in the small watersheds controlled by the check-dams will be absorbed internally, without the need of raising the height of the dams, if some given parameters have reached certain values. A runoff simulation experiment for a single check-dam and a rainfall simulation experiment for the whole check-dam system had been conducted to simulate the induced morphological changes affecting the stability of check dam systems, and develop relationships among the mean slope gradient, mean dam-land elevation and dam-land area.

The results indicated that raising of the dam-land altitude became slower, and the mean gradient of the gully was kept at a constant. The main reasons causing the check-dam to show a good relative stability are the enlargment of the dam-land area, the alleviation of erosion by the check-dam, and the auto-stabilizing mechanism of the gullies.

---

*Corresponding Author: Tel.:+86(0)411-84708946; Fax:+86(0)411-84708946; E-mail address: xzxu@dlut.edu.cn

STABILITY OF CHECK DAM SYSTEMS ON THE LOESS PLATEAU

The fact that the check-dam systems in the Loess Plateau of China always exhibit a good relative stability is really a unique geomorphological phenomenon. It has been found that the *Loess Mesa Ravine Region* and the *Loess Hill Ravine Region*, which cover 200,000 km^2 of the Loess Plateau in China, are among the areas having the highest erosion rates on earth. The most effective way to conserve soil and water in these areas is to build check-dam systems in the gullies. At present, more than 100,000 check-dams have been built, and the amount of sediments retained by the check-dams has been found to be the largest among all measures (Xu. et al., 2004a). Moreover, official documents from the Chinese Ministry of Water Resources have stated that 163,300 check-dams will be built in this area before 2020 (CMWR, 2003). Peasants would prefer to plant crops in the deposited areas behind the dams, which is called the dam-land (Figure 1), owing to their more abundant water content and more fertile soil as compared with the slope land.

According to field investigations on the check-dam systems of the Shanxi Province (Fang, 1996), 6,000~7,500 kg grains could be harvested per hectare of the dam-land, which was 8 to 10 times higher than those on the slope land. At present, the dam-land occupies only 9% of the whole farmland areas in the Loess Hill Ravine Region, but their grain yields comprise 23.5% or higher of the over-all grain production of this region (Xu and Wang, 2000). However, when tens or hundreds of check-dams are built in a small watershed area not larger than 100 km^2, it will present quite complicated hydro-sedimentologic problems. Elements such as the layout of the dam-sites, the dam height and area controlled by the small watersheds should all meet the requirements of relative stability during the design and construction of the check-dam systems.

Figure 1. Dam-land on the Loess Plateau. Photo by X.Z. Xu, 10-09-2006.

The concept of "relative stability of the check-dam system" first emerged in 1960s, which was stemming from the recognition of the key factors that determined the natural balance of the check-dams. Inspired by the natural *Juqiu*, a kind of dam-land resulting from land-slips which never overflows when impounding the floodwater and soil for centuries, people came to recognize the important role of dam height and the ratio of dam-land area to that of the controlled watershed. It was observed that if the above parameters have reached certain values, soil and water in the small watershed could be internally absorbed, without need of raising the height of the dam. Fang (1995) has suggested that check-dam systems exhibiting a good relative stability have to fulfill the following requirements: (1) No flooding will occur. That is, it should be ensured that the check-dam system can withstand rainstorms. (2) Guarantee of harvest from the dam-land, which implies the reducing of losses of the planted crops due to rainstorms. (3) Conservation of floodwater and sediment via impounding. (4) Increasing of the dam height and repairing of the dams after prolonged utilization are unnecessary. To attain relative stability of the check-dam system to meet these purposes, many elements have to be considered: hydrological, geographical and geological conditions of the controlled small watersheds, area of the dam farmland, varieties of the crops, etc. Empirically, the check-dam becomes relatively stable when the ratio of the dam farmland area to that of the controlled watershed is between 1/25 and 1/15. When the impounded water depth is less than 0.8 m and the storage time is shorter than 3~7 days, a dam designed for enduring the biggest rainstorm in a century is found to be relatively stable (Zeng et al., 1995).

The coefficient for relative stability of the dam system I is defined as the ratio of the area of dam farmland to that of the watershed corresponding to flood frequency p. When the flood frequency p is 2% and the depth of impounded water in the dam farmland is equal to the critical value, the coefficient is called the critical coefficient for relative stability of the dam system I_C. If $I > I_C$, the dam system is stable, and if $I < I_C$, the dam system is not stable.

$$I_C = \frac{W_p}{\delta \cdot F} \qquad (1)$$

where, W_p is the amount of impounded floodwater (m^3) when the flood frequency p is 2%; δ is the design depth of water for the dam farmland (m); and F is the area of controlled watershed by the check-dam ($km^{2)}$ (Zeng et al., 1999).

Zeng (1995) has confirmed the feasibility of maintaining the relative stability of the check-dams by comparing the relationship between the dam height and the retention area in the Wangjiagou Watershed of China. Fang (1995 and 1998) and Zeng et al. (1999) have studied the condition, criterion and mechanism for the maintaining of the relative stability by investigating hundreds of typical check-dams on the Loess Plateau. Empirically, in the small watersheds of the Loess Plateau, the check-dam systems will remain relatively stable when the ratio of the dam-land area to the area of the controlled watershed is between 1/25 and 1/15. When the impounded water depth is less than 0.8 m and the storage time is shorter than 3~7 days, a dam designed for withstanding a rainstorm occurring once in a hundred year has been found to be relatively stable (Zeng ,1995). Lei and Zhu (2000) have developed a mathematical model for optimizing the layout of the dams in a small watershed according to the principle of relative stability. Nevertheless, dissensions still exist on the theory of relative stability of check-dams. Li (2004) expressed the idea that no check-dam would be stable if sediment has to be detained in the reservoir when soil and water is inputted continuously from the upper reaches.

CHARACTERISTICS OF THE STUDY AREA

The Loess Plateau is located in the upper and middle reaches of the Yellow River, covering a total area of 624,000 km^2, and over 60% of the land is subjected to soil and water losses (Figure 2). The most crucial area is the *Loess Mesa Ravine Region* and the *Loess Hill Ravine Region*, which cover altogether 30% of the total area. As one of the most severe areas of soil and water losses in the world, the average erosion rate can be as high as 5,000~10,000 t/km^2.a, and sometimes even up to 20,000~30,000 t/km^2.a (Meng, 1996).

Most areas of the Loess Plateau are arid or semi-arid regions with dry air, little clouds and abundant illumination, but short of moisture. The average annual precipitation in the Loess Plateau is only 350 to 550 mm, decreasing gradually from the southeast to the northwest, and most of the precipitation is concentrated in the rainy season of June to September. Usually, a single rainstorm, most of which is in the form of short but very intense rainfalls, can account for more than 60%, even 90%, of the total precipitation in the year, and causing the greatest amount of soil loss (Zhou et al. 2000). Therefore soil given out by the small watersheds for a whole year can be simulated by a single run of artificial rainfall in the laboratory. The rain often lasts 30~120 minutes, which will result in a 10~40 minutes hyperconcentrated flow with average concentration of 200~300 kg/m^3 and a maximum suspended concentration of 1, 000 kg/m^3 (Wang and Jiao, 1996).

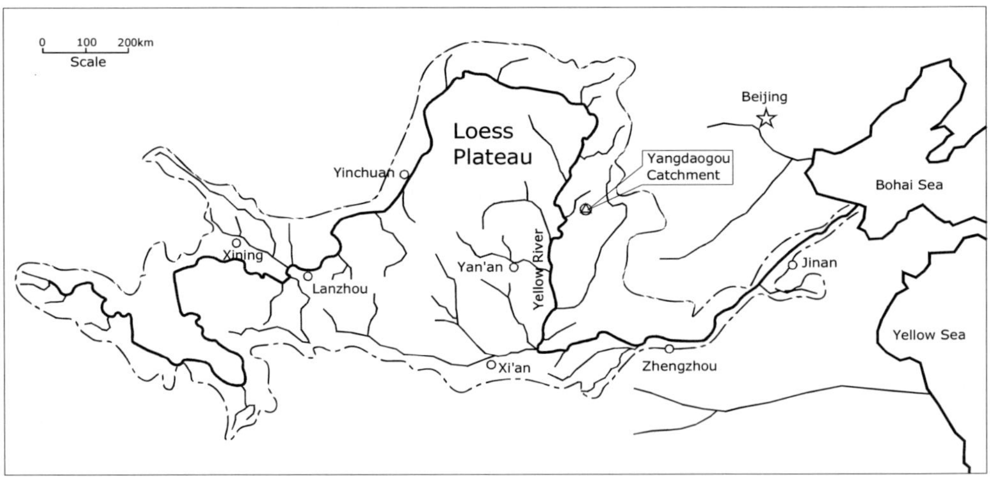

Figure 2. A sketch map of the Loess Plateau, China.

As the prototype catchment or the runoff experiments, the Yangdaogou Catchment (Figure 3) is a typical small watershed located in the *Loess Hill Ravine Region* covering an area of 0.206 km^2. The total length of the gullies is 752 m, and 240 m of which are from the outlet, with a “U”-shaped gradient of 2.3%.

Figure 3. Headwater of the Yangdaogou Catchment.

The ground surface is covered with the Lishi loess and the Malan loess (Cai et al. 1998). The precipitation in this area can be found in the general condition of the Loess Plateau mentioned above. In the 1960s, no dams had been built in this region, so that the average annual runoff rate was 36,700 m^3/km^2, and the average annual soil loss rate was 20,811 t/km^2 (Zhang et al., 1995). Therefore, the annual soil loss, namely, the soil loss of the conceptual prototype in a rainfall event, S_p, could be obtained by the multiplying the watershed area by annual soil erosion rate, which is 4.29×10^6 kg/a. As in engineering practices, the annual erosion rate is adopted as an index when planning the check dam system, and a rainfall event in the recapitulated prototype watershed accounts for soil loss in one year in the prototype watershed.

The prototype catchment of the rainfall experiments is abstracted according to the common erosion characteristics of the Loess Plateau. The conceptual watershed is also located in the *Loess Hill Ravine Region*, but it covers an area of 3.32 km^2. The total length of the gullies is 3 km, and 694 m of which are from the outlet, with a "U"-shaped gradient of 2.3%. Soil of the ground surface, the average annual runoff rate, and average annual soil loss rate in this small watershed are all correspondingly same to those in the Yangdaogou Catchment. The soil loss of the conceptual prototype catchment in a rainfall event, S_p, is 6.87×10^7 kg/a.

EXPERIMENTAL STUDIES

Scaled-down models are scale models of a general class that possess geomorphic features, which can be regarded as the replica of a natural landform, but scaled down in such a way that ratios of significant dimensions and forces are equal to those in nature (Timmons,

1984). Three conditions should be satisfied to ensure physical similarity: (a) geometric similitude; (b) kinematic similitude; and (c) dynamic similitude. While some similarity requirements seemed to have been firmly established, others have not been fixed so far (Albertson et al., 1960; Timmons, 1984). Scale modeling has been widely used and has a long history in the field of hydraulics and river engineering (Zhang, 1994). However, few researchers have published data on the simulating of the process of soil loss by scale model experiments, as it is very complicated with respect to rainfall, soil surface crusting, land-use and vegetated cover, etc. Hancock and Willgoose (2004) have examined the effect of erosion on a back-filled and capped earthen dam wall by constructing an experimental model landscape simulator in the laboratory. However, the designing of a rainfall simulator for directly scaling the processes of rainfall runoff to the fields is rather difficult. Consequently, no attempts have been made to match the rate of gully development on the tailings dam to the field-scale processes. Recently, scale model experiments on soil and water erosion in the small watersheds have been performed, and progress has been made in similitude methodologies. Shi et al. (1997a; 1997b) have observed quantitative erosions in gullies and on slopes of a small watershed model shrunk from the landform of the Xiaofanjiagou Gully of the Shaanxi Province in China (the length scale was 75). Jiang et al. (1994) and Yuan et al. (2000a; 2000b) have carried out a series of scale-down model experiments for varying extents of erosion control in designated small watersheds of the Loess Plateau to investigate the relationship between the runoff and the sedimentation. Rainfalls and soils similar to those of the prototype were applied, and the landforms were scaled down according to the prototype in the above-mentioned experiments. Nevertheless, no explicit scale relationship between the prototype and the model has been reported in the above studies.

Xu et al. (2009) has proposed a semi-scale physical model experimental method to simulate the soil and water process in the small watersheds of the Loess Plateau, and confirmed it with serial model experiments. The key idea of the method is that the ratio of the modeled geomorphological evolvement rate to the corresponding rate of the prototype should be kept constant after the determined runs of the rainfalls, so that the soil and water erosion processes of the prototype could be represented by the result of the model experiment. Four measures have been proposed to ensure the ratio R a constant: (1) A bare land model with a corresponding erosion rate should be applied in the experiments. (2) Dimensions of the landform, including the check-dams, should be normally scaled down according to the prototype watersheds. (3) Soil similar to those of the prototype should be used. (4) The antecedent water content before each rainfall simulation should be kept constant. The proposed method provides the quantitative proportion of soil loss between the prototype and model, using similarity criteria. Nevertheless, it did not strictly meet the similar situation in the conventional scale model experiment and could not provide the scale relationship of the erosion depths between the model and the prototype, etc.

Two experiments were carried out to demonstrate the hydro-sedimentologic balance on the dam-land, one for a single check-dam via runoff simulation, and the other for the check-dam system via rainfall simulation. The objectives of this study are to prove the feasibility of the relative stability idea for check-dams from the viewpoint of hydro-sedimentologic balance, and present a method for designing scale model experiments on soil and water erosions.

Runoff Simulation

Generally, according to the correlative criterion (CMWR, 1996), middle scale check-dams with dam heights of 15~25 m and reservoir capacities of 0.1~0.6 million m^3 should be built in the lower reaches of the main branch gullies. In the present study, the hydro-sedimentologic process and the geomorphological variation for 18 runoffs were investigated with a middle-scale check-dam of 21 m high, which was built near the outlet of the lower reach of the Yangdaogou catchment.

Obviously, deposition on the dam-land is determined by its own reservoir capability as well as the quantity of water-solid mixture from the upper reaches, and is independent of the source of the mixture, as these check-dams are small-sized reservoirs in nature. Thus, the effects of land use, vegetation, rainfall, etc, on soil and water erosions could be embodied in the flux and concentration of the mixture, and the latter could be calculated according to the prototype erosion rate. The Similarity Theory for deposition in reservoirs with hyperconcentrated flow is practicable and has been applied to several practical projects (Zhang et al., 2001; Zhang et al., 2002), so that some of their technical approaches could be used as references in this study.

The experiments were carried out in the Laboratory Hall of the Yellow River Research Center, Tsinghua University, Beijing, China. Due to the limitation of the experimental ground, the length scale λ_L was restricted to 60, which implied that the gully dimension and the dam height were shrunk to 1/60 of the prototype ones. The matrix loess stacked near the experimental spot, which was homologous to the loess in the Yangdaogou Catchment, was used as the erodible material in this experiment. The material had passed through a sieve with 1 cm sieve pores for removing large clods, debris and grass roots. To ensure a regular original microrelief, the soil was prepared by hand-patting for generating a smooth roughness. The diameter of 50% soil particles, D_{50}, was $30.8\ \mu m$, and the dry density γ_0 was $1.44 \times 10^3\ kg/m^3$.

No matter it is a field survey or a laboratory experiment, the gravitational effect is more decisive than the viscous one for the turbulent flow caused by rainfall, so the duration of each rainfall is demanded by Froude's Law of Time Scale (Jiang et al., 1994):

$$t_{m0} = \frac{t_p}{\lambda_t} = \frac{t_p}{\sqrt{\lambda_L}} = \frac{10 \sim 40}{\sqrt{60}} = 1.3 \sim 5.2(\text{min}) \qquad (2)$$

where, t is time, min; λ_t is the time scale; λ_L is the length scale. Since one of the objectives of this study is to validate the trend of check-dams in attaining a relative stability, the reliability of the experimental results would not be degraded if several runoffs were performed continuously. Consequently, the duration of each experimental runoff, t'_m was prolonged to 12 minutes without making any changes in the sediment concentration and the flow rate, so that it was possible to observe the evolution process of the gully landform in greater details and to complete the experiments more quickly.

In practice, the perennial annual runoff rate and erosion rate are usually taken as design indices for soil and water products in programming the sequence of dam constructing for the check-dam system in the small watersheds (Zheng, 2003). The flow rate and concentration of the mixture flowing into the reservoir had been accurately calibrated before the formal experiments, which were 10^{-3} m^3/s and 80~90 kg/m^3, respectively. Thus the soil product pouring into the reservoir in each runoff of the experiment is shown as follows:

$$S'_{m} = Q'_{m} \times C_{m} \times t'_{m} = 10^{-3} \times 85 \times 12 \times 60 = 61.2(\text{kg}) \tag{3}$$

Thus the scale of cumulative sediment for each rainfall can be obtained as follows:

$$\lambda_S = \frac{S_P}{S_m} = \frac{4.29 \times 10^6}{61.2} = 7.01 \times 10^4 \tag{4}$$

Table 1 shows the experimental parameters for the runoff simulation experiments.

Table 1. Experimental parameters for the runoff simulation experiments

Prototype element of the small watershed		Scale	Narration
Length of the gully, L (m)	752	$\lambda_L = 60$	Restricted by the experimental ground
Catchment area, A (km^2)	0.206	$\lambda_A = 3265$	Geometrical similarity
Height of the check-dam, H (m)	21	$\lambda_H = 60$	Geometrical similarity
Duration, t (min)	10~40	——	Deforming erosion rate
Silt concentration, C (kg/m^3)	≈ 200	$\lambda_C \approx 2.5$	$\lambda_C = 1.15 \sim 3$ (Zhang et al., 1994)
Sediment yield (kg)	4.29×10^6	$\lambda_S = 7.01\times10^4$	Deforming erosion rate
Dry bulk density, γ_0 ($10^3 kg/m^3$)	≈ 1.3	$\lambda_{\gamma_0} \approx 1$	Model landform was hand packed

The experimental system was made up of a water circuit and a gully model, as illustrated in Figure 4. Water and loess were added into the round puddling pool and stirred by the spin puddler to become homogeneous enough, and then the mixture was pumped into the constant head tank. To prevent the sediment from subsiding, a reciprocating puddler was set up in the constant head tank. As the sediment-laden fluid flowed into the gully model, part of it was restored in the upstream of the check-dam and the rest arrived at the storing pool. To prevent the runoff from changing the sediment concentration in the puddling pool, a valve was fixed at the tail of the storing pool. Thus, when an experimental run was completed, the valve could be opened, and then the fluid in the storing pool could be released into the puddling pool, which could be reused in the next experiment.

The gully model covered an area of 14 m by 3.5 m. The branch gully at the foreside of the model was "V" shaped, while the main gully at the rear-side was "U" shaped. The length of the branch gully was about 9 m. The dam-site at the main gully was 0.9 m away from the end of the branch, which stood in the entrance of the prototype main gully. The effective height of the dam was 0.35 m before the first runoff. A spillway was located in the left side of the check-dam, which allowed the floodwater to overflow into the storing pool.

Figure 4. Landscape simulator in which the runoff simulation experiments were conducted.

Among the inlet pipes, an electric flow meter was fixed to measure the flux invading the gully. The two sidewalls, which were straddled by a measurement bridge with a height finder, were parallel and their tops were of approximately the same altitude. The bridge could move along the sidewalls and the height finder could slide along the bridge, thus the altitude of any point on the gully bed could be measured. The surface altitude was digitized for both horizontal directions in 0.1 m grids.

Eighteen experimental runs were conducted with the same flux of 1.0×10^{-3} m^3/s and the same duration of 12 minutes. The sediment concentration of the flow running to the gully was kept at approximately 80~100 kg/m^3. There was a 24 hours break following each runoff. The initial microtopography made according to the blueprint of the scale model was designated as "N_m (0)", the one after the first runoff as "N_m (1)", and the one after the second runoff as "N_m (2)", and so on. Each microrelief formed in 24 hours after the runoff became the initial landform of the next run. Before the next runoff, muddy water stored in the reservoir in the upper reach of the check-dam was defecated and pumped out, and the microrelief of the gully bed was measured by the height finder.

In this experiment, the dam area after each runoff could be calculated by the contour line of the gully bed. After the map of the gully bed had been converted into an AutoCAD file with the ratio of 1 unit to 1 cm, the dam-land enclosed by the contour line could be calculated by using the computer command "Area".

Rainfall Simulation

Rainfall simulations were conducted in a small watershed model to study hydro-sedimentologic process and geomorphological evolution in the gullies after the check-dam system was built.

The experiments were carried out in the laboratory of the Yellow River Research Center, Tsinghua University, Beijing, China. The landscape simulator consisted of a rainfall simulator suspended above a flume in the small watershed model. Nozzles with an inner rotor were screwed onto the ends of the pipes in the spot, which could spray downwards to the landform 5.5 m below. The sprinkling intensity could be adjusted by pressure valves to maintain a constant hydraulic head for all nozzles which were located at different elevations, as illustrated in Figure 5. Each basic unit watered a circular area of approximate 5 m in diameter. Two lines of five units each were used to simulate rainfalls in the experimental spot of 6 m by 10.8 m. Clean water was pumped from the nearby reservoir to a constant head tank, and then pumped to the rainfall simulators. A flow-meter and a cut-off valve for each rainfall unit were used to adjust manually the delivery rate of the immersed pump to achieve the desired rainfall intensity. Distribution uniformity of the rainfall was determined using rain-gauges equidistantly spaced over the landscape surface and was measured for three separate periods of ten minutes prior to the experiments. The spatial distribution of the rainfall intensities for the simulator was not uniform, but was constant with time, and was comparable to the sprinkler delivery systems used by others (e.g. Zhou et al., 2000), especially when the rainfall intensity was in the range of 1.0 mm/min~2.5 mm/min. Drops with an average diameter of about 1.2 mm ~ 2.0 mm, which were measured by catching the drops on a sheet of absorbent paper, were produced by the rainfall simulator (Xu et al., 2004b).

Figure 5. Landscape simulator in which the rainfall simulation experiments were conducted.

Brick walls of the sample holders were constructed with an outside facing knife edge (with an angle of 60 degree) on the upper rim. The matrix loess gathered close to the experimental spot, which was homologous to that in the conceptual prototype catchment, was used as the erodible material in this experiment. The same as mentioned above, the materials had passed through a sieve of 1 cm pores for removing large clods, debris and grass roots. The 50% diameter of soil particle D_{50} is $52.2\,\mu m$, and the dry density γ_0 is $1.56 \times 10^3\ kg/m^3$. An open book type catchment was used as the initial conditions for landscape development. To ensure a regular original microrelief, the soil was prepared by hand-patting to generate a smooth roughness. A permeable trough base could allow free drainage of infiltrated water.

Ten rainfalls were applied after the check-dams were constructed according to the sequence list in Figure 6. Simulated rainfalls resembling a characteristic natural rainstorm in the prototype watershed were applied in the model experiments (Figure 7). For each rainfall, the duration was 20 minutes, and the intensity was approximate 1.60 mm/min (with errors $\leq$ 10%). Before the commencement of any rainfall application, the rainfall intensity was verified. The surface of the plot was covered with tarpaulins and the rainfall was collected in a calibrated tank, then measured volumetrically, and was sampled every 2 minutes. Soil water storage was found to be very sensitive to rainfall (Peugeot et al., 1997), and we had made great efforts to keep an equal antecedent soil moisture before the commencement of the experiment in the same series of rainfall applications.

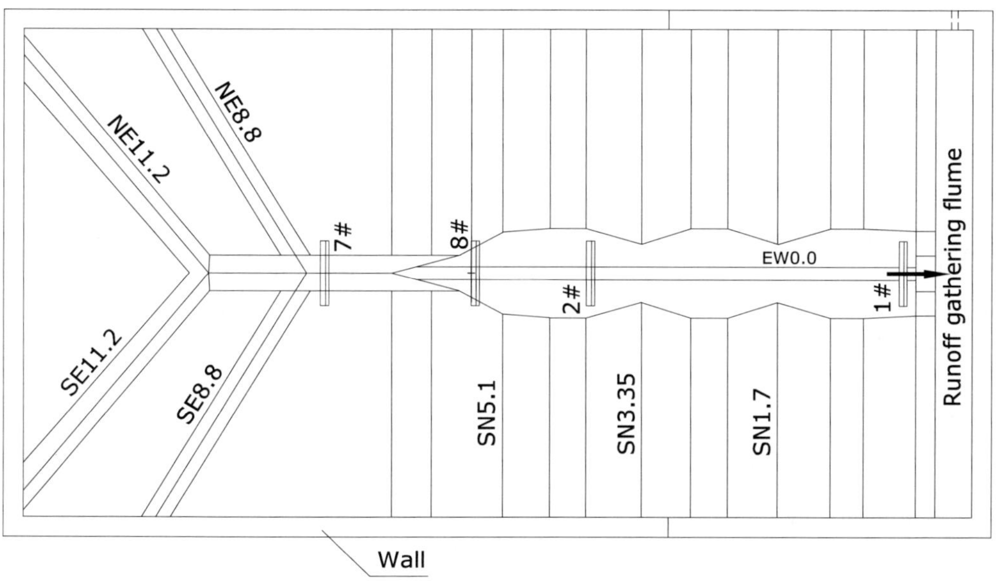

Figure 6. Plane graph of the model relief and arrangement scheme of the check-dams. the check dam arrangement order of schemes A was as follows: the 7# dam was constructed before the 2th rainfall session; the 8# dam was constructed before the 4th rainfall session; the 2# dam was constructed before the 5th rainfall session; and the 1# dam was constructed before the 6th rainfall session.

Figure 7. Simulated rainfall was applied on the model check-dam system.

An equal period (24 hours) was kept after each rainfall, and the water content was kept almost constant, except in the first experiment. Soil samples and test beds were initially protected by plastic sheets until the rainfall intensity was correctly adjusted and the data logging equipment was operating satisfactorily. Then the cover was removed suddenly and the test was started. During each rainfall application, the total water-solid mixture was collected in the calibrated tank and the runoff container at the outlet of the plot. Runoff samplers were also collected with a sampling bottle of 100 ml for determining the sediment concentration. The flow rate was also measured using the calibrated bucket every 2 minutes. After the experiment, the deposited bed load was dried and weighted, and the weight of the suspended load was calculated from the concentration determined gravimetrically and from the volume measured by the bucket gauge. The microrelief of the gully bed, which was digitized in the horizontal directions in 0.1 m grids, was measured by the height finder. The dam-land area after each rainfall was measured by means of a digital camera and a wooden frame with a known dimension. The wooden frame was placed flatwise on the dam-land, which was proportionally deformed with the dam-land in the plan-photo. Then the areas of the dam-land and the wooden box in the photo could be calculated respectively via the AutoCAD program, and the result so obtained was the practical area of the dam-land.

Before the above scenario demonstration as the dams were built orderly, the ratio of the erosion extents between the prototype and the model for each run of rainfall should be verified. An identifying test of simulated rainfall before the constructing of the dams, in which the geometrical scale, the antecedent land cover and the rainfall in the identifying test were homologous to those of the above experiment for the dams, was conducted. Cumulative sediment volume for the experiment is 121.53 kg. Thus the scale of cumulative sediment for each rainfall can be obtained as follows:

$$\lambda_S = \frac{S_p}{S_m} = \frac{6.87\times10^7}{121.53} = 5.65\times10^5 \quad (5)$$

where S is the soil production, kg; the subscript p represents prototype; the subscript m represents model.

The parameters of the rainfall experiments are presented in Table 2.

Table 2. Experimental parameters for the rainfall simulation experiments

Prototype element of the small watershed		Scale	Narration
Name	amount		
Length of the gully, L (m)	3008	$\lambda_L = 240$	Restricted by the experimental ground
Height of the check-dam, H (m)	*	$\lambda_H = 240$	Geometrical similarity
Catchment area, A (km^2)	3.3	$\lambda_A = 57600$	Geometrical similarity
Duration, t (min)	120~480	$\lambda_t = 7.75$	Gravity similarity
Silt concentration, C (kg/m^3)	≈ 200	$\lambda_C \approx 3$	$\lambda_C = 1.15 \sim 3$ (Zhang, et al. 1994)
Sediment yield (kg)	6.87×10^7	$\lambda_S = 5.65\times10^5$	Deforming erosion rate
Dry bulk density, γ_0 ($10^3 kg/m^3$)	≈ 1.56	$\lambda_{\gamma_0} \approx 1$	Model landform was hand packed

*Dam 7# is 35m high, and Dam 1#, 2# and 8# are 68m high.

EXHIBITION OF RELATIVE STABILITY

The decrease in dam-land elevations and the re-stabilizing of the gully gradients after the building of the check-dams represent a hydro-sedimentologic balance of the check-dam or the check-dam system, and the building of the check-dam is an effective way to control soil erosion in the gullies.

Mean dam-land altitude variation. Both in the runoff simulation experiment for a single check-dam and in the rainfall simulation experiment for the check-dam system, the increase in dam-land altitude became smaller and the gully gradient became more stable as the experiment was progressing. Figure 8 (a) illustrates that the increase in dam-land altitude for each runoff became gradually smaller in the runoff experiment. As the water-solid mixture was flowing to the check-dam during the experiment, the thalweg was eroded in the upper reach of the gully, which was out of control of the check-dam, and the gully-bed in the lower reach was filled up, and even a certain land area was gradually formed in the upper reach of the dam. Although the altitudes of the dam-land after each runoff had showed an increasing trend, the increased amount before the N_m (12), especially before the N_m (5), was greater than that after the N_m (13). In the rainfall simulation experiment, reservoirs of Dam 2#, 7#, and 8# were filled up when one or two runs of simulated rainfall had been conducted, due to the too exquisite erosion extent of each rainfall. However, Dam 1# was not filled up and the dam-land was growing up all along owing to its larger capacity and the time of construction was late. The elevation trend of this dam-land was analogous to that of the single check-dam in the runoff simulation experiment, as shown in Figure 8 (b).

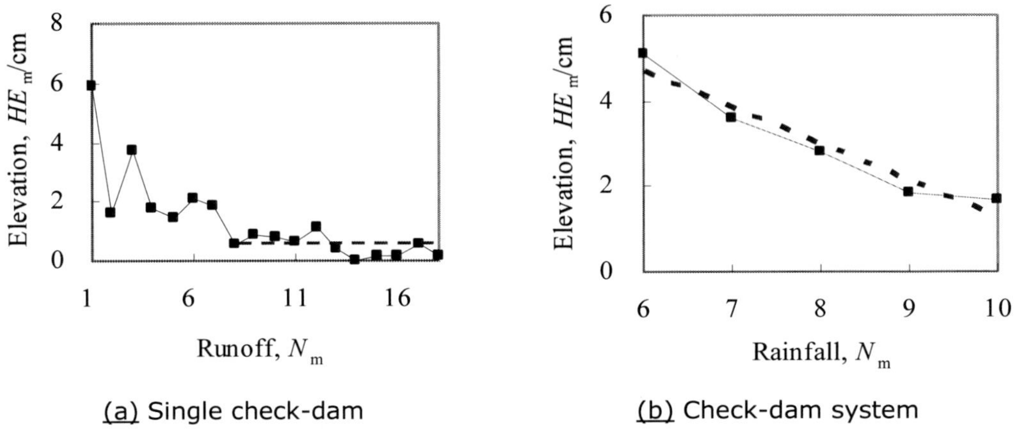

Figure 8. Variation of the mean dam-land altitude with runoff/rainfall.

Mean slope gradient variation. The base level of the gully was raised and the mean gradient was lowered as depositions took place on the dam-land. Referring to ASL (Casalý et al., 1999), we defined PL as the length-weighted average slope gradient:

$$PL = \frac{\sum L_i P_i}{\sum L_i} \tag{6}$$

where L_i is the length of each segment of the gully, and P_i is the slope gradient of this segment with uniform characteristics. In these experiments, L_i and P_i could be calculated based on the coordinates along the gully thalweg. *PL* of the main gully after each runoff/rainfall simulation experiment for a single check-dam and for a check-dam system was listed in Figure 9 (a) and Figure 9 (b), respectively. The gradient was steeper in the initial stages, but became less steep as the experiment went on, since the dam-land area was enlarged. For the check-dam system experiment (Figure 9 (b)), the gradient of the main gully was becoming gentler, as the Dam 1# had not been filled up yet even after 10 simulated rainfalls. However, in the runoff experiment of the single check-dam (Figure 9 (a)), beginning from the N_m (13) runoff, the PL was kept at 7%, and almost did not vary as the experiment during the experiment. This means that the gully had become relatively stable, and the slope had only varied a little, despite of the minute variations in the microtopography.

However, if the runoff experiments were continued, the increasing rate in dam-land elevation would become much slower. It was really difficult to provide strict threshold values for gully slopes and dam-land elevations only by model experiments. In practice, the threshold value of dam-land increase per year which could be borne by the local peasants is decided by the local social and economic conditions (Li, 2004), and generally amounts to 0.3 m/a (Wang and Ma, 2003; Zheng, 2004).

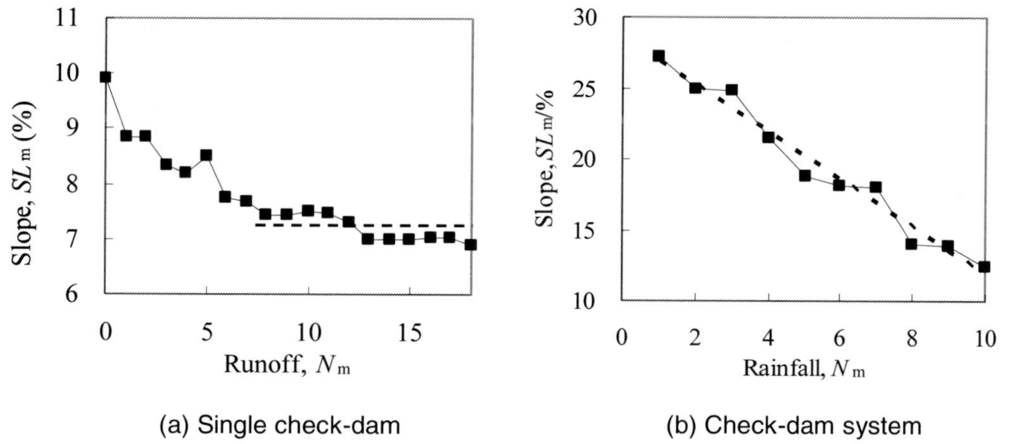

Figure 9. Variation of the mean slope with runoff/rainfall.

In fact, from our experimental results, the increase rate of the dam-land altitude was evidently shrinking as the runoff was carrying on, and the trend was exhibiting a relative stability in the gully in respect of the hydro-sedimentologic balance.

Mechanism of the Developing of Relative Stability

Erosion alleviation by the check-dam. Figure 10 shows the variation in composition of the sediment on the surface of the gully bed after 18 runs of model runoff. Just like any other water-blocking structure such as reservoirs, retention ponds, etc., the check-dam can slow down the velocity of the sediment-laden water, resulting in a lower sediment transport capacity. Thus, some of the sediments would sink down, and the sediment concentration in the outlet would become smaller than that in the inlet. Moreover, before the next runoff, the sediment-laden fluid stored in the reservoir would become clear, and some sediment would be left on the dam-land to prevent the dam-land from erosion.

On the other hand, development of a "shielding" layer composed of relatively heavy soil particles could protect the underlying soil from runoff erosion. When the soil surface was initially inundated with runoffs, the soil particles were detached from the soil surface and were entrained into the gully flow. Lighter particles with low settling rates would move far away from their original locations, whilst heavier particles would settle more quickly near their original positions. If this process was continued, eventually most of the lighter particles would be removed, leaving a shielding layer of heavier particles, which could act to protect the underlying soil.

Augmentation of the alluvium area. As the dam-land rose, its area expanded in virtue of the sloping gully bank. Figure 11(a) and Figure 11(b) illustrate a linear rising trend of the dam-land area. In the rainfall simulation experiment for the check-dam system (Figure 11(b)), Dams 2#, 7#, and 8# had been filled up before the 6th rainfall simulation. The increase in dam-land area in the figure was mainly attributed to Dam 1#. Whether it was in the single check-dam experiment or in the check-dam system experiment, the dam-land areas were expanding linearly, as shown in Figure 11(a) and Figure 11(b). Consequently, even if the bulk of the intercepted sediment was unvaried for every rainfall, the dam-land would rise to a smaller extent owing to the augmenting of the area after each runoff.

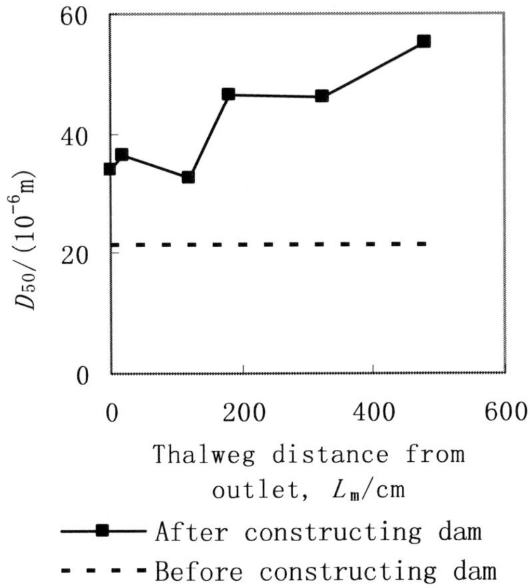

Figure 10. Coarsening of the gully bed after runoffs.

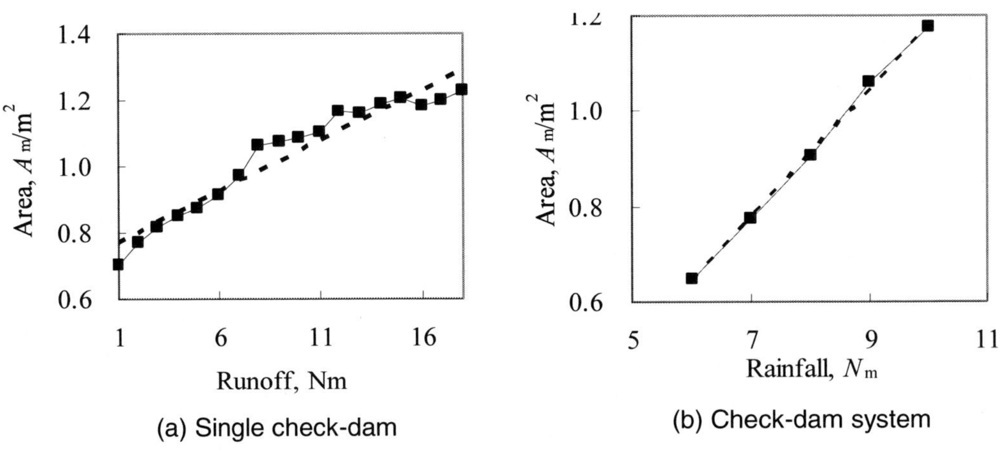

Figure 11. Variation of the alluvium area with runoff/rainfall.

However, other elements can also contribute to the relative stability of the check-dams. For example, as the dam-land rises, the reservoir capacity will decrease, and less sediment-laden fluid can be stored in the upstream of the check-dam. As a result, the amount of deposited sediment will decrease as well. In fact, in the later stage of integrative control of small watersheds of the Chinese Loess Plateau, the dam-land has increased more slowly owing to a great reduction of the amount of sediment originating from the slopes and gullies, thus causing smaller magnitude of floods and smaller sediment concentrations.

Auto-stabilizing mechanism of the gullies. Scour/deposition balance is a common phenomenon of gully evolvement which has been testified by investigations on other watersheds. The rate of gully incision is controlled by the flow rate, depth, turbulence and temperature of the water flow, as well as by soil texture, soil mechanical pattern, and the level of vegetation protection (Sidorchuk, 1999). Foster (1982) has found that gullies were eroded downwards, until a layer of low erosion susceptibility was reached. After this, the gullies would become widened, and the erosion rate would decrease. Eventually, the gullies would "equilibrate" with the overland flow. Novak (1985) has developed equations for calculating the time required to reach equilibrium for rill and channel erosions. Recent studies have indicated that gully formation can be divided into two stages (Sidorchuk, 1999). The initial stage takes up about 5% of a gully's lifetime during which the morphological characteristics of a gully, such as length, depth, width, area and volume, are developing, and stability cannot be attained. After this, a stable stage will be reached, and the gully will attain a maximum in size. In the initial stage, hydraulic erosion is predominant at the gully bottom, and rapid mass movements occur on the gully sides, whilst in the stable stage, sediment transport and sedimentation become the main processes at the gully bottom, and the gully width increases limitedly due to lateral erosion. The mass movements go on slowly in this stage. In our natural world, if the local conditions are permitting, some streams can maintain their relatively stable channels over even a century, despite that there have been great floods in such a long period of time (Jeff, 2002; McEwen, 1994; Warburton et al., 1993).

As check-dams have been built in the small watersheds on the Loess Plateau, the relieves of the gullies were changed. Substantial sediment caused by rainfalls and floods have been stored on the dam-land for decades, and the mean slope gradients of the gullies became gentler. Consequently, erosion in the gullies has been alleviated, and the gullies have attained a relative stability at last. On the other hand, even though the amount of sediments from the upper reaches has not been reduced, the rising rate of the dam-land becomes slower due to the enlargement of the dam-land area. Hence a critical height of a check-dam exists for a given gully having definite conditions of soil, water, geology and physiognomy. If the check-dam in a gully is higher than the critical value, and there are depositions in the upper reach of the check-dam, the mean slope gradient of the gully would be gentle enough to diminish the erosion greatly. As a result, if the annual sedimentation thickness is very little in the later stage of the dam-land formation, the workload for heightening the dam is so little that it can be afforded by the local people (Zeng, 1999). Although other factors also play their roles in the maintaining of the relative stability of the check-dam systems (Fang, 1995), equilibrium of soil and water in the gullies is the most important criterion.

In fact, many natural check-dams, such as those in the Balihe Catchment and the Sanshilihe Catchment in Shannxi Province and the Laobatou Catchment and the Qianqiuzi Catchment in Gansu Province, all have achieved a good relative stability. Huangtuwa in the Zizhou County of Shannxi Province, another famous natural check-dam covering 40 hm^2 of lowlands, has been working favourably for more than 400 years. Some check-dam systems built in the 1960s, such as the Kanghegou check-dam system in Shanxi Province and the Wangmaogou dam system in Shannxi Province, are also typical projects attaining good relative stabilities, and have retained enormous amounts of sediments and yielded plentiful crops. Hence, it is most evident that the attaining of relative stability should be the ultimate goal of build check-dam systems.

FURTHER RESEARCH ON THE THEORY OF RELATIVE STABILITY

The amount of dam-land increase per year. It is supposed that floodwater was evenly spread on the dam-land in the initial hypothesis of relative stability of check-dams. In fact, the water depth near the dam is larger than that far from the dam, due to the longitudinal deposition profile existing in the upper reach of the check-dam, and the wallop of the floodwater pouring into the reservoir. According to the field investigation, the average longitudinal slope gradient of the check-dam lands in Shannxi Province is 0.2% ~ 0.5%, which makes obvious impact on distribution of the impounded floodwater, especially that on the check-dam land with large reservoir capacity (Tian et al., 2004). The phenomenon above mentioned was also widely found in the rainfall simulation experiments. Figure 12(a) shows the slope gradient of the check-dam land gradually became small in the runoff experiments of the single check-dam.

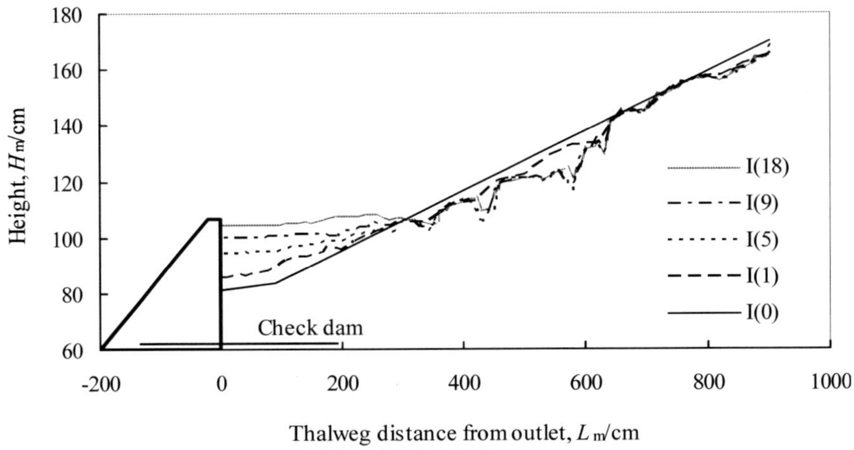

(a) Single check-dam.

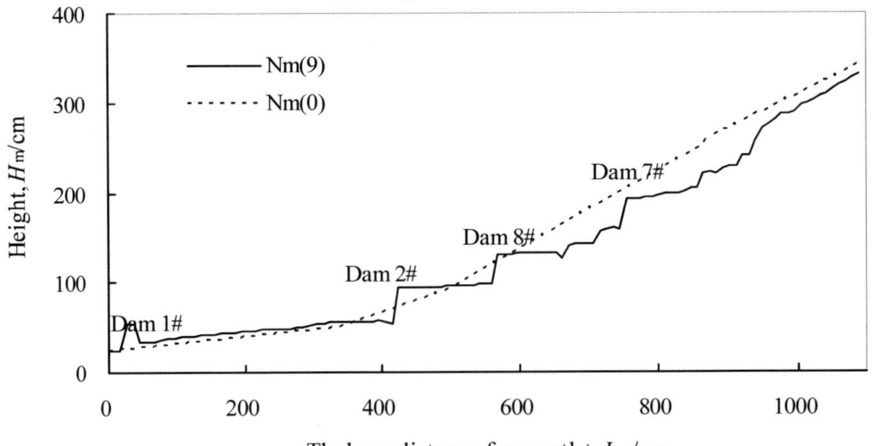

(b) Check-dam system.

Figure 12. Longitudinal deposition profile of the check-dam land.

Figure 12(b) shows the change of the gully bed after 9 events of rainfalls. Both figures illustrates that the longitudinal slope gradient of the check-dam land was larger before the reservoir filled, and it became smaller after the reservoir filled, but the gradient never be equal to 0. Consequently, the annual increment of the check-dam height in the connotation of *Relative Stability* should be the spatial mean value of the dam-land increase per year (Fang, 1995). Experimental data related to the amount in this chapter had been also processed according to the above method. Clarification of the definition is helpful to construction of the check-dam systems in the practice of erosion control.

Relationship between the construction of the check-dam system and the control of the slope erosion. From the previous experimental data it could be found that the main function of the check-dam system is to control soil and water loss in the gullies, other than directly alleviate soil and water loss on the slope. Thus comprehensive treatment including tree planting, terrace constructing and pit digging should be implemented order to improve the eco-environment of the Loess Plateau.

Actually, the relative stability of the check-dam system and the control of the slope erosion may be mutually promoted. Gully is usually the influx of the runoffs, which is also the channel of surface runoffs, therefore, the water resources is comparatively centralized. With the reservoir of the check dam system in the gully, the water resources could be fully impounded to irrigate the plant on the check-dam and the slope, and indirectly reduce soil and water loss on the slope. At the same time, the control of slope erosion could reduce the flood peak to a certain extent, which favors the flood control of the check-dam system. Besides, the control of the slope erosion decreases the amount of soil flowing to the reservoir. Consequently annual increase of the check-dam land becomes smaller, and the equilibrium of soil and water in the catchment would be easy to be realized.

Relationship between relative stability of the dam system and that of the single dam. The relative stability of the dam system was firstly recognized from the natural *Juqiu* on the Loess Plateau, and then the relative stability of the check-dam system was developed to emphasize the safety of the flood preventing and controlling (Fang, 1995). Unfortunately, a big dispute arose on the relationship between the critical coefficient for relative stability of the dam system and the flood preventing and controlling. As stated before, the relative stability of the dam system mainly refers the balance of the soil and water. Relative stability of the check-dam system has positive effects on the safety of the dam system. Nevertheless, it is not the determined factor in the structural safety of a single check-dam.

The relative stability of the dam system reflects the overall development of the check-dam system. The premise is that the soil and water production on the whole slope is homogeneous, and the deposition and inundation on a single dam is also basically identical (Zhu et al., 1997). Therefore, the relative stability of the dam system only represents the relative balance of the soil and water in some limited small and ideal catchment (Shi, 2005). Up to the present, the dam system and single check-dam are not clearly separated in the literatures on the balance of the soil and water. The concept of the relative stability mainly replies the relative balance of soil and water retained by a single check-dam.

In combination with the model experiment of the dam optimization programming, we performed the rainfall simulation experiments with several scenarios of dam construction sequence and found out that the small dams in reservoir were quickly filled up.

The most contribution of the relative stability of the dam system is that the concept of the critical coefficient for relative stability provides a quantitative criterion for the dam system programming in the small watershed, with which the number of dams could be figured out according to the corresponding reservoir capability.

CONCLUSION

The check-dams in the Loess Plateau of China are excellent examples of preventing land degradation and developing deserted lands into agricultural areas. They are so popular in China that more than 100,000 check-dams have been built in the past 50 years, in spite that only few counterparts have appeared in other countries. The amount of sediment retained by these check-dams has been found to be the largest among all other measures, and high crop yields have been obtained in the deposited dam-lands in virtue of the plentiful moisture and nutrients.

Runoff simulation experiments for the single check-dam and rainfall simulation experiments for check-dam systems were conducted to simulate the induced morphological changes affecting the stability of check dam systems, and develop relationships among mean slope gradient, mean dam-land elevation and dam-land area. The essential point of this kind of scale down model experiments is that the ratio between the model geomorphological variables and the corresponding prototype ones keeps a constant after simulated rainfalls, so that the processes of soil and water erosion in the prototype could be represented by the results of the model experiments. Four measures have been proposed to ensure the ratio R a constant: (1) A bare land model with a corresponding erosion rate should be applied in the experiments. (2) Dimensions of the landform, including the check-dams, should normally be scaled down according to the prototype watersheds. (3) Soil similar to those of the prototype should be used. (4) The antecedent water content before each rainfall simulation should be kept constant.

Hydro-sedimentologic balance is the most important element for realizing relative stability of the check-dams, which means that soil and water in the small watersheds controlled by the check-dams could be internally absorbed, without the need of raising the height of the dam, if the above parameters can reach certain required values. Quantitative results from the runoff experiments indicate that the mean altitude of the dam-land and the mean slope gradient of the gully were almost constant after 13 simulated runoffs, which imply that the check-dams were not necessary to be heightened any more. By comparing the quantitative results in the rainfall experiments, it was found that one of the dams in the experiments was becoming stable. The reason for the check-dam to be able to establish relative stability is that the dam-land area expanded in virtue of the sloping gully bank, resulting in higher capacity for storing sediments from the upper reaches, and the dam can also alleviate erosion by slowing down the flow velocity of the sediment-laden water, by clarifying the sediment-laden fluid stored in the reservoir and by capturing sediments for the dam-land, so as to prevent the dam-land from erosion, and the gullies could auto-stabilize themselves after the check-dams were built.

Moreover, the annual increment of the check-dam height in the connotation of *Relative Stability* should be the spatial mean value of the dam-land increase per year due to the longitudinal slope gradient of the check-dam land; the relative stability of the check-dam system and the control of the slope erosion may be mutually promoted, indirectly alleviating soil and water loss on the slope. Relative stability of the check-dam system has positive effects on the safety of the dam system. Nevertheless, it is not the determined factor in the structural safety of a single check-dam. The most contribution of the relative stability of the dam system is that the concept of the critical coefficient for relative stability provides a quantitative criterion for the dam system programming in the small watershed, with which the number of dams could be figured out according to the corresponding reservoir capability.

ACKNOWLEDGMENTS

This study is supported by The Found of Key Laboratory of Yellow River Sediment Research□MWR (Grant No. 2009002) and the Open Research Fund Program of State key Laboratory of Hydroscience and Engineering (Grant no. sklhse-2008-B-01).

REFERENCES

Albertson, N. L., Barton, J. R., and Simons, D. B. (1960*). Fluid Mechanics for Engineers Prentice-Hall.* Englewood Cliff, NJ, 561.

Casalý, J., Lopez, J.J. and Giraldez, J. V. (1999). Ephemeral gully erosion in southern Navarra Spain. *Catena*, 36, 65-84.

Cai, Q. G., Wang, G. P., and Cheng, Y. Z.. (1998). *The Process and Simulation of Soil Erosion for Small Watersheds in the Loess Plateau*, China. Science Press: Beijing, China, 135-137 (in Chinese).

Chinese Ministry of Water Resources (CMWR). (1996). GB/T16453.3. The Technical Criterion for Soil and Water Conservation: Gully Control Method, Beijing: Chinese Press of Hydraulics and Hydropower (in Chinese).

Chinese Ministry of Water Resources (CMWR). (2003). Programming for Check-dams in the Loess Plateau (unpublished technical report), 47-48 (in Chinese).

Fang, R. Y. (1996). Consideration on check-dam system agriculture in Shanxi Province. *Soil and Water Conservation in China*, (7), 47-49 (in Chinese).

Fang, X. M. (1995). Criterion and Condition for the Relative Stability of Check-dam System. *Soil and Water Conservation in China*, (11), 29-33 (in Chinese).

Fang, X. M., Wang, Z. H., and Kuang, S. F. (1998). Mechanism and Effect of Check-dam to Intersect Sediment in the Middle Reach of Yellow River. *Journal of Hydraulic Engineering*, 20, 49-53 (in Chinese).

Foster, G. R. (1982). Modeling the erosion process. In: Hann, C. T., Johnson, H. P., and Brakensiek D. L. Hydrologic Modeling of Small Watersheds. ASAE Monograph, St. Joseph, MI, 297-380.

Hancock, G. R., and Willgoose, G. R. (2004). An experimental and computer simulation study of erosion on a mine tailings dam wall. *Earth Surface Processes and Landforms*, 29, 457-475.

Jeff, W., Melanie, D., and Duncan, W. (2002). Stability of an upland gravel-bed stream, Swinhope Burn. Northern England. *Catena*, 49, 309-329.

Jiang, D. S., Zhou, Q., Fan, X. K., Zhao, and H. L. (1994). Simulated experiment on normal integral model of water regulating and sediment controlling for small watershed. *Journal of Soil and Water Conservation*, 8(2), 25-30 (in Chinese).

Lei, Y. J., and Zhu, X.Y. (2002). Formation, Control Principle and Method for Relative Stability of Check-dam System. Yellow River, (2), 23-26 (in Chinese).

Li, M. (2004). Analyse on relative stability for check-dam system. In: Proceedings of Key Technique to Construct Check-dam System in the Small Watershed on the Loess Plateau, China Sci. and Tech. Spread Center of water resource ministry and Yellow River Research Socity: Xi'an, 5-9 (in Chinese).

McEwen, L. J. (1994). Channel planform adjustment and stream power variations on the middle River Coe, Western Grampian Highlands, Scotland. Catena, 21, 357-374.

Meng, Q. M. (Ed.). (1996). Soil and Water Conservation in the Loess Plateau. Water Resource Press of Yellow River: Zhengzhou, 75 (in Chinese).

Novak, M. D. (1985). Soil loss and time to equilibrium for rill and channel erosion. *TRANSACTION of the ASAE*, 28(6), 1790-1793.

Peugeot, C., Esteves, M., Galle, S., Rajot, J. L., and Vandervaere, J. P. (1997), Runoff generation processes: results and analysis of field data collected at the East Central Supersite of the HAPEX-Sahel experiment. *Journal of Hydrology*, 188-189, 179-202.

Shi, H., Tian, J. L., and Liu, P. L. (1997a). Study on relationship of slope-gully erosion in a small watershed by simulation experiment. Journal of Soil Erosion and Soil and Water Conservation, 3(1), 30-33 (in Chinese).

Shi, H., Tian, J. L., Liu, P. L., and Zhou, P. H. (1997b). Study on spatial distribution of erosion yield in a small watershed by simulation experiment. *Research of Soil and Water Conservation*, 4(2), 75-85 (in Chinese).

Shi, X. J. (2005). Development and suggestions on relative stability of the check-dam system in the small watershed of the Loess Plateau. *Water Conservancy in China* (A), (4), 49-50.

Sidorchuk, A. (1999). Dynamic and static models of gully erosion. *Catena*, 37, 401-414.

Tian, Y. H., Liu, H. J., Yang, M., and Zhang, X. (2004). Relative stability condition and feasibility of the check-dam system in the small watershed of the Loess Plateau. In: Scientific Extension Center of Water Conservancy Department and Association for Yellow River Research, Proceeding of the Workshop on Key Technology Check-Dam System Construction in the Small Watershed of the Loess Plateau (21-24): Zhengzhou, China.

Timmons, M. B. (1984). Use of physical models to predict the fluid motion in slot-ventilated livestock structures. *TRANSACTIONS of the ASAE*, 27, 502-507.

Wang, W. Z. and Jiao, J. Y. (1996). Rainfall and Erosion Sediment Yield in the Loess Plateau and Sediment Transportation in the Yellow River Basin. Science Press: Beijing, 165, 221 (in Chinese).

Wang, Y. S., and Ma, H. (2003). Research and application on the relative stability coefficient of check-dam system. *Water Conservancy in China* (A), (9), 57-58 (in Chinese).

Warburton, J., Davies, T. R. H., and Mandl, M. G. (1993). A Meso-Scale Field Investigation of Channel Change and Floodplain Characteristics in an Upland Braided Gravel-Bed River New Zealand. In: Best, J. L., Bristow, C. S. (Eds.), Braided Rivers. *Geological Society Special Publication*, London: The Geological Society, 75, 241-255.

Xu, M.Q., and Wang, G. (2000). To accelerate the construction of check-dams in the Loess Plateau. Yellow River, 22(1), 26-28 (in Chinese).

Xu, X. Z., Zhang, H. W., Wang, G. Q., Chen, S. Q., and Dang, W. Q. (2008). An experimental method to verify soil conservation by check dam on the Loess Plateau, China. *Environmental Monitoring and Assessment*, 159, 293-309.

Xu, X. Z., Zhang, H. W., and Zhang, O. Y. (2004a). Development of Check-Dam Systems in Gullies on the Loess Plateau, China. *Environmental Science and Policy*, 7(2), 79-86.

Xu, X. Z., Zhang, H. W., and Zhu, M. D. (2004b). Study on measuring method of particle size of raindrop and its improvement. *Soil and Water Conservation in China*, (2), 22-24 (in Chinese).

Yuan, J. P., Lei, T. W., Jiang, D. S., and Zhou, Q. Y. (2000a). Simulated Experimental Study on Normalized Integrated Model for Different Degrees of Erosion Control for Small Watersheds. *Transactions of the CSAE*, 6(1), 22-25 (in Chinese).

Yuan, J. P., Jiang, D. S., and Gan, S. (2000b). Simulated experiment on normal integral model of different control degrees for small watershed. *Journal of Natural Resources*, 15(1), 91-96 (in Chinese).

Zeng, M. L., Fang, X. M., Kang, L. L., and Wang, E. P. (1995). Relative steadiness of development of valley-dam system is absolutely feasible. *Yellow River*, (4), 18-21(in Chinese).

Zeng, M. L., Zhu, X. Y., Kang, L. L., and Zuo, Z. G. (1999). Effects of Sediment Reduction and Erosion Control and Development Prospects of Warping Dam in Water and Soil Loss Areas. *Research of Soil and Water Conservation*, 6(2), 127-132 (in Chinese).

Zhang, H. W., Jiang, E. H., Bai Y. M., Shao, X. M., Zhang, Q., and Lu, X. D. (1994). Similarity Law for Physical Model of the Hyperconcentrated Flow in Yellow River. *Henan Science and Technology* Press: Zhengzhou, 80, 115, 156-162 (in Chinese).

Zhang, H. W. (1994). The Study of the Law of Similarity for models of Flood Flows of the Lower Reach of the Yellow River (Unpublished PhD dissertation), Tsinghua University, Beijing, China (in Chinese).

Zhang, J. H., Zhang, H. W., Jiang, C. B., Wang, G. D., and Chen, S. K. (2001). A primary study on physical model similarity for reservoir on the Yellow River. *Journal of Hydroelectric Engineering*, (3), 52-58 (in Chinese).

Zhang, H. W., Zhang, J. W., Wang, G. D., Chai, Q. F., Zhang, L. R., and Yao, W. Y. (2002). Design of the movable bed for the Dongzhuang reservoir on the Jinghe River. *Journal of Sediment Research*, (2), 71-77(in Chinese).

Zhang, Z. G., Wang, G. P., Jia, Z. J., Wang, X. P., and Niu, S. P. (1995). *Sediment source of Wangjiagou Gully in a well-controlled way.* Science and Technology on Soil and Water Conservation in Shanxi, *(2), 6-8 (in Chinese).*

Zheng, B. M. (2004). Discussion on key technology to construct check-dam systems in the small watershed. In: *Proceedings of Key Technique to Construct Check-dam System in the Small Watershed on the Loess Plateau,* China Sci. and Tech. Spread Center of water resource ministry and Yellow River Research Socity: Xi'an, 16-20 (in Chinese).

Zheng, X. M. (2003). Discussion on dam-system construction in the Loess Plateau of China. *Water Conservancy in China* (A), (9), 19-22 (in Chinese).

Zhou, P. H., Zhang, X. D., and Tang, K. L. (2000). Rainfall installation of simulated soil erosion experiment hall of the state key laboratory of soil erosion and dryland farming on the Loess Plateau. *Bulletin of Soil and Water Conservation*, 20(4), 28-30 (in Chinese).

Zhu, X. Y., Lei, Y. J., and Liu, L. B. (1997). Cognition on several important problems about relative stability of dam system. *Soil and Water Conservation in China*, (7), 53-56 (in Chinese).

In: Check Dams, Morphological Adjustments…
Editors: C. C. Garcia and M. A. Lenzi
ISBN: 978-1-62618-856-3

Chapter 12

Local Scouring at Check-Dams in Mountain Rivers

F. Comiti*[1], M. A. Lenzi[2] and L. Mao[3]
[1]Faculty of Science and Technology,
Free University of Bozen-Bolzano, Italy
[2]Dept. Land and Agroforest Environments,
Università di Padova, Legnaro (Padova), Italy
[3]Department of Geography,
University of Hull, Cottingham Road, Hull, UK

Abstract

Check-dam stability is heavily influenced by local scouring processes, such that structure undermining and failure can occur if scour depth is not adequately accounted for. Therefore, the foundation depth of grade-control structures is generally assigned to equal the maximum scour depth in clear-water conditions.

The present chapter will first present an overview of the physical processes taking place in local scour phenomena due to overfalls and then a state-of-the-art of the prediction of the maximum scour depth at single and sequence-forming check-dams. Differences between the two cases will be highlighted, and the complex interaction between closely-spaced structures will be addressed.

Introduction

Steep (>2-3%) mountain streams, because of their high stream power and shear stress due to their high gradient, are prone to intense bedload transport phenomena. Being these channels usually supply-limited of sediments, the large sediment transport capacity tends to erode the alluvial boundaries, causing vertical incision and bank failure. Large sediment

* Corresponding author: francesco.comiti@unibz.it

volumes are then delivered to downstream reaches at milder slopes where aggradation occurs, leading to avulsion and flooding hazards.

In the European Alps, humans have attempted to control excessive bed erosion and bedload transport for centuries by using transversal structures. When bed incision is to be prevented, grade-control works (check-dams and bed sills) are built in staircase-like sequences (Figure 1a), and incision is controlled both by fixing the bed height at several points and by reducing stream transport capacity through the lowering of the initial longitudinal channel slope S to a lower value S_{eq} (ultimate or equilibrium slope). On the other hand, in order to trap the flood-transported sediment upstream of sensitive locations, tall retention check-dams are used (Figure 1b). Over the two past decades this type of structures has increasingly become "open" or "filtering", i.e. with openings to achieve a self-cleaning mechanism for the finer sediments at lower flows (see D'Agostino, chapter 9, this book).

Figure 1. Examples of check-dams in the European Alps: a) evenly-spaced sequence of 1.5m high structures (Pramper River, Italy) for preventing bed incision; right, single 3.5m high structure for sediment retention upstream (Biois River, Italy). A subsidiary check-dam is visible at the foreground.

The stability of grade-control structures can be heavily impaired by the local scouring process occurring downstream, which may lead to their failure through foundations undermining. In fact, the foundation depth is generally assigned at least equal to the maximum depth of the scour hole formed downstream of the structure. A reliable prediction of the maximum scour depth is therefore needed to design adequate foundations of the control works, usually against flood events with return periods of 50-100 yr.

Therefore, a proper check-dams design cannot overlook the estimation of maximum scour depth, which entails complex relationships among hydraulic, geometric and sediment variables.

HYDRAULIC CHARACTERISTICS

From a hydraulic perspective, a check-dam is fundamentally a weir, broad- or sharp-crested depending on head and crest length, issuing a jet which causes a hydraulic jump in alluvial material downstream. Right after the completion of the structure, a certain height above the bed can be present at its upstream side, and thus the approaching flow velocity can be very low (subcritical conditions, Fr<<1). Eventually, once the impoundment becomes filled with sediment over the years, the approaching flow will feature a steeper energy gradient depending on the equilibrium slope established upstream. Actually, uniform flow conditions can be attained only if the channel presents a sloping bed for a significant length, and this condition is satisfied when the distance between the works is sufficiently large. In the case of short-spaced structures, the approaching flow will still present an accelerating profile downstream of the hydraulic jump, and the bed slope will be lower than the equilibrium value (Martin-Vide and Andreatta, 2003).

For check-dam designers, the evidence (Grant, 1997; Comiti et al., 2007, 2009) that supercritical flows are very uncommon in alluvial channels even at steep slopes (at least up to 10-14 %) is very convenient. In fact, the flow over the check-dam crests can be correctly assumed to feature critical conditions ($Fr \approx 1$), such that flow depth (i.e. the initial jet thickness) can be evaluated as critical flow depth h_c ($=(q^2/g)^{2/3}$). Only when the uniform flow region features a rocky or concrete bed or exceptionally deep and steep flows, supercritical ($Fr>1$) conditions may establish and thus characterise the flow at the structure's crest.

Flow hydrodynamics at drop structures can vary substantially depending on the flow regime, and several regime classification exist. After the early works on the hydraulics of rectangular drops by Moore (1943) and Rand (1955), Little and Murphey (1982) had described the transition between a direct jump (the so called B-jump) to an undular jump at low grade-control structures when the drop ratio z/h_c (drop height over critical flow depth, Figure 2) becomes approximately 1. Leutheusser and Birk (1991), Ohtsu and Yasuda (1997) and Mossa et al. (2002) schematized different conditions that can occur below a weir, depending on the tailwater level. For low tailwater depths, a free hydraulic jump forms, where the jet impinges onto the bottom (B-min jump).

Figure 2. Role of grade-control structures' spacing in determining the hydraulic pattern and local scouring processes (Padova Hydraulic Labs, Italy). Uniform flow conditions establish only if a sufficient bed length is available (right side of the picture, widely spaced structures), whereas when the weirs are close the entire spacing is occupied by the hydraulic jump (left side).

An increase in the tailwater depth pushes the jump against the weir and the nappe plunges into the water cushion, the jump becoming submerged with the onset of strong vortexes with horizontal axis (B-jump). A further raise of the downstream level leads to the disappearance of the surface roller, the nappe stays on the surface forming standing waves, as in an undular jump.

Wu and Rajaratnam (1996, 1998) defined two "jet regimes" below sharp- and broad-crested weirs based on the flow pattern, the impinging jet and the surface flow regimes. In turn, the latter comprises three sub-regimes, named breaking wave, surface wave and surface jet, in order of increasing submergence conditions. The transition between jet regimes at drops depend both geometric and kinematic parameters (Wu and Rajaratnam, 1996, 1998). However, due to the difficulty in accurately predicting flow velocity in coarse alluvial bed, the use of one simple geometrical parameter (i.e. the drop ratio z/h_c, Figure 4) was proposed by Comiti and Lenzi (2006) to discriminate among the two basic regimes, i.e. impinging vs surface (Figures 3 and 4), in accordance with the previous study by Little and Murphy (1982): an impinging regime is generally established for drop ratios $z/h_c > 1$, but its existence was observed even with values about 0.4-0.5 (Comiti, 2003; Comiti and Lenzi, 2006).

Recently, it was found that the drop ratio can be effectively used in predicting the reach-scale flow resistance on a stream stabilized by check-dams (Lenzi et al., 2004a). Indeed, the disappearance of rollers in step-pool streams associated to the drowning out of steps (i.e. the onset of a surface regime with skimming characteristics due to steep slopes) was observed to coincide with an abrupt decrease in flow resistance (Comiti et al., 2009). Similarly, it can be expected that reach-scale flow resistance would sharply decrease (i.e. mean flow velocity increase) at the onset of surface regime along a check-dam reach during a intense flood.

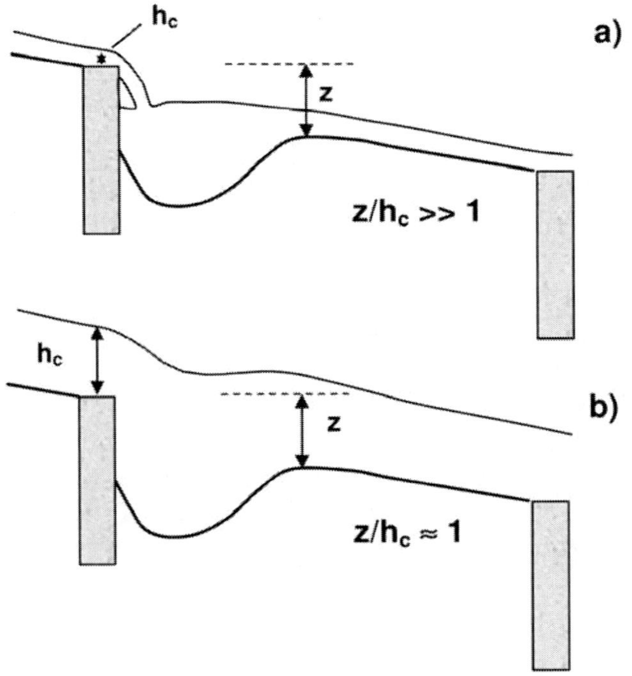

Figure 3. Jet regimes at check-dams: a) impinging and b) surface regime.

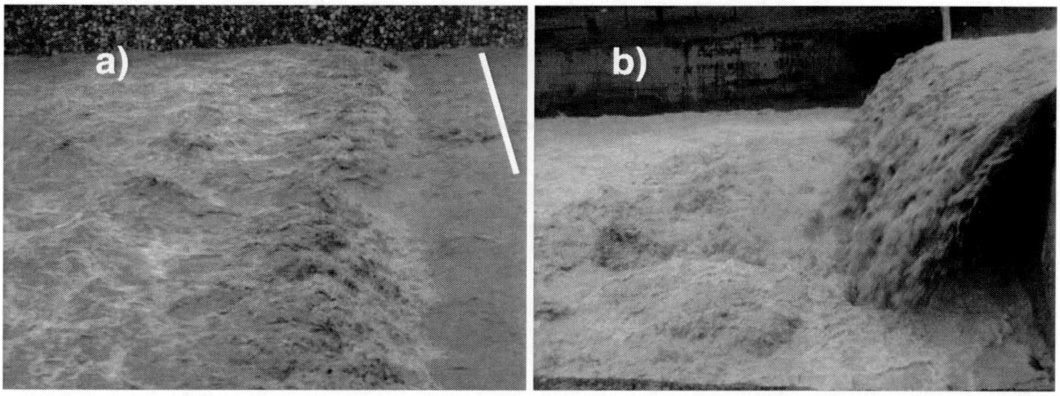

Figure 4. The degree of submergence influence the jet hydrodynamics: a) breaking surface wave (surface regime) generated by a bed sill (white line indicates its crest); b) impinging jet issued from a check-dam.

Jet hydraulics are very complex. For a comprehensive presentation of the abundant literature about hydrodynamic characteristics (impact velocity, air entrainment and concentration, jet disintegration, pressure distribution, velocity decay, etc..), the reader should see Breusers and Raudkivi (1991), Vischer and Hager (1995), Schleiss and Bollaert (2002). However, several important findings are worth to be mentioned here.

Jet impact velocity is directly proportional to the issuance velocity and to the square root of the drop height, but a corrective – reducing – factor accounting for aerodynamic drag should be used for large heights when disintegration occurs (Bohrer and Abt, 1996). In fact, jet aeration reduces the available pressures at the bottom (Ervine et al., 1997) because its total energy is expanded into a larger area before impingement; also, the air concentration of the jet at the tailwater surface influences the centerline velocity decay (Bohrer and Abt, 1996).

Jet centerline velocity decreases as the jet transfers energy to the surrounding fluid by shearing and is inversely proportional to the square root of the diffusing distance and directly proportional to the square root of jet original thickness (Kuroiwa, 1999). As to the tailwater effects, hydrodynamic pressures and their fluctuations on a plate impinged by a jet diminish as the tailwater level increases (Fiorotto and Rinaldo, 1992) but only up to a certain depth (Ursino et al., 2003).

GEOMETRICAL FACTORS

The jet potential energy depends on the difference in water level H (Figure 5), i.e. on the acceleration along the air-surrounded travel length. In a sequence of grade-control structures, such potential head can be effectively substituted by the drop height z (Figure 5), called "morphological jump" by Gaudio et al. (2000), and which can be approximated by the relationship

$$z = (S - S_{eq})L \qquad (1),$$

where S is the mean channel slope measured between crests, S_{eq} is the equilibrium (or compensation) slope, and L is the horizontal distance between two structures (i.e. the spacing, Figure 5).

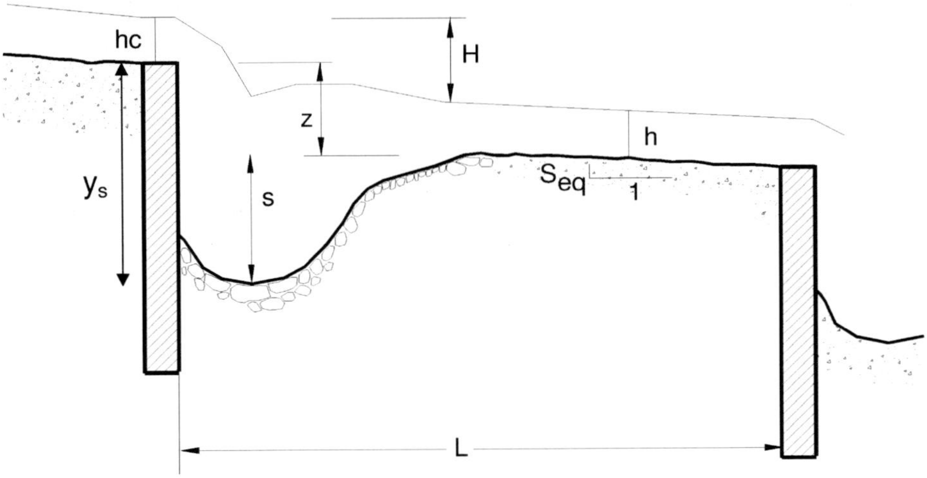

Figure 5. Sketch of the main geometric variables used for analysing local scouring below check-dams: hc critical flow depth, z drop height between bed elevations, H difference in level between upstream and downstream water surfaces, h uniform flow depth, s residual scour depth, ys the total scour depth, L structure spacing, Seq equilibrium slope.

The mean channel slope is imparted by the topographic valley slope, and in most cases does not represent a design variable. In contrast, the equilibrium slope depends on water and sediment discharge, channel cross-section, and sediment characteristics. Its determination, based on bed stability or equilibrium conditions, is subject to a large degree of uncertainty, especially in torrential mountain stream. The reader can found details in Porto and Gessler (1999). The spacing L is instead a fundamental design variable (Wu et al., 2002), tightly associated to drop height.

It is important to highlight that the drop height z (i.e. and thus the head drop H) is not a fixed dimension, but changes over time. For bed sills (i.e. structures built with their crests not protruding significantly from the channel bed), it develops from approximately zero up to the value predicted by Equation 1. For check-dams, it can start from several meters (i.e. drop height just after the completion of the works) and then will increase or decrease depending on the actual equilibrium slope. However, drop height variation associated to equilibrium slope adjustments are relatively more important for bed sills than for check-dams, but in any case the wider the spacing the larger the maximum potential drop height variation (Equation 1). Spacing is therefore a key variable for structures' stability.

Furthermore, recent findings by Lenzi et al. (2003_a) introduced evidence of the 'interference' on scour development due to closely spaced structures. If the downstream boundary condition, i.e. the non erodible point at the subsequent crest position, is closer than a certain critical distance, the size of the scour hole can be significantly different compared to the potential scour that may form in an unrestricted geometry. This effect, termed 'geometrical interference' (Lenzi et al., 2003a) invalidates existing formulae for scour depth prediction when sills are closely-spaced (see later). However, the difference in scour hole dimensions seems to respect a geometric constraint, and it was therefore hypothesised that scour length controls the maximum scour depth. In fact, the part of scour length downstream of the point of jet impact is likely to be determined primarily by the length of the turbulent roller associated to the hydraulic jump: closely-spaced sills would then constrain the roller.

This control appears to be mainly acting by fixing the scour length, and because of the inherent scour hole invariant shape (Comiti, 2003; Tregnagi et al., 2007) the maximum scour depth is constrained not to go under a certain level, physically depending on scour hole's sides stability. In turn, this stability depends on sediment characteristics (sediment size and gradation).

Sediment Sorting and Upstream Sediment Supply

The evaluation of the degree of armoring of pool bottom should help identify which grain size of the original bed mixture is the limiting one in the scouring process. Hallmark (1955) and Kuroiwa (1999) studied in a physical model scour holes below dams, and analyzed sediment sorting in the scour hole. Hallmark found that the material constituting the bottom surface of scour holes was approximately D_{85} whereas Kuroiwa reported median grain size at the bottom ranging from D_{77} to D_{85}, and selected D_{85} to be used in his maximum scour depth equation.

Recently, longitudinal grain size sorting occurring in a sequence of grade-control structures was studied in the field in two channels of the Italian Alps (Comiti et al., 2004), showing that the mean diameter at pool bottom ranges between the D_{80} and D_{87} of the surface grain size distribution found on the equilibrium slope bed, the berm-forming clasts to be on average the D_{90}-D_{95} and finally particles at exit slope and pool end to correspond to the D_{30}-$D_{38.}$

As to the impact of upstream sediment supply, Whittaker (1987) presented the first evaluation of the reducing effect brought about by a certain bed load rate, q_s, coming from upstream. He considered the scour depth with reference to the sill crest, $y_{s.}$ (=s+z, Figure 5). The ratio between the maximum scour depth for live-bed conditions ($y_{s\ qs}$) and the correspondent clear-water case (y_s) is given by:

$$\frac{y_{s\,q_s}}{y_s} = \ln\left(\frac{0.8662 S^{0.1097} q^{0.2164}}{D_{90}^{\ 0.0579} q_s^{\ 0.1778}}\right) \quad (2),$$

where q = unit water discharge (m^2s^{-1}), S = initial slope (m/m), D_{90} = grain size for which 90% of the sediments are finer (m), q_s = unit sediment discharge (m^2s^{-1}).

Recently, Marion et al. (2006) demonstrated that upstream sediment supply can be accounted for in the variable which describes the morphological jump (i.e. the drop ratio) via its influence on equilibrium slope, so that another variable is not required to describe the impact of sediment transport on the development of the local scouring phenomenon.

TEMPORAL EVOLUTION AND TUMBLING FLOW INSTABILITIES

Bateman et al. (2000) described in detail the different phases characterizing scouring development at bed sills. At the beginning of the process, a standing wave establishes when acceleration is experienced by the flow at sill crest as soon as a difference in elevation appears. The jet shapes the bed scouring a trough and depositing a mound downstream, below the wave crest. The mound grows up until the wave breaks when its characteristics achieve the critical threshold for wave stability. According to Bateman et al. (2000), this stage is the most effective as to the erosive action, and an indirect confirmation was observed by Comiti and Lenzi (2006).

After the breaking wave stage onsets, the scouring action decreases and eventually the roller is pushed upstream towards the sill, drowning the hydraulic jump and thus altering the flow pattern inside the scour hole (i.e. onset of impinging jet regime) Scouring proceeds, but at slower pace due to the higher energy dissipation due to turbulent mixing.

However, in some cases a flow instability is observed, characterized by a regular periodical shift between a standing wave (surface wave regime) and a drowned roller (impinging jet regime). Bateman et al. (2000) attributed this phenomenon to the propagation of a small sediment accumulation in the scour hole, noted also by Gaudio and Marion (2003).

A detailed study on the temporal evolution of scour holes at bed sills has later been carried out by Gaudio and Marion (2003). Scour formation showed to develop rapidly and conditions close to equilibrium were achieved in a short time. The introduction of a morphological time permitted the non-dimensional description of the development of scour depth over time through a unique exponential decreasing curve for different sediment and flow conditions. In this curve a short- and a long-term local scour regions were differentiated.

Once pools are scoured below the structures, a regular pattern of supercritical-subcritical flow transition forms, analogous to the so-called tumbling flow in natural step pool systems (Whittaker, 1987). In the case of sequences of grade-control structures, the assigned spacing between the works will give more or less hydrodynamic and energetic relevance to such an analogy (Lenzi and Comiti, 2003; Comiti et al., 2005). In fact, if the distance between two drop structures is large enough – as is usually the case for relatively low gradients – a fairly long stretch of the stream bed will achieve an equilibrium slope. However, in steep streams short spacings are usually assigned , so that the next drop structure locates just at the end of the upstream pool, thus making the profile very similar to a natural step pool sequence with a typical tumbling flow pattern.

Two types of tumbling flow were reported by Whittaker (1987). He carried out several flume tests with fixed weirs and movable bed, and observed that the scour hole develops as flow rate increases only up to a certain point. After this the downstream step begins to limit the scour development, distorting the scour process. As flow rate is increased beyond this point, the flow attains an instability limit and slug waves form. Whittaker called this regime unstable tumbling flow, to differentiate it from the stable tumbling flow occurring before the instability is reached. The same pool deformation process was previously described by Volkart (1972).

Whittaker and Davies (1982) described the formation of water and sediment unsteadiness in similar experiments on tumbling flow characteristics. They reported the occurrence of roll waves, attributing their formation when "...the discharge coefficient for flow over a baffle increased due to the water surface in the pool upstream of the baffle rising beyond a certain level. This increase in discharge resulted in washing out of the hydraulic jump immediately upstream of the baffle into the pool downstream. The water level in the upstream pool thus fell, while that in the inundated pool rose. The same process subsequently occurred with greater amplitude at the next baffle downstream….". Recently, field evidences of instability phenomena both in a natural step pool stream and in a mountain river trained with check-dams have instead been reported and subsequently investigated in the laboratory (Ganz, 2003; Ganz and Schöberl, 2003). Wave properties and flow stability fields have been determined, even though the exact nature of these wave trains is not totally clear yet.

Scouring Process under an Impinging Jet Regime

Check-dams are mostly characterised by drop height much larger than critical flow depth, hence the impinging jet regime is more important to be focused on in this chapter, i.e. they function hydraulically as real dams. However, it should be borne in mind that the same structure may feature both impinging and surface regime depending on the flow rate.

Three distinct phases or zones can be observed when an impinging jet scours an erodible bed (Kuroiwa, 1999; Figure 6). At first (zone 1), the jet is issued above the tailwater depth. Gravitational acceleration transforms potential into kinetic energy, and the jet impinges with a certain angle with respect to the vertical at a higher velocity than the initial, unless it disintegrates. This happen if the trajectory path is large in comparison to the disintegration length, which depends on unit discharge and turbulence level at issuance (McKeogh and Elsawy, 1980; Ervine et al., 1997). If the jet disintegrates, its velocity might be less than that at issuance.

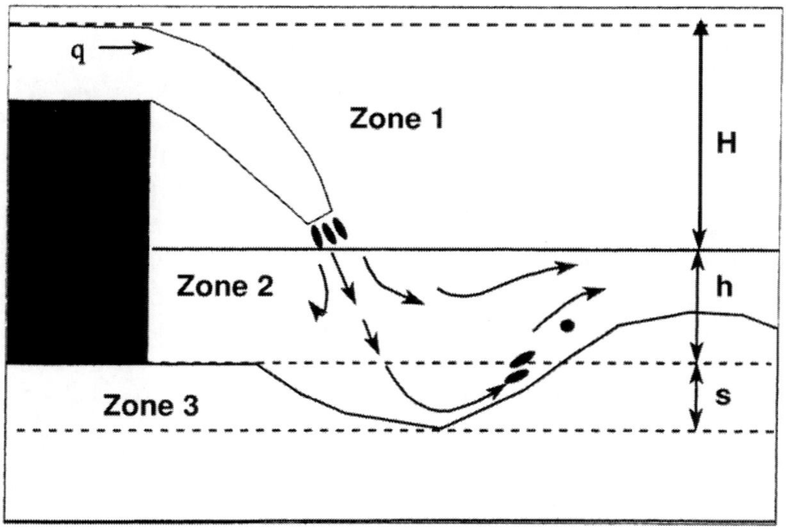

Figure 6. The three distinct zones in a jet scouring process: 1) gravity-driven accelerating jet; 2) jet diffusion and deceleration; 3) sediment entrainment and transport (scour hole formation). Modified after Kuroiwa (1999). See Figure 5 for symbols.

In the zone 2, the jet dissipates energy after it impinges on the tailwater, and its velocity undergoes decay due to momentum transfer to the surrounding slower fluid. Therefore jet residual velocity hitting the bed depends on the length of this diffusion path. In zone 3, if the residual power of the jet is sufficient to entrain and transport bed particles a scour hole forms. Because the forces available at the beginning on the bed surface are higher than those acting afterwards when the scour depth increase lengthens the dissipating path in the tailwater, the temporal rate of scour decreases as time passes (Breusers and Raudkivi, 1991; Stein and Julien, 1994; Gaudio and Marion, 2003). However, jet diffusion inside the scour hole can be expected to be different from that occurring above in the tailwater cushion, where the flow is less restricted.

As the scour hole starts to form, deposition occurs downstream of it. Depending on the tailwater depth (D'Agostino, 1994; Kuroiwa, 1999), material will either deposit immediately downstream of the hole (relative deep tailwater) or will be carried away (shallow tailwater). In the first case, a typical mound or dune – analogous to a flow separation berm (Carling, 1995) – will buildup, whose presence strongly influences scouring processes and dimensions (D'Agostino, 1994).

During the process of approaching the equilibrium dimensions, lighter (i.e. smaller) particles are transported away whilst heavier (i.e. coarser) remain at the bottom of the scour hole, determining armoring of its surface. If the downstream channel width is approximately the same as the width of the jet, downstream flow streamlines will be straight and no recirculation (i.e. backflows) will occur. If recirculation takes place, the scour hole will appear three-dimensional, with the presence of crosswise side slopes along with the streamwise one, and the maximum scour depth will be smaller (D'Agostino, 1994; see later). The gradient and the roughness of the scour hole slopes will in fact affect scouring by inducing different energy losses.

Models for Predicting Maximum Scour Depth

Taking into account the three zones described above, an equation to predict scour depth – and scour length too – must include terms which consider the jet travel in the air (acceleration, aerodynamic drag and jet aeration, impinging angle), the effect of energy dissipation in the tailwater depth (deceleration, turbulence generation, pressure fluctuations), and the resistance of the bed to jet scouring (sediment relative density, characteristic diameter, interlocking factor). Since most of the phenomena and factors cited above are so closely linked to each other, some complications arise when attempting to generalize and simplify.

For example, keeping fixed the other conditions, the higher the drop the deeper the scour, but only to a certain value when aeration and disintegration decelerate the jet (Ervine et al., 1997); or, the deeper the tailwater the smaller the scour depth, but only above a threshold under which an increase of tailwater level leads to deeper holes (Thomas, 1953; Hallmark, 1955; D'Agostino, 1994); or again, the more vertical the jet, the steeper the downstream side of the scour hole (Rajaratnam, 1981; Bormann and Julien, 1991) only up to a given jet angle (Kuroiwa, 1999, around 18 degrees to the vertical, corresponding to a scour slope of 26 degrees to the horizontal), and the deeper the scour hole (Mossa, 1998) but again up to a certain jet inclination only (Mirtskhulava et al., 1967, around 15 degrees to the vertical).

As to the sediment characteristics influencing scouring processes, no general agreement has been reached yet. Given the armoring processes taking place in the scour hole, the majority of the authors use D_{90}, thus the effect of sediment gradation has been so far reduced to the selection of a coarse armoring size, neglecting other possible effects on sediment transport inside the scour hole. To the authors' knowledge, only Ghodsian and Faradonbeh (2001) and Marion et al. (2004) proposed equations taking explicitly into consideration the degree of sorting of the mixture. Indeed, Lenzi et al. (2003b) found a non significant role of sediment size on scour depth, similarly to the results by Marion et al. (2004) which showed a much higher relevance of sediment gradation.

Analytical and Numerical Models

The complexity of factors and the highly turbulent hydrodynamics determining scour dimensions strongly inhibit the adoption of a purely analytical approach for the derivation of predictive formulae. However, some attempts have been made.

Bormann and Julien (1991) developed an equation based on an analysis of jet diffusion and sediment incipient motion, later modified by Stein and Julien (1994). Also Hoffmann (1998) derived an equation using a dynamic approach using jet momentum However, all these methods rely on coefficients (i.e. for diffusion, friction, incipient motion) which are of difficult prediction.

A different approach was adopted by Chen and Hong (2001), who presented a complex geometrical analysis of the shape of scour holes formed below check-dams. These provided equations for scour dimensions and volume, then validated by experimental tests with uniform and graded sediment.

As to numerical models, Jia et al. (2001) simulated computationally the scouring process in a plunge pool with uniform bed material due to a two-dimensional plane impinging jet. A finite-element-based unsteady three-dimensional model with k-epsilon turbulence closure was employed to solve the flow field. The Authors considered the effects of both shear stress and lift force on sediment particles due to pressure fluctuations. The latter was taken into account by using empirical relationships. Both of these effects were incorporated in a non-equilibrium sediment transport model consisting of sediment pickup rate and step length adopted for the jet scour problem. In practice, applying complex numerical models – whose parameters must be calibrated for each field site – for designing check-dams is not worth while and still subject to a high degree of uncertainty.

Empirical Models

Historically, empirical equations, i.e. dimensional formula, where variables are not organized in non-dimensional groups, for predicting maximum scour depth were firstly developed for the case of dams. With some limitations, such formulas can be used for check-dams too; actually, these "old" equations are still widely applied. Here only a general summary will be presented, given the abundance of published reference available (e.g. Brausers and Raudkivi, 1991).

The most common general form of these equations can be generalized as (see Figure 5):

$$s + h = a\frac{H^{\alpha} q^{\beta} h^{\gamma}}{D_x^{\ \lambda}} \tag{3}$$

where s = maximum scour depth (m), h = tailwater depth (m), H = drop height (m), q = unit water discharge (m^2s^{-1}) and D_x = quantile of the sediment grain size distribution (m); coefficient a and exponents α, β, γ, λ are determined empirically (most often by laboratory experiments).

Examples of formulae following the model represented by Equation 3 are numerous (among others, Schoklitsch, 1932; Veronese, 1937; Jaeger, 1939; Eggenberger and Müller, 1944; see also Breusers and Raudkivi, 1991). Very similar are Falciai and Giacomin's (1978) – derived from field data – and Mason e Arumugam's (1985) formula which do not contain the tailwater depth at the left side of the equation. The following – rather wide – ranges have been reported by the different authors: $a = 3.27\text{-}12.4$, $\alpha = 0.05\text{-}0.5$; $\beta = 0.54\text{-}0.64$, $\gamma = 0\text{-}0.15$, $\lambda = 0.10\text{-}0.42$. It is worth to point out how the majority of studies found a weak influence of sediment size on scour depth (i.e. $\lambda \approx 0.10$).

Semi-Empirical Models

A second group of equations can be identified as semi-empirical, because they derive from a dimensional analysis carried out by applying the Buckingham's Pi-Theorem to the variables involved in the phenomenon. Non-dimensional groups are thus obtained, and then experiments follow to allow the determination of coefficients and exponents of the dimensionless expression. Here, only formulations specifically developed for grade-control structures (bed sills and check-dams) will be presented. The reader can find information of semi-empirical relationships for dam scour in manuals (e.g. Breusers and Raudkivi, 1991; Vischer and Hager, 1995; Schleiss and Bollaert, 2002).

To the writer's knowledge, the work by Volkart (1972) and Volkart et al. (1973) represent the first investigations providing a formula for scour dimensions at bed sills forming a sequence in a sloping channel. After a dimensional analysis and several clear-water flume tests, the equation for the maximum scour depth with reference to the sill crest y_s (y_s=s+z; Figure 5) was proposed as (Volkart, 1972):

$$y_s = 1.25\frac{q^{1/2} S^{1/2} L^{2/3}}{D_{90}^{\ 5/12}} \frac{\gamma_w}{\gamma_s g^{1/4}} \tag{4},$$

where q = unit water discharge, S = initial slope, L = sill spacing, D_{90} = grain size for which 90% of the sediments are finer, γ_w = specific weight of water, γ_s = specific weight of sediment, g = acceleration due to gravity. The slope range was $0.0075 < S < 0.07$.

Gaudio et al. (2000) were the first to introduce the concept of morphological jump for bed sills (Equation 1), which bring together spacing, initial and equilibrium slope, and defines a proxy for the drop height H. They carried out laboratory experiments on low gradient channels ($S < 0.015$) and large sill spacings. Subsequent investigations (Lenzi et al., 2002) broadened the slope range (up to 14%) and the flow regimes (i.e. both surface and impinging) of such approach. Lenzi et al. (2002) derived the following formula:

$$\frac{y_s}{H_s} = 0.44 + 1.45\left(\frac{z}{H_s}\right)^{0.86} + 0.06\left(\frac{z}{\Delta D_{95}}\right)^{1.49} \qquad (5),$$

with y_s = maximum scour depth; H_s = specific critical energy (=1.5h_c); D_{95} = grain size for which 95% of the sediment in weight is finer; z = morphological jump (i.e. drop height) given by Equation 1 and Δ = sediment relative submerged density defined as:

$$\Delta = \frac{\rho_s - \rho}{\rho}$$

with ρ_s and ρ = sediment and water density respectively. The experimental range was $0.225 < a_1 / H_s < 1.872$ and $0.161 < a_1 / \Delta D_{95} < 1.150$, and equation 5 showed satisfactory capabilities with field data on scour holes below check-dams and bed sills in Alpine mountain stream. (Lenzi et al., 2003b; Lenzi et al., 2004b).

A further step in the formulation was to include the cases of closely-spaced structures (i.e. leading to interference phenomena) and the effect of sediment gradation. Starting from the following dimensionless groups:

$$\frac{y_s}{H_s} = \phi_2\left[\frac{z}{\Delta D}, \frac{z}{H_s}, \frac{L}{H_s}, SI\right],$$

where SI is the sorting index:

$$SI = \frac{\dfrac{D_{84}}{D_{50}} + \dfrac{D_{50}}{D_{16}}}{2}$$

Marion et al. (2004) combined flume datasets of Lenzi et al. (2002) and Lenzi et al. (2003a), 13 field data presented in Comiti et al. (2004), plus new flume experiments on slopes ranging from 4% to 8% and with variable spacing for a total of 108 data, and obtained:

$$\frac{y_s}{H_s} = 2.68\left(\frac{z}{H_s}\right)^{0.43} SI^{-0.19}\left(1 - e^{-0.14\frac{L}{H_s}}\right) \qquad (6).$$

The experimental range was $0.188<z/H_s<1.872$ and $1.48<SI<5.95$. It is interesting to note that the dimensionless group including the sediment size was not found to be statistically significant if the sorting index was used. The exponential part (in parenthesis) of equation 6 is a corrective factor accounting for the reduction in scour depth due to the geometrical interference (see above), and is potentially applicable to other predictive equations developed for non-interfering scours.

Marion et al. (2006), after performing 48 new flume tests with and without sediment feeding, reanalysed the complete dataset (i.e. including all data in Marion et al., 2004) obtaining slightly dfferent parameters compared to equation 6:

$$\frac{y_s}{H_s} = 2.68\left(\frac{z}{H_s}\right)^{0.60} SI^{-0.19}\left(1-e^{-0.25\frac{L}{H_s}}\right) \quad (7).$$

The effect of sediment feeding, as already mentioned above, is here included in the different (i.e. lower) value attained by the drop heigh z because of the higher equilibrium slope. This calibration is valid for $0.07<z/H_s<1.87$.

It is important to point out that equations 5, 6, 7 refer to the total scour depth ys, i.e. measured from the structure crest (Figure 5). Such approach, valid from a practical perspective, contains nevertheless a degree of spurious correlation, because y_s is expressed as a function of the drop height z, which is part of y_s itself (Figure 5). However, dimensionless equations directly addressing the residual scour depth s (Figure 5) have been proposed as well. For example, D'Agostino and Ferro (2004) proposed an equation calibrated on laboratory data on single structures with null bed slope downstream, in contrast to the previously described cases of sloping beds. The dataset included experiments by Veronese (1937), Bormann and Julien (1991), D'Agostino (1994) and Mossa (1998). The obtained equation was:

$$\frac{s}{z} = 0.65\left(\frac{b}{z}\right)^{0.593}\left(\frac{h}{H}\right)^{-0.126}\left(A_{50}\right)^{0.544}\left(\frac{D_{90}}{D_{50}}\right)^{-0.856}\left(\frac{b}{B}\right)^{-0.751} \quad (8),$$

where $A_{50} = \frac{Q}{bz\left(\Delta g D_{50}\right)^{0.5}}$, b = crest width (m) and B = channel width (m).

They also proposed a simplified formula:

$$\frac{s}{z} = 0.975\left(\frac{H_s}{z}\right)^{0.863} \quad (9).$$

Comiti et al. (2006) developed further the approach by D'Agostino and Ferro (2004), including the case of sloping channels. The initial functional relationship was:

$$s = f(z, H, h, H_s, D_{50}, D_{90}, D_{30}, S_{eq}, \rho_s, \rho, b, B)$$

They arranged the variables into the following dimensionless groups:

$$\frac{s}{z} = \Phi\left(\frac{h}{z}, \frac{H}{z}, \frac{H_s}{z}, \frac{\Delta D_{50}}{z}, \frac{D_{90}}{D_{30}}, \frac{b}{B}, S^*\right)$$

where S* was introduced as a more robust way to parameterize the equilibrium slope for the cases of nearly horizontal beds based on the equilibrium bed inclination α_{eq} expressed in degrees:

$$S^* = \tan(45° - \alpha_{eq})$$

This relationship was calibrated using 186 clear-water flume data referring to grade-control structures where their spacing was long enough to avoid any geometric interference as described above, using the data of D'Agostino ad Ferro (2004), Gaudio et al. (2000), Gaudio and Marion (2003), Lenzi et al. (2002), Marion et al. (2004). The following equation was obtained (R^2=0.95):

$$\frac{s}{z} = 0.65\left(\frac{h}{z}\right)^{0.23}\left(\frac{H}{z}\right)^{0.32}\left(\frac{H_s}{z}\right)^{1.06}\left(\frac{\Delta D_{50}}{z}\right)^{-0.64}\left(\frac{D_{90}}{D_{30}}\right)^{-0.48}\left(\frac{b}{B}\right)^{2.27}\left(S^*\right)^{-10.25} \quad (10).$$

However, not all the groups are statistically significant (i.e. h/z) whereas others are of difficult determination (H/z, S*, D_{90}/D_{30}). Therefore, a simpler yet less precise equation (R^2 = 0.85) was proposed as most suitable for practical predictions, where D_{90} is used instead of D_{50}:

$$\frac{s}{z} = 2.00\left(\frac{H_s}{z}\right)^{0.59}\left(\frac{b}{B}\right)^{2.34}\left(\frac{\Delta D_{90}}{z}\right)^{-0.09} \quad (11).$$

Despite the good performance of equation 11 for such a wide range of laboratory conditions, its comparison against field data (i.e. 73 scour holes, Lenzi et al. 2003b) showed errors in predicted s/z values up to ±50% (Comiti et al., 2006). However, equations 10 and 11 illustrates how relevant can be the three-dimensionality of the scouring process (i.e. the ratio b/B) on reducing the scour depth, as first demonstrated by D'Agostino (1994).

Other Approaches: Artificial Neural Networks

Recently, artificial neural networks were successfully implemented for the prediction of maximum scour depth in local scouring processes. Liriano and Day (2001) applied this method to scouring at culvert outlets, Azmathullah et al. (2006) to ski-jump spillways, and recently Guven and Gunal (2008) to grade-control structures. The latter authors trained an explicit neural networks formulation on the dataset presented by D'Agostino and Ferro

(2004), and claimed a higher predictive performance of such method compared to traditional regression-based equations.

CONCLUSION

The local scouring process downstream of grade-control structures is an extremely complex macroturbulent phenomenon which defies present numerical modelling techniques and hardware capabilities. The empirical and semi-empirical equations presented in this chapter are still the most reliable tools to predict the maximum scour depth and thus to adequately design their foundations. Furthermore, once a relationship between maximum scour depth and discharge (or critical flow depth) is assumed, the degree of safety (e.g. ratio scour depth/foundation depth) of each structure for different recurrence intervals can be estimated. Particular caution must be used for relatively low structures featuring a surface jet regime, where some empirical equations may fail.

Designers should be aware that the spacing between grade-control structure – or otherwise the distance to the first non-erodible section downstream in the case of single structures – is the key factor determining the dimensions of scour holes, being primarily involved in the formation of the drop height. In addition, if the spacing is short enough (approximately $L/H_s < 15\text{-}20$), interference processes reducing scour depth will onset. Another important advantage of having relatively short spacing is that the unavoidable errors in the prediction of the equilibrium slope can be made less severe, because the associated error in determining the drop height will be smaller. Also, shorter spacings imply smaller structure heights, which can be favourable for both fish passage and aesthetic purposes.

Therefore, when a sequence of grade-control structures is built using different spacings, the foundation depth of each structure must be calculated considering its actual distance from the next downstream. If all the works feature the same foundation depth, the stability of the overall sequence is determined by the one having the longest spacing, assuming the equilibrium slope to be the same along the whole reach. Once a structure fails for undermining, all the upstream ones are very likely to collapse in a "domino" way for the absence of a downstream fixed bed level.

In several cases (e.g. when a single check-dam is very large or is the downstream most structure of a sequence), small secondary transversal structures (i.e. subsidiary dams) are effectively utilized in order to limit scouring by fixing the bed level downstream. Other countermeasures such as concrete aprons or "paving" with large blocks the scour holes are commonly implemented. However, it is safer to design foundations assuming that the scour hole will attain the maximum scour depth even if aprons are provided, because these structures have shown a high susceptibility to heavy damages and therefore cannot be completely relied on.

REFERENCES

Azmatullah, H. M., Deo, M. C., and Deolalikar, P. B. (2006). Estimation of scour below spillways using neural networks. *J. Hydr. Res.*, 44 (1), 61–69.

Bateman, A., Andreatta, A., and Muñoz, N. (2000). Estudio experimental de la erosión local producida por la presencia de umbrales de fondo en lecho vivo, *Proc. XIX Congreso Latinoamericano de Hidráulica*, Córdoba, Argentina, (in Spanish).

Bohrer, J. G., and Abt, S.R. (1996). Plung pool velocity prediction of rectangular free falling jets, *Report,* Dept. Civil Engineering, Colorado State University, Fort Collins, Colorado.

Bormann, N., and Julien, P.Y. (1991). Scour downstream of grade-control structures. *Journal of Hydraulic Engineering*, 117(5), 579-594.

Breusers, H.N.C, and Raudkivi, A.J. (1991). Scouring. *Hydraulic Structures Design Manual*, IAHR, A.A. Balkema, Rotterdam, The Netherlands.

Carling, P.A. (1995). Flow-separation berms downstream of a hydraulic jump in a bedrock channel. *Geomorphology*, 11, 245-253.

Chen, J.Y., and Hong, Y.M. (2001). Characteristics of check dam scour hole by free over-fall flow, *J. Chin. Inst. Eng., 24(6)*, 673-680.

Comiti (2003). Local scouring in natural and artificial step pool sytems. PhD dissertation, University of Padova, 211 pp.

Comiti, F., and Lenzi M.A. (2006). Dimensions of standing waves at steps in mountain rivers, *Water Resour. Res.*, 42, W03411, doi:10.1029/2004WR003898.

Comiti, F., Andreoli A., Lenzi M.A. (2004) Dimensions of scour holes and effectiveness of grade-control structures in two cobble-gravel bed rivers, *Proc. 2nd Intl. Conf. on Scour and Erosion*, IAHR, Stallion Press, vol. 1, 358-365.

Comiti F., Andreoli A., Lenzi M.A. (2005). Morphological effects of local scouring in step-pool streams, *Earth Surf. Process. Landforms,* 30(12), 1567-1581.

Comiti, F., Mao, L., Wilcox, A., Wohl, E., and Lenzi M.A. (2007). Field-derived relationships for flow velocity and resistance in high-gradient streams. *Journal of Hydrology* 340, 48-62.

Comiti, F., D'Agostino, V., Ferro, V., and Lenzi M. A. (2006). Fenomeni di erosione localizzata a valle di opere trasversali: integrazioni tra indagini di campo e di laboratorio, *Quaderni di Idronomia Montana,* 25, 217-225 (in Italian).

Comiti, F., Cadol, D., Wohl, E. (2009) Flow regimes, bed morphology and flow resistance in self- formed step-pool channels. *Water Resources Research.* 45, W04424, doi:10.1029/2008WR007259.

D'Agostino, V. (1994). Indagine sullo scavo a valle di opere trasversali tramite modello fisico a fondo mobile. *L'Energia Elettrica*, 71(2), 37-51 (in Italian).

D'Agostino, V., and Ferro, V. (2004) Scour on alluvial bed downstream of grade-control structures. *J. Hydr. Engrg.*, ASCE, 130(1), 24-37.

Eggenberger, W., and Müller, R. (1944). Experimentelle und theoretische Untersuchungen über das Kolkproblem, *Mitteilung aus der Versuchsanstalt für Wasserbau*, 5, Verlag A.G. Gebr. Leemann and Co., Zürich, 59-79 (in German).

Ervine, D. A., Falvey, H.T., and Withers, W.A. (1997). Pressure fluctuations on plunge pool floors, *J. Hydr. Res.*, IAHR, 35(2), 257-279.

Falciai, M., and Giacomin, A. (1978). Indagine sui gorghi che si formano a valle delle traverse torrentizie, *L'Italia Forestale e Montana*, 33, 111-123 (in Italian).

Fiorotto, V., and Rinaldo, A. (1992). Turbulent pressure-fluctuations under hydraulic jumps, *J. Hydr. Res.*, 30 (4), 499-520.

Ganz, T.F. (2003). Development of Flow Instabilities in Steep Stepped Channels - Experimental Analysis of the Influence of Channel Geometry, Bottom Roughness and

Suspension Concentration. *PhD thesis*, Water Resources Institute, University of Innsbruck, Austria (in German).

Ganz, T., and Schöberl, F. (2003). Limits of flow instability in steep stepped channels. *Proc. XXX IAHR Congress*, Thessaloniki, Greece, 24-29 August, 141-148.

Gaudio, R, and Marion, A. (2003). Time evolution of scouring downstream of bed sills. *J. Hydr. Res.*, IAHR, 41(3), 271-284.

Gaudio, R, Marion, A., and Bovolin, V. (2000). Morphological effects of bed sills in degrading rivers, *J. Hydr. Res.*, IAHR, 38(2), 89-96.

Ghodsian, M., and Faradonbeh, A.A. (2001). Effect of sediment gradation on scour below free over fall spillway. *Proc., 3rd Int. Symp. Environmental Hydraulics*, ISEH, Tempe, USA.

Grant, G. E. (1997). Critical flow constrains flow hydraulics in mobile-bed streams: a new hypothesis, *Water Resourc. Res.,* 33 (2), 349-358.

Guven, A., and Gunal, M. (2008) Prediction of scour downstream of grade-control structures using neural networks. *J. Hydr. Engrg.*, ASCE, 134 (11), 1656-1660.

Hallmark, D. E. (1955). Influence of particle size gradation on scour at base of free overfall, *MSc thesis*, Colorado State University, Fort Collins, USA.

Hoffman, G. (1998). Jet scour in equilibrium phase, *J. Hydr. Engrg.*, ASCE, 124(4), 430-437.

Hoffmans, G.J.C.M., and Verheij, H.J. (1997). Scour manual. A.A. Balkema, Rotterdam, The Netherlands.

Jaeger, C. (1939). Über die Ähnalichkeit bei flussbaulichen Modellversuchen. *Wasserwirtschaft und Wassertechnik*, 34, N° 23/27 (in German).

Jia, Y., Kitamura, T., and Wang, S. S.Y. (2001). Simulation of scour process in plunging pool of loose bed-material. *J. Hydr. Engrg.*, ASCE, 127 (3), 219-229.

Kuroiwa, J.M. (1999). Scour caused by rectangular impinging jets in cohesionless bed. *PhD thesis*, Colorado State University, Fort Collins, USA, 365 pp.

Lenzi, M.A, and Comiti, F. (2003). Local scouring and morphological adjustments in steep channels with check-dam sequences, *Geomorphology*, 55, Special Issue, 97-109.

Lenzi, M.A., Marion, A, Comiti, F., and Gaudio, R. (2002). Local scouring in low and high gradient streams at bed sills, *J. Hydr. Res.*, IAHR, 40(6), 731-739.

Lenzi, M.A., Marion, A., and Comiti, F. (2003a) Interference processes on scouring at bed sills, *Earth Surf. Process. Landforms*, 28(1), 99-110.

Lenzi, M.A., Marion, A., and Comiti, F. (2003b) Local scouring at grade-control structures in alluvial mountain rivers, *Water Resourc. Res.,* 39(7), 1176-1188.

Lenzi, M.A, Mao, L., Comiti, F., and Lucchini G. (2004a). La resistenza al flusso in corsi d'acqua sistemati con sequenze di briglie. *Quaderni di Idronomia Montana*, 26, 375-386 (in Italian).

Lenzi, M.A., Comiti F., Marion A (2004b). Local scouring at bed sills in a mountain river: The Plima River, Italian Alps, *Journal of Hydraulic Engineering, ASCE*, 130(3), 267-269.

Leutheusser, H.J., and Birk, W.M. (1991). Downproofing of low overflow structures. *J. Hydr. Engrg.*, ASCE, 117(2), 205-213.

Liriano, S. L., and Day, R. A. (2001). Prediction of scour depth at culvert outlets using neural networks. *J. Hydroinform.*, 3(4), 231–238.

Little, W.C., and Murphey J. B. (1982). Model study of low drop grade control structures, *J. Hydr. Engrg.,* ASCE, 108(10), 1132-1146.

Marion, A., Lenzi M. A., Comiti, F. (2004). Effect of sill spacing and sediment size grading on scouring at grade-control structures, *Earth Surf. Process. Landforms,* 29(8), 983-993.

Marion, A., Tregnaghi, M., Tait, S. (2006). Sediment supply and local scouring at bed sills in high-gradient streams. *Water Resour. Res*., 42, W06416, doi:10.1029/2005WR004124.

Martin Vide, J. P., and Andreatta, A. (2003). Bed sill disturbance on the slope of steep streams. *Proc. XXX IAHR Congress*, Thessaloniki, Greece, 24-29 August, 149-157.

Mason, P. J, and Arumugam, K. (1985). Free jet scour below dams and flip buckets. *J. Hydr. Engrg.*, 111(2), ASCE: 220-235.

McKeogh, E. J., and Elsawy, E. M. (1980). Air retained in pool by plunging water jets, *J. Hydr. Div*., ASCE, 106(10), 1577-1593.

Mirtskhulava, E., Dolidze, I. V., and Magomedova, A.V. (1967). Mechanism and computation of local and general scour in non-cohesive, cohesive soils and rock beds. *Proc. Twelfth Congress of Intl. Ass. Hydr. Res*., Colorado State University, Fort Collins, USA, C20 (1-8).

Moore, W.L. (1943). Energy loss at the base of free overfall. *Transaction*, ASCE, 108, 1343-1360.

Mossa, M. (1998). Experimental study on the scour downstream of grade-control structures. *Proc. of XXVI Convegno di Idraulica e Costruzioni Idrauliche*, Catania, Italy, 581-594.

Mossa, M., Petrillo, A., and Chanson, H. (2002). Tailwater level effects on flow conditions at an abrupt drop. *J. Hydr. Res*., IAHR, 40(4), 1-13.

Ohtsu, I., and Yasuda, Y. (1997). Characteristics of flow passing over drop-structures. *Proc. Conf. on Management of Landscapes Disturbed by Channel Incision*, Wang, S.S.Y, Langendoen, E.J., Shields F.D. Jr (eds.), 217-222.

Porto P., and Gessler J. (1999). Ultimate bed slope in calabrian streams upstream of check dams: field study. *J. Hydr. Engrg.,* ASCE, 125(12), 1231-1242.

Rajaratnam, N. (1981). Erosion by plane turbulent jets. *J. Hydr. Res.*, IAHR, 19(4), 339-358.

Rand, W. (1955). Flow geometry at straight drop spillways. *Proceedings* ASCE*,* 81, 1-13.

Schleiss, A.J., and Bollaert, E. (2002). Rock scour. Balkema Pub., Rotterdam, The Netherlands.

Schoklitsch, A. (1932). Kolkbildung unter Überfallstrahlen. *Die Wasserwirtschaft*, 341 pp. (in German).

Stein, O. R, and Julien, P. Y. (1994). Sediment concentration below free overfall. *J. Hydr. Engrg.,* ASCE, 120(9), 1044-1053.

Tregnaghi, M., Marion A., and Gaudio R. (2007). Affinity and similarity of local scour holes at bed sills. *Water Resour. Res.*, 43, W11417, doi:10.1029/2006WR005559.

Thomas, R. (1953). Scour in a gravel bed at the base of a free overfall. *MSc thesis*, Colorado State University, Fort Collins, CO, USA.

Ursino, N., Salandin, P., and Da Deppo, L. (2003). Fluctuating pressures at the bottom of a plunge pool, *Proc. XXX IAHR Congress*, Thessaloniki, Greece, 24-29 August, 191-198.

Veronese, A. (1937). Erosioni di fondo a valle di uno scarico. *Annali dei Lavori Pubblici*, vol. LXXV, 9, 717-726.

Vischer, D., and Hager, W.H. (1995). Energy dissipators. *I*AHR Hydraulic Structures Design Manual 9, *A.A. Balkema, Rotterdam, 206 pp.*

Volkart, P. (1972). Die Stabilisierung von Flussläufen mittels einer Folge von Querschwellen. *Mitteilungen der Versuchsanstalt für Wasserbau, Hydrologie und Glaziologie*, No. 6, Zürich, Switzerland, 59 pp.

Volkart, P., Tshopp, J., and Bisaz, E. (1973). The effect of sills on river bed. *Proc. of International Symposium on River Mechanics*, IAHR, 1, Bangkok: 167-178.

Whittaker, JG. (1987). Sediment transport in step-pool streams. In: *Sediment Transport in Gravel Bed Rivers,* Thorne CR, Bathurst JC, Hey RD (eds.). Wiley and Sons; 545-570.

Whittaker, J. G., and Davies T. R. H. (1982). Erosion and sediment transport processes in step-pool torrents. In: *Recent developments in the Explanation and Prediction of Erosion and Sediment Yield*, IAHS Publ. No. 137, 99-104.

Wu, S., and Rajaratnam, N. (1996). Submerged flow regimes of rectangular sharp-crested weirs, *J. Hydr. Eng,* ASCE, 122 (7), 412-414.

Wu, S., and Rajaratnam, N. (1998). Impinging jet and surface flow regimes at drops. *J. Hydr. Res*., IAHR, 36(1), 69-74.

Wu, C. C., Hou, C. H., Chen, C. N., and Shih, C.Y. (2002). Determination of groundsills interval for stream training, *Proc. River Flow 2002 Conf*., 4-7 Sept., Louvain-La-Neuve, Belgium, Balkema Pub., 1109-1115.

In: Check Dams, Morphological Adjustments…
Editors: C. Conesa and M. A. Lenzi

ISBN: 978-1-62618-856-3

Chapter 13

Bed Scouring-Sedimentation Balance Induced by Check Dams in Semiarid Catchments with Different Lithology

C. Conesa-García[*1] *and R. García-Lorenzo*[2]
[1]University of Murcia, Dep. of Physical Geography,
Campus de la Merced s/n. 30001 Murcia, Spain.
[2]Autonomous Community of Murcia Region,
Environmental Integration and Information Service.
C/ Catedrático Eugenio Ubeda, s/n. Murcia, Spain

Abstract

This study has as its objective to show the bed scouring-sedimentation balance due to the gabion check dam construction in ephemeral channels submitted to Projects of Hydrological Forest Restoration (PHFR) under different lithological conditions. For such purpose, two semiarid catchments have been chosen in South-East Spain: the Cárcavo catchment, which is characterized by Miocene marls and Quaternary pediments, and the metamorphic Torrecilla catchment. A comparison of bed scouring-sedimentation balances in both study catchments has been made from variables on geometric features of the sedimentary wedges and scour wedges, bed *slope*-induced *changes* and bed particle sizes variations between upstream and downstream reaches. Volumetric variation curves of the erosion and sediment wedges according to the distance from the check dams are also shown. The field information has been simplified and the relations between variables explained through factorial analysis of principal components (APC), applying an orthogonal varimax rotation. This method contributes to a better understanding of the geomorphological behavior of the reaches affected by check dams.

Keywords: Check dams, ephemeral channels, bed scouring-sedimentation balance, volumetric variation curves, South-East Spain.

[*] E-mail: cconesa@um.es

1. Introduction

Channel constitute the last path of sediment released in a drainage basin. This is the most important vehicle for direct transfer in the whole hydrographic system. In this system waters and sediment originate from the rest of the watershed, which might or might not flow depending on their geometric features and fluvial dynamics. The insertion of check dams along the channel creates a series of water and sediment collectors which produces infiltration and attenuates the general bed slope. In successive floods, the materials retained by check dams refill the nearest channel reach upstream, thus raising the level of the bed until it reaches the compensation slope which is lower than that of the natural channel (MMA, 2005). The elevation of the channel bed throughout the sediment-filled reach is accompanied by the formation of a flat surface on the bed, cross-sections that are wider than the original ones, and a decrease in the hydraulic radius, which tends to be like the flow depth. All of this causes the flow velocity to slow down and reduces bed-load discharge and erosion capacity.

As well as slowing down the movement of the water and acting as barriers, the check dams also interrupt the normal transit of downward flowing sediment, causing bed erosion and reducing the distribution of nutrients to lower reaches. However, unlike large dams, fine sediment is not completely interrupted, and therefore there is a less nutrient loss. The volume of retained sediment is comparatively greater than that due to erosion downstream from the check dam, but variations in the proportions of each are very sensitive, depending mainly on the channel slope and the bed-sediment grain size. The sequence of several consecutive check dams causes an alternation in sedimentary wedges and erosion which gives the bed a terraced shape. This has been a frequent solution adopted in countries such as Japan, Austria, Spain, Switzerland, Italy, Brazil, Bolivia, Nigeria and the United States, among others. In such cases seried gabion check dams have often been constructed. These are relatively permeable and after being filled up with sediment they form a longitudinal system of steps which has a lower gradient than the original bed (Suárez, 1993). This structural system is quite effective in ramblas and gullies, which typically have great transport capacity and hillslope instability. The correction of *ramblas* (ephemeral channels) and gullies in the South East of Spain is often based on actions of this nature, even when they follow design and location criteria that are not very homogenous.

Works for hydraulic correction and stabilization of torrential channels are intended to control their hydrological and geomorphological downstream effects (Conesa García and García Lorenzo, 2007). Check dams are constructed in order to control these processes in the bed and banks, preventing the incorporation of large solid discharges to floodwaters. By installing several consecutive check dams an attempt is being made to reduce the magnitude of erosive processes and the transport capacity of floods in reaches of greater instability. Indeed, indirectly, this system could be relatively effective for stabilizing adjacent hillslopes, given that it involves the formation of flat sedimentary wedges that are not as steep as the original bed. However, in the end its overall effectiveness depends on the spatial distribution of the check dams, their location, number and the type of works. The importance of these factors is becoming increasingly recognized by environmental management and planning agencies responsible for erosion (The European Environmental Agency, 2000; CMA de la Junta de Andalucía, 2005; Secretaria de Estado de Aguas y Costas, 2003, among others). The Basin Management Programs and Forestry Hydrological Planning Policy developed in Spain has an essential task that is often difficult to carry out: to establish criteria for the selection and optimum use of correction systems in torrential streams by integrating cost and effectiveness (Martínez de Azagra y Navarro Hevia, 1996; Del Palacio et al., 1999; Dirección General de Conservación de la Naturaleza, 2002).

Many of the geomorphological impacts caused by check dams are usually fairly common in torrential channels (bed agradation upstream into the sedimentary wedges, deepening and narrowing of the bankfull channel downstream, etc) (García Ruiz and Puigdefábregas, 1985; Martínez Castroviejo et al., 1990; Conesa García, C. et al., 2007). Nevertheless, the extent of these impacts and consequently the degree of effectiveness of the cross-structures is very different depending on the environmental characteristics of the catchments and on the hydrological rectification systems (spatial distribution of check dams, location, number and type of work). A clear indicator of such an influence is the channel response to the deficit and overfeeding of sediments (Brandt, 2000; Lenzi et al., 2003; Marion et al., 2006; Comiti et al., 2005). In such a way, this study has as its objective to show the influence of staggered check dams of gabions on the scouring-sedimentation balance in Mediterranean ephemeral channels submitted to Projects of Hydrological Forest Restoration (PHFR).

2. Study Area

For areas of study, two semiarid gullied catchments with a strong tendency to dry up have been chosen: Torrecilla and Cárcavo catchments (South-east Spain) (Figure 1).

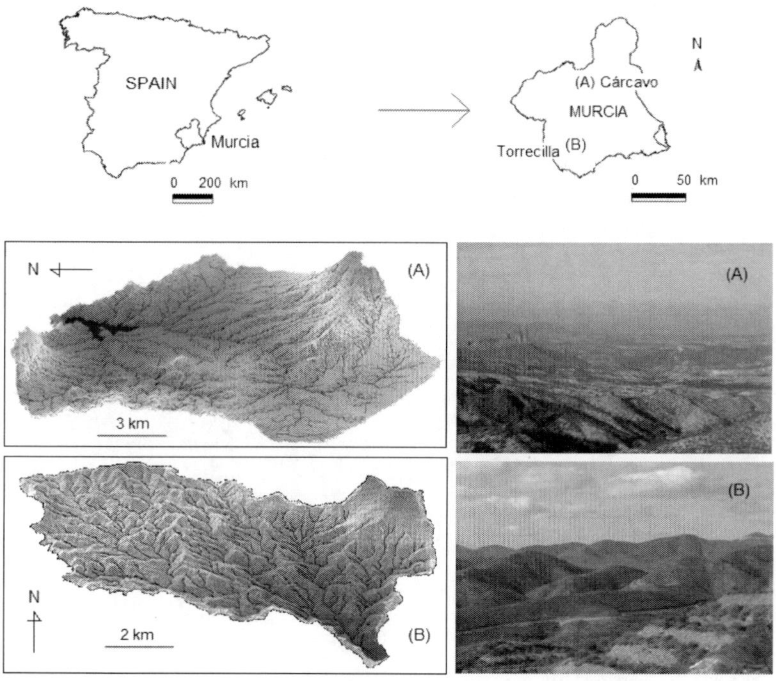

Figure 1. Location of the study area (Murcia Region, Spain). General view and digital elevation model of the Cárcavo (A) and Torrecilla (B) catchments.

These catchments have a semi arid climate with a mean annual precipitation of 260 to 275 mm, annual potential evapotranspiration higher than 850 mm and torrential precipitations in the autumn and at the end of the summer. Rainfall intensities can reach more than 100 and 200 mm in a few hours, causing as a result extreme flash floods (Conesa García, 2005).

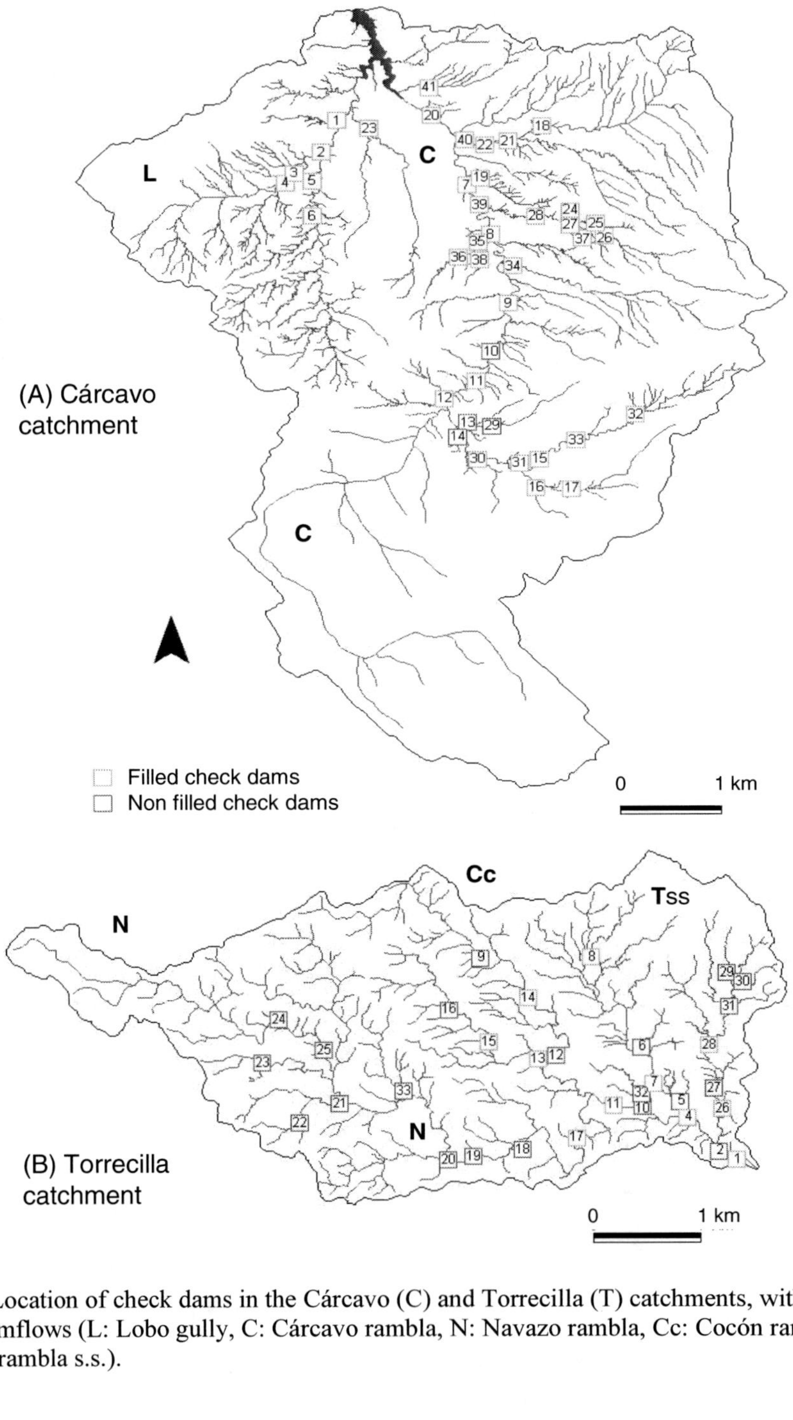

Figure 2. Location of check dams in the Cárcavo (C) and Torrecilla (T) catchments, with indication of main streamflows (L: Lobo gully, C: Cárcavo rambla, N: Navazo rambla, Cc: Cocón rambla and Tss: Torrecilla rambla s.s.).

Both areas have a dense drainage network, characterized by a torrential behaviour, depending on the relief, lithology and land use changes (López Bermúdez et al. 1998). The Torrecilla catchment (15.5 km^2) shows a gullied area developed on metamorphic materials (slates, phyllites, schists and quartzites), while the Cárcavo catchment (34.8 km^2) is drained by ephemeral channels and gullies that dissect deeply the Miocene marls and Quaternary pediments. Channel bifurcation, soil erodibility and high mean slopes give a strong erosive potential in these areas. To combat such erosion the Dirección General de Montes, Caza y Pesca del Ministerio de Agricultura approved and checked a forestry-hydraulic restoration Project, which included the repopulation of 1500 ha with pine (Pinus halepensis) and the construction of check dams (33 and 40 check dam series were constructed respectively during the 1970s, the majority of them with gabions) (Figure 2).

3. Methodology

The alluvial-fill thickness in each point of the sedimentary wedge, behind the check dam, has been calculated using the values of elevation of a three-dimensional image generated from the surfaces of the original and current beds. The surface area of the wedge was measured with GPS (Geoexplorer3C Trimble©) and the bottom area (bed previous to filling) deduced from trigonometrical calculations and from the topographical plans included in the memory of the PHFR (Ministry of Agriculture, 1972). For the cubing of the erosion wedges, consecutive channel cross-sections have been topographied, with intervals of separation of 10 to 20 m depending on each case. Taken from these have been obtained by interpolation other intervening cros-sections setting intervals of 5 m. As much as the spatial interpolation of the cross-sections as the estimation of their geometrical data, in particular of the area, is carried out via the HEC-RAS software. The field information has been simplified and the relations between variables explained through factorial analysis of principal components (APC), applying an orthogonal varimax rotation. Initially, the variables that were handled in this study are related to geometric features of the sedimentary wedges (area -A-, length -L-, mean width -W-, máximum thickness -H-, volume -Vs-), the volume of the scour wedges (Ve), the bed-scouring-sedimentation balance (B), changes in the bed slope (decrease in the gradient above check dams, Δs-up; decrease in the gradient below check dams, Δs-dw), and variations in the bed particle sizes between upstream and downstream reaches (D_{84}dw – D_{84}up).

The main component analysis has made it possible to relate these variables to factors, providing better understanding of their effect on the geomorphological behavior of the reaches affected by check dams.

The structural matrix, specifically, contains the correlation coefficients between factors and original variables that clarify some aspects that are confusing a priori. For ortogonal factors this coincides with the load matrix. In order to evaluate whether a load expresses a relationship or not (it is sufficiently elevated), sample size and the number of variables used has been taken into account, in addition to adopting a rotation which could eliminate some ambiguities. In this case the Varimax ortogonal rotation has been used, which is less realistic conceptually, but which maximizes the "explanation" of variance and could be useful for clearly defining the most influential groups of relationships among the variables. In this analysis these variables are associated to volume retention and final sediment remotion by check dams. The Varimax rotation is an ortogonal type of rotation of the factors that tries to minimize the number of variables with high saturations in a factor. The objective of the rotation of factors is to obtain a solution that is easier to interpret, in the sense that variables that are strongly correlated have high saturation levels over the same factor and are low over the rest. Varimax normalization has been carried out using Kaiser's test (1958), an easy ratio between simple and partial correlations.

4. Results and Discussion

4.1. Retention and Remotion of Sediments Induced by Check Dams in the Marly Cárcavo Catchment

The sedimentary wedges developed upstream of the check dams reach considerably large sizes along the main channel of the Cárcavo rambla, with average values very superior to the ones observed in their tributaries (surface: 3,678 m^2, length: 152 m and width: 16 m). Such large sizes are due to the low bed slope (<0.025) (the result of a deep channel incision from its headwater) rather than the mean height of the check dams (2.9 m). However, in the lower reach, the combination of lower gradient values (0.01-0.023) and higher check dams (> 4.5 m) has brought about the formation of very significant sedimentary wedges. Most notable are the cases of the wedges originating upstream from check dams C7 and C8, which have a surface area of about 11,600 and 5,100 m^2, respectively, and their longitudinal development is greater than 230 m (Figure 3). Obviously, there are less variations in the gradient induced by check dams than in the rest of the channel (Δs-up = 0.008; Δs-dw = 0.004) and changes in the bed texture are much more significant ($\Delta t \approx 0.180$). This last fact is undeniably relevant in the analysis of bed stability, given that it reflects a marked increase in coarse material, and therefore roughness, downstream from the check dams, making these geomorphologically effective. Stabilization is sometimes apparent, given the large proportion of retained sediment that is saturated in the flood events and the subsequent increase in hydrostatic pressure.

Figure 3. Sedimentary wedge behind check dams C12 (a) and C7 (b). Main channel of the Cárcavo rambla.

In fact, the eroded base on which the sedimentary wedges rest on the lower reach of the Cárcavo rambla is usually carved out of limy-marly bedrock, which can act as an aquiclude as well as a slippery surface under conditions of hydric saturation, and thus the stability of the structure is put at risk.

Most of the tributary streams have a gullied form, closely reflected in the geometric features of the sedimentary wedges. Two types of channels should be differentiated: i) those belonging to the largest torrential sub-basins with a higher drainage density; and ii) the gullied channels of the smaller headwater catchments. In the first type, the surfaces of sedimentary wedges retained by check dams are between 1,000 and 1,500 m^2, and the lengths are between 90 and 150 m, whilst in steep headwater channels the wedges are smaller having extensions of less than 1,000 m^2 and are no longer than 90 m. The latter have a limited capacity as sediment storage devices; however, they play an important role in controlling erosion inside the channel bed as well as in the adjacent hillslopes which acquire greater stability. In these cases the formation of the sedimentary wedge involves vertical bed accretion and a flatter channel bed. On this new bed, homogenous talus with smaller angles than in non rectified sections are deposited at the sides. This brings about more intensive vegetation colonization that strengthens the general stability of the system.

The readjustment of the bed in reaches affected by check dams also involves the excavation of the channel bed downstream from its base, giving rise to a new channel entrenched into the preexisting bed ("*scour wedge*"), associated with a progressive undercutting in the same direction. The shape of the resulting incision shows an initial width-depth *ratio* which is consistent with the geometry of the check dam, its position in the context of the watershed area and the bed texture; but in any case, this *ratio* increases progressively in the downstream direction, as the distance from the check dam increases. The scour wedges have a length that varies according to the reach, the longitudinal channel gradient, the height of the check dam and its filling level. In geometric terms, the longitudinal section of the scour wedge is the equivalent of a rectangular triangle, the largest side of which forms the profile of the rectified bed, the shortest side, the depth of local erosion at the foot of the check dam, and the hypotenuse, the profile of the original bed.

In any of the beds analyzed the greatest geomorphological impact occurs in its lower reaches, as a consequence of the greater volume of concentrated run-off and the susceptibility of the bed to erosion processes, due to the large amount of sediments that can be stored by the check dam system installed upstream. In these cases, the rectangular triangle, that represents the longitudinal cross-section of the wedge of the lowered bed, shows an elongated shape, with a base-side of the triangle 100 times longer than the maximum height of the wedge. In the case of the low reach of the Torrecilla rambla this relationship persists in spite of the insertion of recent check dams that have led to a new morphological adjustment of channel *bed.*

In the main channel of the Cárcavo rambla the scour wedges have a 150 to 440 m longitudinal development (Figure 4), which is greater the lower the position of the reach. In fact, the longest eroded reaches are located in the lower part of the catchment, downstream from check dams C7 and C20 (the latter destroyed), whereas the shortest belong to the headwater sectors, downstream from check dams C14 and C33. Obviously, the development of these "*scour wedges*" slows as the bed slope increases and the height of the check dam decreases, given that in this case the adjustable slope segment tends to be less also. With a slope of 0.041 and a 5 m drop from the check dam, C14 has produced a 178 m long "*scour wedge*"; however, with a similar slope (0.049) and a fall height of 2 m the incision wedge developed from check dam C33 hardly reaches 90 m.

Figure 4. Scour wedge downstream from check dam C7 (lower reach of Cárcavo rambla).

Given the greater length of the scour wedges in the lower reach of the main channel, the intersection point of the new bed excavated on the primitive bed is, in these cases, consistent with the finalization of the incision wedge and, therefore, with progressive erosion. The high local erosion rates observed at the foot of these check dams, together with the flatness of the bed, produce a significant morphological adjustment and particularly a *lowering* of *bed level* in relatively long segments. Local erosion at the base of check dam C7 (4.7 m high), for example, has caused the bed level at this point to decrease by more than 1.5 m and the longitudinal gradient to reduce by about 0.004, which is quite significant for this type of reach. It should be noted however that the channel depth and the dimensions of the erosion wedges in this reach do not currently seem to be progressing, given the stability of the limy-marly bedrock.

In accordance with their longitudinal development, the "*scour wedges*" located in the lower reach of the Cárcavo rambla also cause a considerably greater amount of erosion than those at the headwater: for example, the volume of the bed excavated downstream of check dam C7 increases to 7,030 m^3, versus 972 m^3 at check dam C14 located in the upper reach. Another significant fact is the progressive decrease in scoured bed volume depending on the distance to the check dam. Even in some reaches, as in those downstream from check dams C10 and C7 (middle and lower reaches of the Cárcavo rambla, respectively), the limy-marly outcrops limit the incision depth by shortening the length of the scour wedge. The dip of stratified marly limestones, occasionally against the direction of flow, forms a flaky bedrock that provides great mechanical resistance to the tangential flow. This results in a wider scour wedge and an increase in lateral instability that could end up affecting steep talus of Miocene marls and Quaternary silts around the main channel. Currently, there are frequent fallen rock deposits due to base landslides caused by a strong lateral erosion during the flood of October 2000. What is more, due to the destruction of the closing check dam (C20), in the mid 90's, its sedimentary wedge was swept away, revealing the erosive base that the runoff had carved on the bedrock before the deposition induced by these structures (Figure 5a).

Figure 5. Appearance of the bed surface after the destruction of the closing check dam C20 (lower Cárcavo rambla) (a) and thickness of the dismantled sedimentary wedge (b).

The fast and rough evacuation of the sediments produced by the breakage of the dam provoked a new readjustment of the bed slope upstream over the one that had already been modified by previous filling. This readjustment, linked to the local base level generated at the foot of the broken check dam, is shown through a small *bankfull* channel that advances upstream due to regressive erosion. The same erosive surface on which check dam C20 was installed, formed by less permeable substrate materials, could have served as a slippery surface for sliding away the water-saturated sediment retained by this dam. In sum, the total sediment volume, moved firstly because of the construction of this check dam (considered as the catchment closure check dam), and later due to its destruction, is estimated to be about 3,924 m^3, which is a mass of about 10,400 tm.

As shown in Figure 6, the profiles for the accumulative volumes of the sedimentary wedges are very similar: steep in the reach closest to the check dam and only slightly sloping at the end of the wedge. However, profiles representing net accumulated scouring volumes in downstream direction beyond dams are quite different.

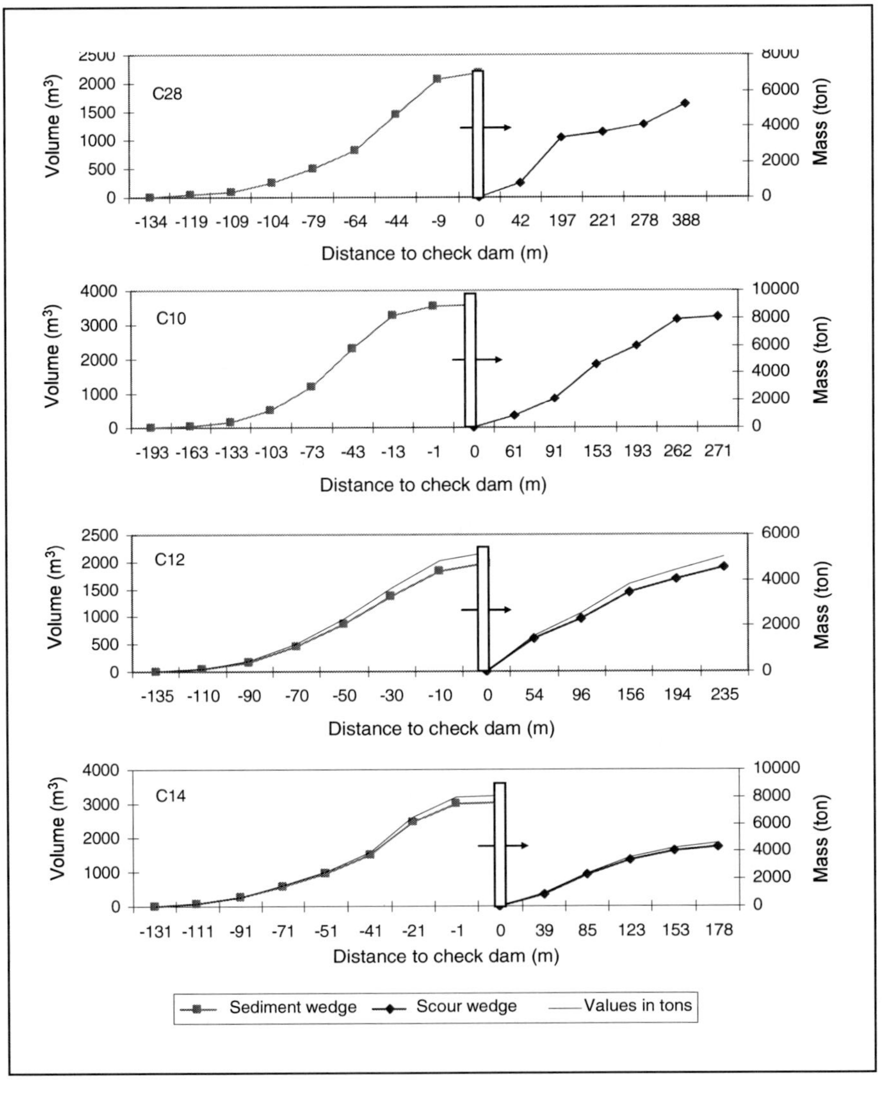

Figure 6. Profiles of accumulative volumes and mass for several sediment and scour wedges versus distance to check dams. Cárcavo catchment.

They have varied shapes, from a convex facies which is typical of a relatively large incision on the bed in the segment right at the structure (C12), to a concave appearance that is more characteristic of progressive downstream erosion (C10) or a broken profile linked to structural geological controls (C28). As distance from the check dam increases upstream, so there is a decrease in the thickness, width and the volume of the sediment wedge. Therefore it is close to upstream dams where there is the greatest relative sediment volume. The same does not occur downstream, in connection with the scour wedge, where many factors cause it to decrease in volume, including geometric features and the design of the structures.

Check dams built in the main channel of the Cárcavo rambla retain the greatest amount of sediment, around 47,060 m^3 which is about 124,700 tm, approximately 49.3 % of the total stored by the check dams in the whole catchment. Similarly, these have the greatest eroded volume downstream (28,834 m^3), in this case accounting for 47 % of total erosion induced in the catchment channels. If we substract the part eroded downstream of the check dams from the quantity retained upstream, the balance is positive, reaching 53.6 % of the total. Next in importance, due to its retained material, is the series of check dams installed in the main tributary gully (Lobo gully). These amount to 15,600 m^3 (41,320 tm), accounting for 16.3 % of the total retained in the basin. The amounts dislodged downstream from the check dams have more relative importance than those deposited behind them, given that they account for 18.4 % of the total volume of erosion induced by all the channels. This fact also explains why the scouring-sedimentation balance rate relative to the overall catchment (27.4 %) is less than in the previous case. The other torrential channels, in spite of having most check dams full, contain scarce amounts of sediment, given their reduced storage capacity, and this fact leads to less development of the incision processes downstream. In nearly all of these the volume of the scour wedge accounts for between 60 and 72 % of the sediment retained upstream.

4.2. Retention and Remotion of Sediments Induced by Check Dams in the Metamorphic Torrecilla Catchment

In contrast to the Cárcavo catchment, the storage capacity of the check dams installed in the Torrecilla catchment are quite small and therefore, the size of the sedimentary wedges upstream from these are also quite small. In addition, the depth and narrowing of the channel, normally produced downstream during the filling phase of these check dams, are reduced once the scoured reach is refilled and the bed load is reestablished. This mainly consists of coarse material (gravels and pebbles), that immediately forms a surface armoured surface layer on the bed preventing its excavation. Normally such armouring is static, as it usually occurs because of scouring or washing away of fine sediment or coarse particles remaining after successive layers of mixed materials have been stirred up. Consequently, the modification of the channel slope upstream from the gabion structures affects short reaches, whilst downstream the *bed level lowering*, that could make this slope disminish to a large degree, is controlled by the rhythm of reposition of coarse alluvial material from upper reaches. Basically, until now, the geomorphological effectiveness demonstrated by these check dams is much lower than has been seen in the Cárcavo catchment: they retain a smaller volume of sediment, they cause a surface washing action that is shallower and they produce less overall adjustment to the longitudinal slope. Unlike in the Cárcavo catchment, here there is a high heterogeneity in wedge sizes which is not really related to the position of each check dam in the catchment. Other factors are more decisive such as the age of the check dam, the spacing between check dams and the geomorphological dynamics of the channel.

The most insignificant sediment retention occurs in two of the check dams located in the lower part of the catchment (T2 and T5), which is demonstrated by the small size of the wedges behind them (< 250 m^2). Both have existed for less than 10 years and they are practically empty of sediment. Downstream from the check dams, the tributary channels of the Torrecilla rambla behave in a different way, as a consequence of the uneven reestablishment of the bed load and the importance of this for filling the "*scour wedge*".

Figure 7 represents several examples in this respect, showing the volumetric variation of the erosion wedges according to the distance from the check dams. In general, the total volume of these wedges is much less than the amount estimated in the Cárcavo rambla, barely 1,200 m^3 in the reaches of greatest development (e.g. T7 and Cc14). The volume of incision also decreases progressively as distance from the check dam increases, but the tendency is not homogenous.

The magnitude of the bed-incision downstream of each check dam is very much conditioned by the effect of the latter on the bed load. The complete interruption of bed transport causes the formation of a pool at the foot of the check dam which involves the dislodgement a large amount of sediment and, generally, a greater volume of erosion compared to the whole of the excavated wedge. On the other hand, the check dams refilled with sediment form steps that do not prevent the drag of material. On the whole, they provide particles that are not as thick as those moved in previous slope conditions. As a result, the "*scour wedges*" previously excavated by clearer water flows undergo a gradual refilling process which extends downstream, from the base of the check dam right to the end of the incision wedge. This is the case of check dam Cc14, located in the Cocón rambla. At the food of this dam a local pool has been partially refilled with gravel and pebbles, removing a lot of capacity to the head of the "*scour wedge*". This process is shown by means a very slightly sloped segment in its initial part in the accumulated erosion volume curve (Figure 7).

The scour wedges induced by check dams in the upper sectors, with steep gradients, are small (< 700 m^3), whereas those that originate in the lower part of the basin have a high rate of filling (T7 and Cc11) or are associated with subsurface flows that are not able to carry out much surface washing (T2, T5 and Cc10).

The general considerations made about the Cárcavo rambla in connection with the profiles of accumulated volumes of retention and erosion are equally valid in the reaches with check dams in the Torrecilla rambla. However, there is a clear difference between both ephemeral channels in relation to the slope of the incision profile and the contrast between this and the sediment retention profile. Often, the "*scour wedges*" developed at the foot of the check dams of the Torrecilla rambla have (according to the distance from the check dams) curves of accumulated volumes with a very low slope and a slightly convex profile (Figure 7).

In accordance with the geometric features, the sedimentary wedges of most volume belong to the check dams already mentioned Cc14 (rambla del Cocón) and T7 (rambla de la Torrecilla s.s.), that retain 3,454 and 3,320 m^3, respectively. Except for these two cases, the mean quantity of sediment contained in the check dams is quite modest (between 500 m^3/check dam and 1,464 m^3/check dam depending on the tributary gullies) compared to estimates for the Cárcavo catchment. The lower amount of filling, the higher channel gradient and the coarser bed texture lead to the formation of smaller sized sedimentary wedges and therefore, lower retention volumes.

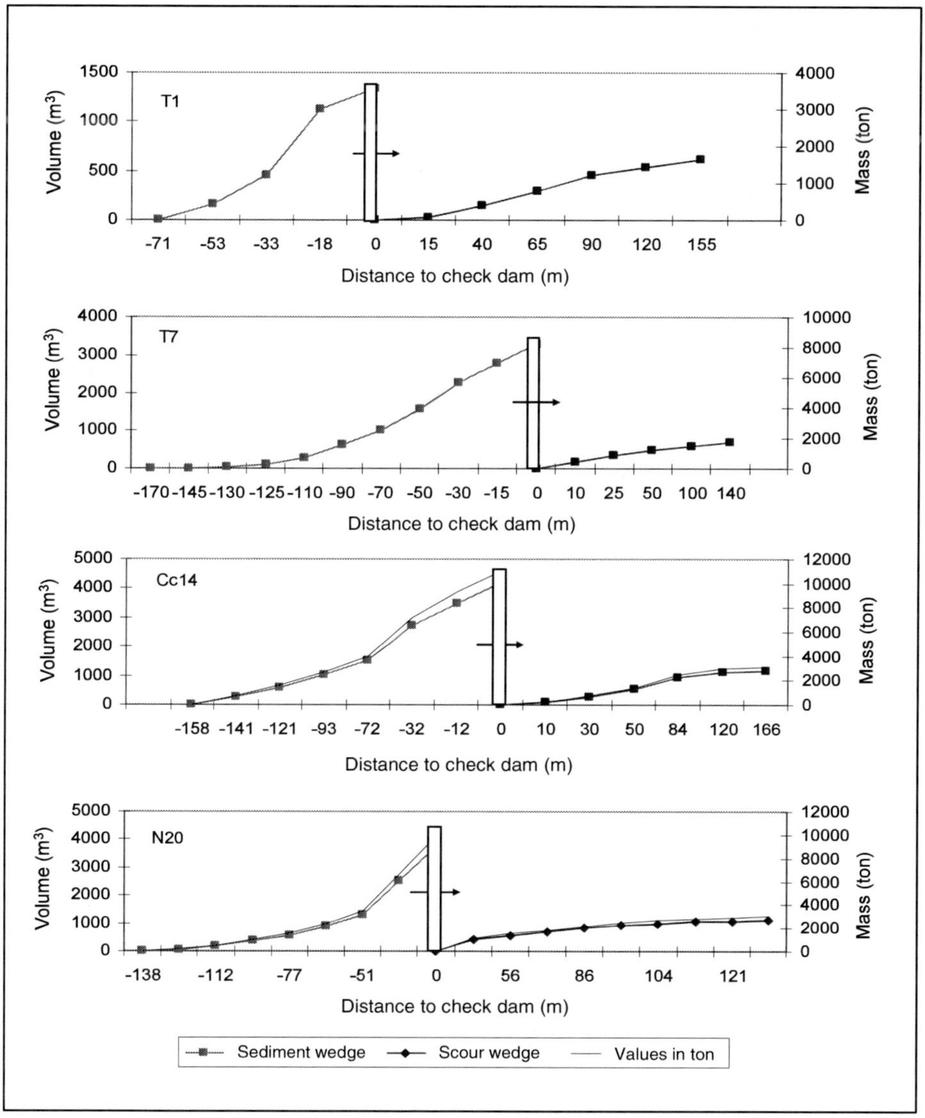

Figure 7. Profiles of accumulative volumes and mass for several sediment and scour wedges versus the distance to check dams. Torrecilla catchment.

However, the most outstanding feature of its geomorphological influence is the low proportion of volume of the scour wedges downstream of these check dams in comparison to the volume of sediment stored upstream. Compared to 69.6 % in the Cárcavo rambla, the Torrecilla rambla and its tributaries have a ratio of only 31.6 %. The most elevated percentages are for those check dams, which have fillings that are close to full capacity and have still not been completely refilled. This is the case, for example, in check dams Cc9 and N20, where the quoted proportion reaches 62 and 51.7 %, respectively. In general, the volumes of these erosion wedges hardly reach 600 m^3/check dam (1,600 tm/check dam). Only the check dams of greatest geomorphological impact, such as Cc14 (Cocón rambla) and N20 (Navazo rambla) are able to cause "*scour wedges*" of more than 1,100 m^3 that involve the dislodgement of approximately 3,000 ton of material. Consequently, the sedimentation-scouring balance of the reaches affected by check dams is quite significant compared to what normally happens in natural conditions. The gains are more than 700 m^3 in most of the reaches affected by the check dams, reaching levels of more than 2,000 m^3 in some cases.

4.3. Comparison of Bed Scouring-Sedimentation Balances Obtained in Both Study Catchments

The comparison of the sediments retained upstream of the check dams and those eroded downstream shows that there are clear differences between the ramblas studied (Table 1 and Figure 8). For the case of the Cárcavo rambla, most retention, undercutting and, therefore, total remotion of the sediments, are found in the lower reach of the main channel, as expected; but this does not occur in the Torrecilla catchment, characterized by a webbed drainage network composed of three main arteries (Navazo, Cocón and Torrecilla s.s. ramblas). Their lower and middle reaches contain most of the sediments at the same time as having recently constructed check dams that are practically empty.

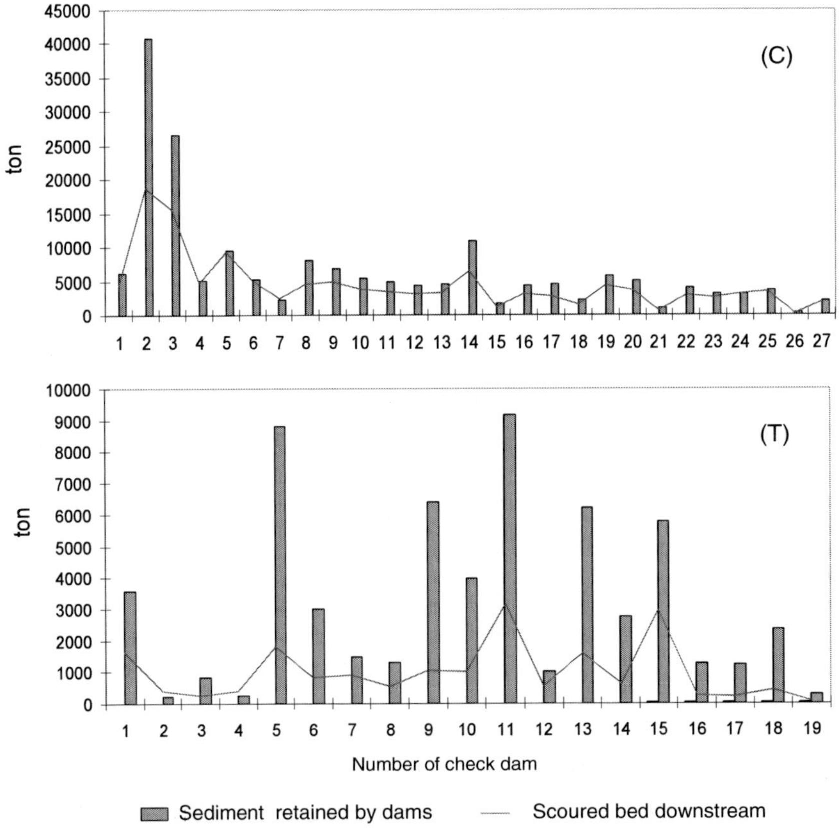

Figure 8. Comparison of sediments retained by check dams and those eroded downstream in the Cárcavo (C) and Torrecilla (T) catchments. Period 1970-2005. Please note: the reach is numbered in ascending order according to its position within the basin. The low numbers of each rambla correspond to the lower reaches and the highest to upper reaches.

Table 1. Total values and sediment retention, scouring and remotion rates calculated for the Cárcavo and Torrecilla ramblas

Rambla	Sedimentary wedges		Scouring wedges		Total remotion (channel)	
	(ton)	$t \cdot ha^{-1} \cdot year^{-1}$*	(ton)	$t \cdot ha^{-1} \cdot year^{-1}$*	(ton)	$t \cdot ha^{-1} \cdot year^{-1}$*
Cárcavo	253007.4	4.03	162924.9	2.59	415932.3	6.61
Torrecilla	85269.1	2.40	28864.6	0.81	114133.7	3.21

*Sediment remotion rates in channel reaches affected by check dams.

In the Cárcavo catchment the mean mass of excavated material downstream from the check dams (4,793.6 ton/check dam) accounts for 70.9 % of that stored upstream (6,765.7 ton/check dam). However, in the Torrecilla catchment the mean eroded material represents only a 31.2 % of the 3,146 ton/check dam which is the average mass of the sedimentary wedges. In absolute terms, the sediment mass retained by check dams in the Cárcavo catchment (253,007.4 ton) is much greater than the amount estimated for the Torrecilla catchement (85,269.1 ton), about three times greater (whilst the total drained surface is only actually twice this amount). In fact the retention rate in the Cárcavo catchment is 4.04 $t \cdot ha^{-1} \cdot year^{-1}$ compared to 2.4 $t \cdot ha^{-1} \cdot year^{-1}$ in the Torrecilla.

The total erosion induced by the check dams downstream is about 162,900 tm in the drainage network of the Cárcavo, whilst in the whole of the Torrecilla channels it is no more than 29,000 tm. Expressed as rates of sediment removal they amount to 2.59 and 0.81 $t \cdot ha^{-1} \cdot year^{-1}$, respectively. Consequently, the retention balance *versus* induced erosion due to the construction of check dams is greater in the Torrecilla catchment (characterized by metamorphic terrains), but total bed remotion turns out to be considerably higher in the Cárcavo catchment (mainly marly): 415,932 ton, which means there is a remotion rate of 6.61 $t \cdot ha^{-1} \cdot year^{-1}$, more than twice the rate of the Torrecilla basin.

4.4. Results of the Factorial Analysis

The study of the geomorphological effects of the check dams involves working with many variables. After describing separately the many variables that are affected by the hydrogeomorphological response of these gullied channels, we should make use of the advantages of factorial analysis for better defining its behavior. To carry out this procedure the number of entry variables has been reduced (10 in each basin), converting them into factors to clarify their meaning.

For each of the data samples (the Cárcavo catchment and the Torrecilla catchment) two main factors or components have been found, as well as original variables that are correlated with these. As a starting point, matrix correlations will be generated between the initial variables (Tables 2 and 3). Some variables refer to the geometry of the sedimentary wedge upstream from the check dam (A, L, W, H, Vs); other are related to bed erosion (Ve and B) and finally there is a group of variables that indicate changes in the channel gradient (Δs-up and Δs-dw) and sediment grain size variations (Δt).

Among the extracted factors the first two were selected as main components, given that these can account for 82 % of the accumulated variance in the Cárcavo rambla and 81.3 % in the Torrecilla rambla (Table 4). By adding a third factor it would only be possible to explain another 6.4 % of the original variability.

Table 2. Correlation matrix of the variables involved in the sedimentation-scouring balance for rectified channels. Reaches affected by check dams in the Cárcavo rambla

Vari-able	A	L	W	Hs	Vs	Ve	B	Δs-uo	Δs-dw	Δt
A	1.000	0.874	0.774	0.479	0.929	0.913	0.892	-0.491	-0.723	0.734
L	0.874	1.000	0.653	0.536	0.884	0.906	0.817	-0.501	-0.795	0.630
W	0.774	0.653	1.000	0.267	0.688	0.756	0.591	-0.500	-0.775	0.729
H	0.479	0.536	0.267	1.000	0.693	0.651	0.693	-0.131	-0.583	0.357
Vs	0.929	0.884	0.688	0.693	1.000	0.968	0.975	-0.510	-0.743	0.683
Ve	0.913	0.906	0.756	0.651	0.968	1.000	0.888	-0.466	-0.796	0.746
B	0.892	0.817	0.591	0.693	0.975	0.888	1.000	-0.521	-0.655	0.591
Δs-ar	-0491	-0.501	-0.500	-0.131	-0.510	-0.466	-0.521	1.000	0.510	-0.431
Δs-ab	-0.723	-0.795	-0.775	-0583	-0743	-0.796	-0.655	0.510	1.000	-0.606
Δt	0.734	0.630	0.729	0.357	0.683	0.746	0.591	-0.431	-0.606	1.000

Geometric variables of the sedimentary wedges (area -A-, length -L-, mean width -W-, máximum thickness -H-, volume -Vs-), another volume variables (volume of the scour wedge -Ve-, bed-scouring-sedimentation balance -B-), variables related to changes in the bed slope (decrease in the gradient above check dams, Δs-up; decrease in the gradient below check dams, Δs-dw), and variations in the bed particle sizes between upstream and downstream reaches ($D_{84}dw - D_{84}up$).

Table 3. Correlation matrix of the variables involved in the sedimentation-scouring balance for rectified channels. Reaches affected by check dams in the Torrecilla rambla

Vari-able	A	L	W	H	Vs	Ve	B	Δs-up	Δs-dw	Δt
A	1.000	0.882	0.571	0.286	0.873	0.858	0.802	0.876	0.579	0.359
L	0.882	1.000	0.449	0.456	0.896	0.771	0.868	0.817	0.538	0.077
W	0.571	0.449	1.000	0.156	0.541	0.667	0.441	0.545	0.439	0.285
H	0.286	0.456	0.156	1.000	0.573	0.334	0.621	0.505	0.679	-0.213
Vs	0.873	0.896	0.541	0.573	1.000	0.845	0.975	0.924	0.791	0.148
Ve	0.858	0.771	0.667	0.334	0.845	1.000	0.707	0.787	0.578	0.465
B	0.802	0.868	0.441	0.621	0.975	0.707	1.000	0.898	0.809	0.004
Δs-ar	0.876	0.817	0.545	0.505	0.924	0.787	0.898	1.000	0.795	0.238
Δs-ab	0.579	0.538	0.439	0.679	0.791	0.578	0.809	0.795	1.000	0.001
Δt	0.359	0.077	0.285	-0.213	0.148	0.465	0.004	0.238	0.001	1.000

Geometric variables of the sedimentary wedges (area -A-, length -L-, mean width -W-, máximum thickness -H-, volume -Vs-), another volume variables (volume of the scour wedge -Ve-, bed-scouring-sedimentation balance -B-), variables related to changes in the bed slope (decrease in the gradient above check dams, Δs-up; decrease in the gradient below check dams, Δs-dw), and variations in the bed particle sizes between upstream and downstream reaches ($D_{84}dw - D_{84}up$).

Figure 9 shows the auto values related to the components for the two ramblas analyzed. In the case of the Cárcavo there is only one component (the first main component) with a value of more than 1 (auto value near to 7) whilst the second component adopts a value of 1, the rest of the auto values are very small compared to the first two. For the reaches of the Torrecilla the first two components were greater than auto value 1, to such an extent that the rest were below 0.65.

Table 4. Total variance explained by the components extracted and rotated in the Cárcavo and Torrecilla ramblas

C	Cárcavo rambla				Torrecilla rambla			
	(a)		(b)		(a)		(b)	
	% of variance	% accum.	% of variance	% accum.	% of variance	% accum.	% of variance	% accum.
1	*71.65*	*71.65*	*41.17*	*41.17*	*65.58*	*65.58*	*53.77*	*53.77*
2	*10.31*	*81.95*	*40.78*	*81.95*	*15.74*	*81.31*	*27.55*	*81.31*
3	6.33	88.28			6.42	87.73		
4	4.61	92.89			5.81	93.54		
5	3.56	96.45			3.22	96.75		
6	2.18	98.63			1.37	98.13		
7	0.73	99.36			0.92	99.05		
8	0.39	99.75			0.60	99.65		
9	0.25	100.00			0.35	100.00		
10	0.00	100.00			0.00	100.00		

C: Component. (a) Sum of squared saturations of extraction. (b) Sum of the squared saturations of rotation.

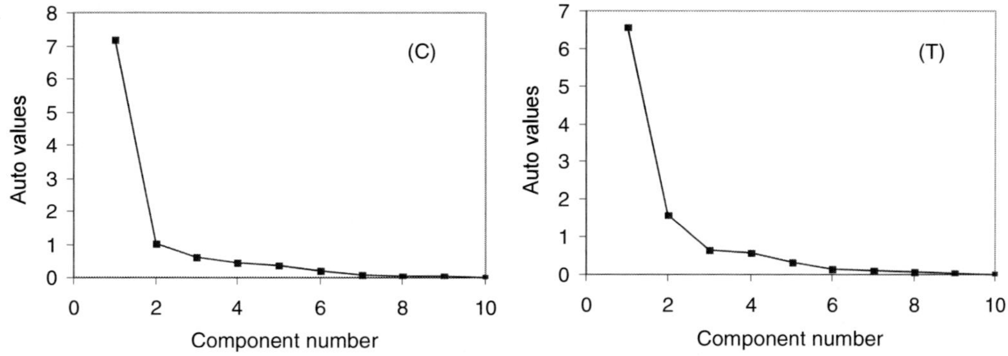

Figure 9. Sedimentation graphs for the components obtained with samples from the Cárcavo (C) and Torrecilla (T) catchments.

Finally, the correlations are shown between each main component and each value (Table 5). In the case of the Cárcavo, it was seen that CP1 is more positively correlated with: (1) the filling height of the check dams (Hs); (2) the volumes of the sedimentation wedges (Vs) and erosion (Ve); (3) and with the sedimentation-erosion balance (B). CP1 also has a certain relationship with length (L) and area (A) of the sedimentary wedges. The association between this type of variables supports two facts: i) the effects of bed scouring downstream from the check dams are consistent with the quantity of sediment retained upstream and the erosion–sedimentation balance inside the channel is a reliable reflection of this; ii) the volume of the sedimentary wedges seems to be defined more by its length and surface area than by its width. In fact, variations in the width of the wedge are hardly correlated with volumetric ones. To conclude, this component groups together a lot of reaches that have a similar geomorphological response downstream of the dams. This response is coherent with the filling process upstream and, therefore, with the geometric features that define it, which include sections of varying width due to geological constrictions. Generally, these reaches are more noted for the impact of the check dams on erosion control than for their effects on the slope variations and the bed texture.

Table 5. Matrix of components rotated for each of the variables. Rotation method: Varimax normalization with Kaiser

Rambla del Cárcavo			Rambla de la Torrecilla		
Variable	Componente		Variable	Componente	
	1	2		1	2
H	0,931	-0,030	B	0,938	0,246
Vs	0,801	0,568	Vs	0,896	0,412
B	0,799	0,493	Δs-ab	0,862	0,106
Ve	0,755	0,619	Δs-ar	0,829	0,466
L	0,674	0,618	H	0,806	-0,270
W	0,293	0,846	L	0,792	0,406
Δs-ar	-0,051	-0,776	A	0,668	0,667
Δt	0,370	0,727	Δt	-0,210	0,833
A	0,631	0,700	Ve	0,601	0,727
Δs-ab	-0,568	-0,640	W	0,368	0,625

The second main component (CP2) is very highly correlated with variables W (width of the sedimentary wedge) and Δt (bed texture changes), and it also has a notable negative correlation with the variables that define the effects of the check dams on the bed slope (Δs-up) (Δs-dw). Therefore, unlike in the previous case, CP2 could be associated with those reaches that in spite of having suffered notorious textural change in their beds, have not experienced a significant reduction in the slope, especially upstream from the check dams (for Δs-up, r = -0.78). Several reaches of great stability belong to this component characterized by wide sedimentary wedges, with limited control over the longitudinal slope (especially upstream) and a radical change of texture, evidenced by a sudden increase in the roughness of the bed immediately downstream from the aforementioned structures. The lower reaches of the main channel of the Cárcavo rambla are probably included in this subgroup. With regard to the reaches studied in the Torrecilla rambla the matrix of rotated components allows us to also establish several categories of differentiated responses (Table 5).

In view of the solution offered by SPSS, it is possible to confirm that the results offered by the APC are also significant in this case, given that the two selected dimensions make it possible to discriminate between enough groups of variables in order to be able to explain the main types of geomorphological response. With regard to the interpretation of the components, it has been seen that CP1 has a very positive correlation with variables B, Vs, H, L, Δs-up and Δs-dw, having factorial coefficients of more than 0.8. It has thus been proven, using this component, that there are a large number of reaches (65.6 % of the total explained variance) with sedimentary wedges and scouring-sedimentation balances that are closely related to the readjustment of the bed slope upstream and downstream from the check dams. This fact differentiates them from modified lower reaches of the Cárcavo rambla, in which the changes in the gradient seem to have a greater link with the grain size variations in the bed (CP2). With regard to the geometric features of the sedimentary wedges, the factorial coefficients of L (length) and H (maximum filling height) are greater than those estimated for the mean width (W) and the surface area (A), suggesting the presence of geostructural controls in these cases and the alternation of V-shaped and trapezium shaped transverse sections. This is however, a logical situation if we take into account the level of fitting in and constriction presented by the gullied channels of this catchment.

Contrary to CP1, CP2 is most positively correlated with the Δt variable (texture difference) and Ve (wedge erosion volume). Its correlation with the measurements of mean width (W) and surface area (A) of sedimentary wedges is also worth noting. This component is associated with reaches in which the scoured volume downstream from the check dams has exhumed the coarser material that was deposited in the previous bed, revealing greater textural contrast compared to the retained sediment. These are reaches in a maximum perturbation phase of the bed in which there has still not been any refilling of the check dam. However, this situation is usually linked to wide wedges with a relatively wide surface and variable length, although this is generally less common than in the previous group.

Graphs of the dimensions of CP1 and CP2 (Figure 10) show the position of the variables compared to the rotated factorial axes. It can be seen that all of these are well represented on the plan, because they are all near the edge of the circle with radius = 1, and none of them are near to the origin. Most of the factorial coefficients obtained in both samples (Cárcavo and Torrecilla) fall into the upper right quadrant, where they represent central values and high positives for the first component and also positive values for the second component. However, there are certain differences regarding the position of the first factor variables related to the abscisas axis (first factorial axis), and the second factor variables with regard to the ordinate axis (second factorial axis).

In the case of the Cárcavo rambla there is a great concentration of variables related to the "control of channel erosion", which reflects a special link between magnitude of both types of wedge (sedimentation and surface washing). The elevated positive values of both factorial coefficients (Vs and Ve), and their proximity in the graph to the balance between these variables, indicate a significant remotion of materials on the bed. The variable H (height of the sediments retained by check dams) is next to the abscisas axis of the first component with a very high coefficient and is some distance away from the rest of the elements that form this subgroup, giving it a more independent character. This is the variable most directly related to the height of the check dams: here there is a clear component of structural influence. In the negative lower quadrant, but separate from the coordinates of origin, we can find the coefficients represented by the variables relative to bed slope reduction.

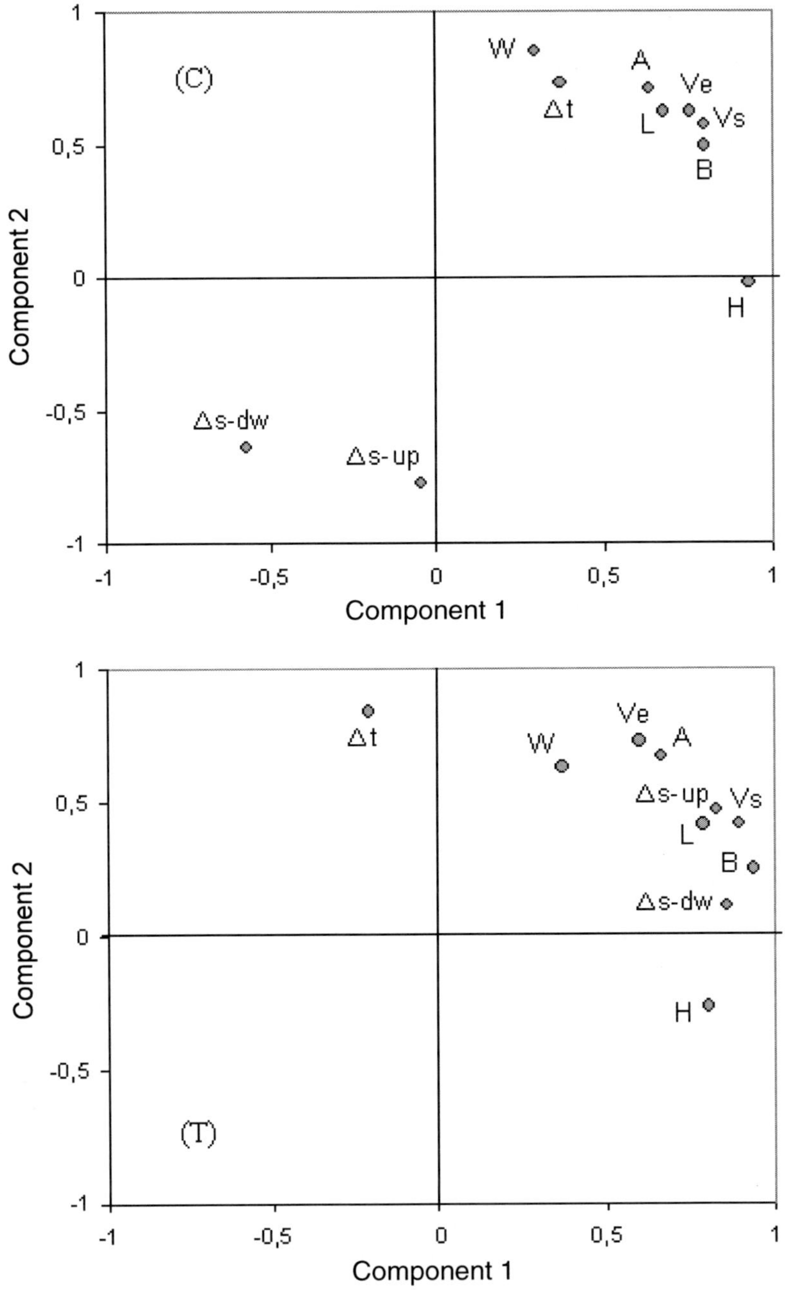

Figure 10. Component Plot in Rotated Space for the Cárcavo (C) and Torrecilla (T) catchments.

The most significant correlation within this group is variable Δs-up, located in a position that is closer to the abscisas axis (figure 10). If we compare the size variables of the sedimentary wedges with the position that they occupy within the opposite quadrant, particularly A and W, it is easy to see that in these types of reaches, the check dams base their greatest geomorphological efficiency on the capacity to retain sediments, although this does not necessarily imply a substantial reduction in bed slope upstream.

The relative position of the factorial coefficients in the graph of rotated components of the Torrecilla catchment highlight the effect of the check dams on the volume of retained sediment and, depending on this, on the bed gradient, which is reduced as the trapped sediment disminishes. However, bed texture has a different behavior, distinguishing it from the rest of the variables (upper left quadrant, second main component). This position supports the fact already described that the influence of the check dams on bed particle size is not really related to the other correlated factors. In any case, the increase in D_{84} downstream from the check dam, compared to the reach upstream, could be related to incision volume downstream (Ve), given that the surface wash leaves coarser materials bare at this position (*static bed armouring*), which contrast with the fine material deposited upstream.

Such considerations, together with the coarser texture of the detritic materials retained by the check dams of the Torrecilla rambla, are a clear indicator of the different type of efficiency exercised between one case and the other. Effectively, the check dams of the Cárcavo rambla have proven to be more efficient as sediment collectors and controllers of erosion, providing greater bed stability and consolidation of talus and adjacent hillslopes. On the other hand, the check dams of the Torrecilla rambla seem to have carried out, and are still carrying out, a more efficient function of bed stabilization, flood lamination and flood wave speed reduction, and, above all, they can be utilized for recharging aquifers. The following reservations should be made however: the check dams located in gullied areas, such as that of the Cárcavo catchment, are very much affected by the direct contribution of rills and landslides, so that refilling is relatively fast. As soon as the check dams are refilled they lose their ability to retain sediment, but they are able to reduce the longitudinal bed gradient and with this the erosive capacity of the streamflow. The immediate upstream effect is the stabilization of the bed and the partial consolidation of unstable hillslopes (i.e. badlands). As a parallel, downstream there are significant scour wedges that have a volume not very different from the volume of sediment retained upstream, and that, ultimately, contribute to guaranteeing the reduction in the bed slope along the affected reach. However, the effects of check dams constructed in metamorphic catchments, arranged around several main channels, such as that of the Torrecilla, are very uneven, depending on their state of filling and the reestablishment of the bed-load. The filled check dams soon replace the transport of coarse sediments downstream, refilling the scour pools and stopping the incision processes. With the replacement of bed load the rhythm of local erosion decreases and moves further away from the theoretically established level for maximum equilibrium erosion.

Conclusion

The comparison between the sediments retained upstream (Up) of check dams and the eroded downstream (Dw) shows marked differences between the two catchments under study, associated with its environmental characteristics, especially lithology and slopes. Scouring-sedimentation balance caused by check dams is greater in the Cárcavo catchment, where an important volume of bed fine material is removed. This is due to a high soil erodibility, scanty influence on the hillslope stability, low reduction in bedload transport rate and strong local bed scour. Torrecilla catchment is composed of metamorphic materials, mainly slates, phyllites and schists, that provide a large quantity of gravels and pebbles to the channels, giving major stability to the beds. In the Cárcavo catchment the average mass of excavated material downstream from check dams (4,794 t/check dam) represents 70.9 % of that stored upstream (6,765.7 t/check dam). On the contrary, in the Torrecilla catchment such a proportion is reduced to 31.2 % of the 3,146 t/check dam which makes up the average mass of the sedimentary wedges. The absolute balance of retention versus bed scour deduced from the construction of check dams is considerably greater in the marly Cárcavo catchment. In addition, the total valuation of removal caused by check dams in the Cárcavo (6.61 $t \cdot ha^{-1} \cdot year^{-1}$) is more than double that which is reached in the Torrecilla catchment.

In the case of the Cárcavo, most retention, scouring, and total removal of sediments is concentrated in the lower reach of the main channel, as it was expected; but it did not happen in the Torrecilla catchment, distinguished by a drainage network palmated with three main arteries (streams of Navazo, Cocón and Torrecilla s.s.), whose lower and middle reaches hoard the main proportion of sediments at the same time as having recently built check dams which are practically empty. Such considerations, together with the coarser texture of detritical materials retained by the Torrecilla check dams are a clear indication of the different type of efficiency affecting each case. Effectively, the check dams of the Cárcavo have proved to be more efficient as sediment collectors and erosion controllers, providing greater stability to the bed and consolidating the adjacent talus and hillslopes. However, the check dams of the Torrecilla rambla seem to have carried out a more efficent function for controlling floods and refilling the aquifers. The following reservations should be made: the check dams located in gullied areas, such as those of the Cárcavo, are very affected by direct contributions from the badlands and therefore they are filled relatively quickly. As soon as the check dams are full they lose their ability to retain sediment, but they are able to reduce the longitudinal be gradient and the consequent erosive capacity of flow. The immediate upstream effect is the stabilization of the bed and partial consolidation of the gullied hillslopes. At the same time, downstream there are considerable erosion wedges with a volume not unlike the volume of sediment retained upstream, and that, finally, contributes to decreasing the bed slope in the affected reach. On the other hand, the effects of check dams constructed in metamorphic catchments, arranged around main channels, such as that of the Torrecilla, are very uneven, depending on the state of the filling and the reestablishment of the bed load. Filled check dams quickly replace the transport of coarse material downstream, filling the scour pools and stopping the incision process. With the reposition of the bed-load the rhythm of local erosion decreases and moves away from the theoretically established limit for maximum equilibrium erosion. The scouring-sedimentation balance caused by check dams is greater in the Cárcavo marly catchment, where an important volume of bed material is removed. This is due to a high soil erodibility, scanty influence on the hillslope stability, low reduction in the bed-load rate and a strong local bed scour. However, the combined action of the check dams with gravel beds in the Torrecilla rambla, fed by large amounts of metamorphic material, has provided greater morphological stability to the system.

ACKNOWLEDGMENTS

This study was carried out within the research project PI/13 framework, "Hydrological and geomorphological response to the fluvial torrential systems affected by forestry-hydrological restoration works in semiarid catchments in South east of Spain", financed by the *Fundación Instituto Euromediterráneo de Hidrotecnia,* European Council, and *Comunidad Autónoma de la Región de Murcia.*

REFERENCES

Brandt, S.A. 2000. Classification of geomorphological effects downstream of dams. *Catena*, 40, pp. 375-401.

CMA de la Junta de Andalucía (2005): Estadística *de obras de corrección hidrológica en Andalucía. http://www.juntadeandalucia.es/medioambiente/red_ambiental/proyectos/obrasde-correcionhidrologicaenandalucia.pdf.*

Comiti F., Andreoli, A. and Lenzi, M.A. 2005. Morphological effects of local scouring in step-pool streams, *Earth Surf. Process. Landforms,* 30(12), 1567-1581.

Conesa García, C., (2005): "Les 'ramblas' du Sud-est Espagnol: Systèmes hydromorphologiques en milieu méditerranéen sec". *Zeitschrift für Geomorphologie*, N° 49, 2. Gebrüder Borntraeger Berlin Stuttgart, pp. 205-224.

Conesa García, C. and García Lorenzo, R. (2007). *Erosión y diques de retención en la Cuenca Mediterránea. Efectividad hidrogeomorfológica de los diques de retención en cuencas torrenciales del Sureste Español.* Fundación Instituto Euromediterráneo del Agua, Murcia, 669 p.

Conesa-García, C.; García Lorenzo, R. and López Bermúdez, F. (2007). "Bed stability variations after check dam construction in torrential channels (South-East Spain)". *Earth, Processes and Land Forms*, 32, 2165-2184. Published online in Wiley InterScience (www.interscience.wiley.-com), DOI: 10.1002/esp.1521.

Del Palacio *et al.* (1999). La restauración hidrológico-forestal en España. Gestión sostenible de los recursos suelo, agua y vegetación. Ministerio de Medio Ambiente. Madrid.

Dirección General de Conservación de la Naturaleza, (2002): Plan Forestal Español. Ministerio de Medio Ambiente. [access on August 8, 2007] *(http://www.mma.es/conserv-nat/planes-/planifor/pfeindex.htm)*

European Environment Agency (2000): *Environmental signals 2000. European Environment Agency regular indicator report.* Environmental assessment report No 6. EEA, Copenhagen, *http://www.eea.eu.int.*

García Ruiz, J.M. y Puigdefábregas, J. (1985): Efectos de la construcción de pequeñas presas en cauces anastomosados del Pirineo Central. *Cuadernos de Investigación Geográfica*, Publicaciones del Colegio Universitario de La Rioja, Logroño, T. XI, fasc. 1 y 2, pp. 91-102.

Kaiser, H.F. (1958). The Varimax criterion for analytic rotation in factor analysis. *Psychometrika*, 23, 187-200.

Kim, J. and Mueller, C.W. (1978). *An introduction to factor analysis: What it is and how to do it.* Beverly Hills, CA: Sage.

Lenzi, M.A.; Marion, A. and Comiti, F. 2003. Local scouring at grade-control structures in alluvial mountain rivers, *Water Resources Research* 39(7), 1176-1188.

Marion, A.; Tregnaghi, M. and Tait, S. 2006. Sediment supply and local scouring at bed sills in high-gradient streams, *Water Resour. Res.*, 42.

Martínez Castroviejo, R.; Inbar, M.; Gómez Villar, A. y García Ruiz, J.M. (1990): Cambios en el cauce aguas abajo de una presa de retención de sedimentos. *I Reunión Nacional de Geomorfología.* Vol. 2. SEG, Teruel, pp. 457-468.

Martínez de Azagra, A.; Navarro Hevia, J. (1996): *Hidrología forestal. El ciclo hidrológico.* Ed. Universidad de Valladolid. 286 pp.

Ministerio de Agricultura, 1972. *Proyecto de Corrección Hidrológico-Forestal de la cuenca del Cárcavo.* Dirección General de Montes, Caza y Pesca Fluvial. Memoria técnica.

Secretaría de Estado de Aguas y Costas (2003): Resolución de 9 de abril de 2003, por la que se dispone la publicación del *Protocolo General de Colaboración* entre el Ministerio de Medio Ambiente, la Comunidad Autónoma del Principado de Asturias y la Confederación Hidrográfica del Norte para *el desarrollo de actuaciones del Plan Hidrológico-Forestal y de restauración de cauces y enclaves naturales.*

In: Check Dams, Morphological Adjustments… ISBN: 978-1-62618-856-3
Editors: C. C. Garcia and M. A. Lenzi

Chapter 14

UPSTREAM AND DOWNSTREAM EFFECTS OF CHECK DAMS IN BRAIDED RIVERS, CENTRAL PYRENEES

***José M. García-Ruiz*[*,1], *Luis C. Alatorre*[1], *Amelia Gómez-Villar*[2] and *Santiago Beguería*[3]**

[1]Pyrenean Institute of Ecology, CSIC, Campus de Aula Dei, Apartado 13.034, 50080-Zaragoza, Spain

[2]Dept. of Geography and Geology, Campus de Vegazana, University of León, 24071-León, Spain

[3] Aula Dei Experimental Station, CSIC, Campus de Aula Dei, Apartado 13.034, 50080-Zaragoza, Spain

ABSTRACT

Check dams constructed in braided rivers in the Flysch Sector of the Central Spanish Pyrenees were studied in order to assess the main geomorphic consequences upstream and downstream the dams. A sudden decrease of sediment transport downstream the check dams resulted in the development of an incised and confined channel with a meandering pattern, leaving the old alluvial plain perched and inducing plant colonization.

In the case of alluvial fans, the entrenchment of the ravine was the most outstanding consequence, favouring sediment transport to the distal area of the fan or even directly to the main river. Upstream the check dam, sedimentation was well organized depending on the distance from the dam, the gradient and the episodic arrival of debris flows during large floods.

Keywords: Fluvial adjustments, check dams, incised channels, channel stabilization, Flysch, Central Pyrenees.

[*] humberto@ipe.csic.es and lalatorre@ipe.csic.es

1. Introduction

Braided rivers are a common morphology in mountain areas, due to the high capacity of the hillslopes to transfer coarse sediment toward the valley bottoms. As many papers have stressed, braided rivers are characterized by the presence of several channels changing frequently in position, a low sinuosity, a high width to depth ratio (over 40), a relatively high longitudinal gradient, and a predominance of bedload material covering the alluvial plain (Leopold and Wolman, 1957).

In general, braided rivers are a consequence of particular environmental conditions in their basins. These conditions were identified by Leopold *et al.* (1964), Schumm (1977, 2005), Richards (1982) and Leopold (1994) as:

- (i) An intense geomorphological activity in the hillslopes (active headwaters, occurrence of debris flows and landslides, scarce plant cover in some sectors of the basin, etc.), allowing for high soil erosion rates and the arrival of large volumes of sediment to the alluvial plain.
- (ii) Little cohesive river banks, in such a manner that channel wandering can increase the availability of bedload.
- (iii) Strong variability in the discharge. Steep slopes and a high drainage density result in a fast hydrological response and the corresponding capacity to carry coarse sediments.
- (iv) Existence of strong contrasts among high and low energy stretches, resulting in sudden decreases in the sediment carrying capacity.

In the case of the Central Spanish Pyrenees, the Flysch Sector is mostly drained by braided rivers, with relatively wide alluvial plains infilled with coarse sediment in which several channels change in position during flood events. These braided channels had historically problems of lateral erosion and sediment transport to the main rivers (Aragón, Gállego, Ara and Cinca), thus contributing to their remarkable instability (including the development of large bars) and shortening the useful life of reservoirs (mainly the Yesa and Mediano reservoirs) and infrastructure such as bridges and roads. A number of check dams were constructed in the most active and relevant tributaries in order to reduce coarse sediment transport and to stabilize the channels.

Some of the geomorphological consequences of check dams were studied more than twenty years ago (García-Ruiz and Puigdefábregas, 1985; Martínez-Castroviejo *et al.*, 1990). Since then, several works were focused on fluvial morphology and sediment transport in relation to the construction of checkdams, both in mountain areas (i.e., Arnáez-Vadillo and Gómez-Villar, 1990) and in semi-arid environments (Castillo *et al.*, 2007; Conesa-García and García-Lorenzo, 2007; Romero-Díaz, 2007).

This paper studies the upstream and downstream effects of check dams in tributaries of the Aragón and Gállego rivers and informs about their role at the catchment scale.

2. The Study Area

The study was carried out in the Ijuez and Estarrún rivers (tributaries of the Aragón River), the Aurín River and the Escuer and Arás ravines (tributaries of the Gállego River), all of them draining the Flysch Sector of the Central Spanish Pyrenees (Figure 1).

The Flysch Sector is one of the most important structural units of the Spanish Pyrenees, between the Inner Sierras and the Inner Depression. It is composed of alternating thin beds of sandstones and marls, intensively folded and faulted by the Alpine tectonics. Some coarser limestone layers are integrated within the flysch, helping to understand the general organization of the structure. The relief is characterized by a series of smooth divides and straight hillslopes, ranging from 700 to 1200 m a.s.l. The divides represent different old erosion levels sloping towards the south. Three geomorphic belts can be identified in the hillslopes (García-Ruiz and Puigdefábregas, 1982):

(i) The sub-alpine environment (over 1600-1800 m a.s.l.), affected by active torrential headwaters, solifluction, shallow landslides and dense networks of parallel rills.
(ii) The forest environment, dominated by infiltration processes, moderate hydrological responses (Serrano-Muela *et al.*, 2008) and, locally, the activity of soil creeping.
(iii) The sub-Mediterranean environment, intensively affected by human activities (abandoned fields and scrubs, particularly in the south facing slopes), with the development of incised gullies, deep landslides and debris flows (Lorente *et al.*, 2002), all of them well connected to the fluvial network.

The climate is sub-mediterranean, somewhat continental. Average annual precipitation ranges from 700 mm (in the southern, lowest part of the study area) to about 2000 mm in the highest divides. The rainy season runs from November to May, with a short, dry season in summer. Local, intense rainstorms are relatively frequent in the study area, mainly from August to December (García-Ruiz *et al.*, 2000).

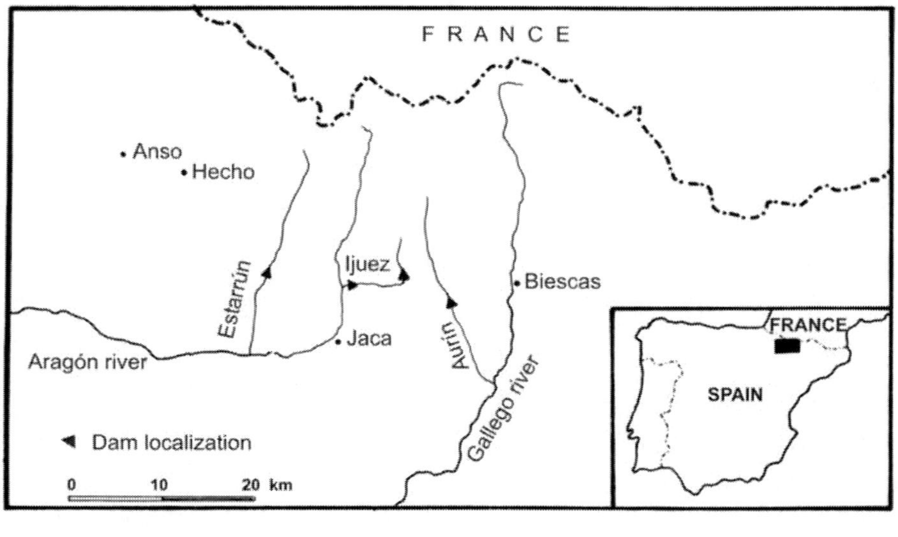

Figure 1. The study area.

Plant cover has been strongly impacted by human activity. Historically, cultivated areas were located below 1600 m a.s.l. in the valley bottoms, perched flats, and on the steep, south-facing hillslopes, which were cultivated even under shifting crop systems. The forests have remained relatively well preserved on the north facing slopes and everywhere between 1600 and 1800 m on the south facing slopes. The upper divides are covered by grasslands of the sub-alpine belt.

Most cultivated fields were abandoned during the 20th century, except in the valley bottoms. Abandoned fields, representing about 25% of the total area, have been affected by a natural process of plant recolonization (mainly by dense shrub communities, but also woodland in the most favourable locations) or have been reforested for land reclamation and to reduce reservoir silting (Ortigosa *et al.*, 1990). Moreover, the use of fire for land management has been abandoned in the last decades, favouring plant recolonization. As a consequence, a decrease in the river discharge and in the frequency and magnitude of floods and the volume of sediment transport has been reported (Beguería *et al.*, 2003; López-Moreno *et al.*, 2006. These hydrological changes resulted in fluvial adjustments that represent the colonization of most of the sedimentary structures and the narrowing of the active sectors of the alluvial plains (Beguería *et al.*, 2006).

3. Methods

Three braided rivers with check dams were selected (Figure 1): Estarrún, Ijuez and Aurín rivers. Sinuosity, width and depth of the channels were measured downstream the check dam. A tape was located along the channel upstream the check dam in order to measure the size of the sediments each half a meter. A check dam constructed in 1966 in the Aurín River was studied in detail, above all the temporal evolution of the channel downstream the check dam before (1956) and after (1978 and 1988) its construction. Besides, the gravel size in the river banks was measured in 12 points along 1,100 m stretch of a channel. In each of the sample points 100 gravel particles were selected following the Wolman's (1954) method.

Finally, the characteristics of the entrenchment in alluvial fans near Biescas (Gállego valley) were also studied, as well as the consequences of a catastrophic hydrological event on the behaviour of almost 40 check dams in the Arás.

4. Results

Table 1 shows the characteristics of the check dams studied, as well as their upstream and downstream influences. The check dams were constructed between 1958 and 1966, their height ranging between 6 and 7.40 m. Two of the check dams were infilled by sediment already in 1985.

Upstream, the construction of the check dams represents a sudden, local change in the longitudinal profile of the river, with a clear decrease of the gradient. The immediate consequence is the settlement of a large part of the sediment load, mainly the coarser sediment. Figures 2 and 3 show the frequency distribution of the grain size upstream the check dams. Gravels of less than 10 cm prevail in the proximities of the dams, and their proportion decreases progressively upstream, though such a trend is not linear. Besides, as the sampling points are located more and more away upstream the dam, there is a higher trend to heterometry, with a greater importance of gravels between 10- and 19 cm and an increasing frequency of boulders between 30 and 39 cm and even more, particularly in one of the check dams of the Ijuez River.

Table 1. Characteristics of the check dams studied

Dams	Construc-tion	Height dam (m)	Height full dam (m)	Sediment level (m)	(1)	(2)	(3)	(4)	(5)	(6)
Estarrún	1966	7.40	6.30	4.50	120	410	435	1.08	35.50	2.20
Ijuez (low)	1958	6.65	6.00	4.65	165	515	4600	1.09	15.00	2.00
Ijuez (middle)	1966	6.50	5.75	5.75	20	195	1750	1.02	13.00	1.20
Aurín	1966	6.00	4.80	4.80	77	600	480	1.16	15.00	1.00

1. Maximum reach of fine sediment upstream of the dam (m).
2. Maximum reach if the dam influence, upstream (m).
3. Maximum reach if the dam influence, downstream (m).
4. Sinuosity of the channel in the section influenced by the dam (downstream).
5. Width of the channel 60 m downstream of the dam (m).
6. Depth of the channel 60 m downstream of the dam (m).

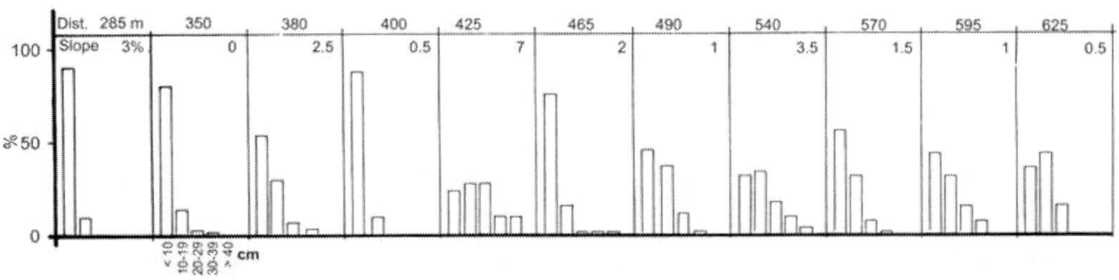

Figure 2. Frequency distribution of the gravel size upstream of the lower check dam in the Ijuez River. Source: García-Ruiz and Puigdefábregas (1985).

This spatial organization of sediment can be considered as "expected", because the most immediate effect of the dam is to simplify the sediment distribution, in such a manner that the coarser gravels sediment first, and then the smaller ones, reducing progressively the heterometry close to the dam (Figure 4). This can be observed in all of the check dams analyzed, particularly in that of the Aurín River.

The existence of steps that break the linear decreasing trend towards the dam is explained by the local gradient. In fact, the proportion of small gravels decreases in the steeper stretches, substituted by coarser blocks. In order to evaluate the relative importance of the distance to the dam and the gradient, a step by step multiple regression analysis was performed, using both variables as predictors, and the percentage of gravels smaller than 10 cm as dependent variable. Table 2 shows that the distance to the dam is not the only factor in explaining sediment sorting upstream the dam, and that the role of the gradient increases with the distance from the dam.

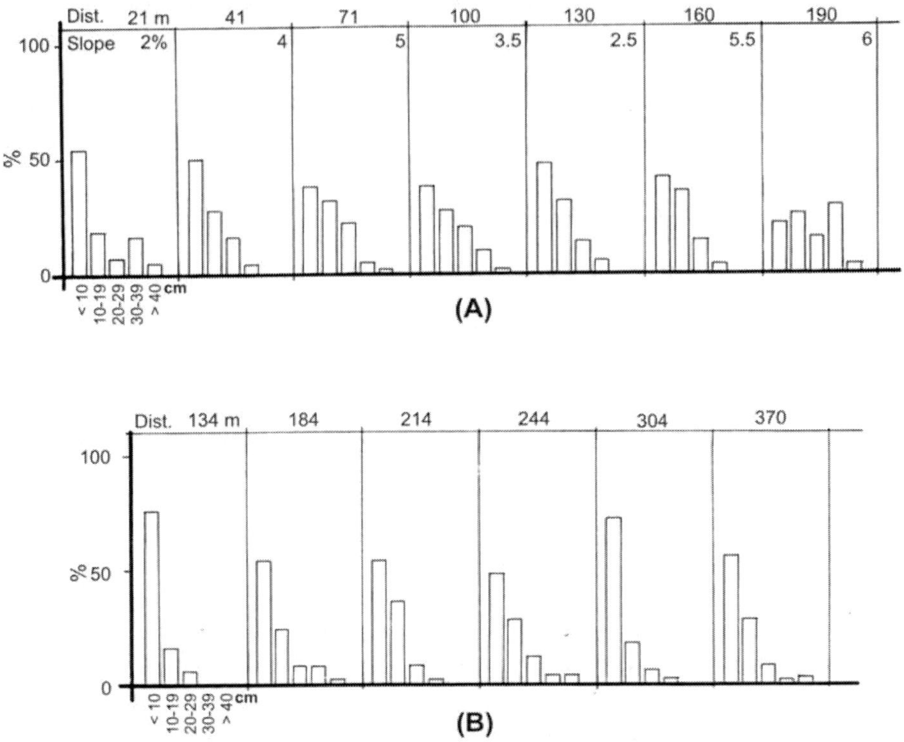

Figure 3. Frequency distribution of the gravel size: (A) upstream of the upper check dam in the Ijuez River; and (B): upstream of the check dam in the Estarrún river. Source: García-Ruiz and Puigdefábregas (1985).

Figure 4. Check dam in the Aurín River, showing the presence of massive fine sediments immediately upstream the dam.

Table 2. Stepwise multiple regression

Location	Predictors	Explained variance
Ijuez (low)	1° x_1: distance to the dam 2° x_2: slope (%) $\psi = -7.152\xi_1 - 0.16\xi_2 + 147.3$	0.403* 0.360** 0.763*
Ijuez (middle)	1° x_1: distance to the dam 2° x_2: slope (%) $\psi = -0.071\xi_1 - 3.842\xi_2 = 65.02$	0.605* 0.100 0.705

* Signification level = 0.01.
** Signification level = 0.05.

A simplified, ideal pattern of sediment structures upstream a dam is illustrated in Figure 5. In the upstream extreme the channel shows the typical aspect of a braided river, with several winding channels separated by longitudinal bars with coarser materials. When the river reaches the area of influence of the check dam, the channels almost disappear, and the sediment becomes finer. At the same time, partially-buried old bars with coarser sediment, appear between the sediments of the dam. During extreme events, like debris flows or hyperconcentrated flows, some coarse sediment can penetrate the area of influence of the check dam (Figure 6). This coarse sediment obligates to increase locally the gradient, disturbing the linear pattern in the grain size organization. Close to the dam the gravels, the more and more smaller, are organized as small lobes between a matrix of fine sediments. In consequence, the burying of old bars, the absence of new, well-developed gravel bars, the organization of sediment according to grain sizes are the main consequences upstream the check dams.

All the bed load and a part of the suspended sediment load is retained immediately upstream the dams. For this reason, there is an increase in the power carrying capacity of the stream downstream the dam, which can be used to erode the channel itself and the river banks, affecting occasionally bridges and roads. Clear processes of degradation were observed in the four check dams studied. A new, narrow channel was opened, incising the braided alluvial plain (Figure 7). In the fist one hundred meters downstream the dams all the small and medium size gravels (lower than 40 cm) were removed, resulting in an armoured structure of large blocks and boulders (Figure 8). The width of the new channel oscillated between 13 and 35.5 m, limited by abrupt banks, suggesting a recent process of incision.

Other important aspect is that the pattern of the channel also changed. The braided pattern was substituted by a straight, unique channel, with an incipient trend toward meandering. Sinuosity was still low, though meanders were developing. The gradient of the channel was slightly lower than the previous one, though the classical riffle-pool sequence was still evident. The downstream effects of check dams reached variable distance. In the check dams of the Estarrún and Aurín rivers (as well as in one of the Ijuez River) the influence reached a distance close to 500 m. Further away, the channel was not confined anymore and appeared subdivided into several channels, adopting the original braided pattern. In the other check dam of the Ijuez River the influence was lost only 180 m downstream the dam, due probably to the steeper gradient of the channel stretch.

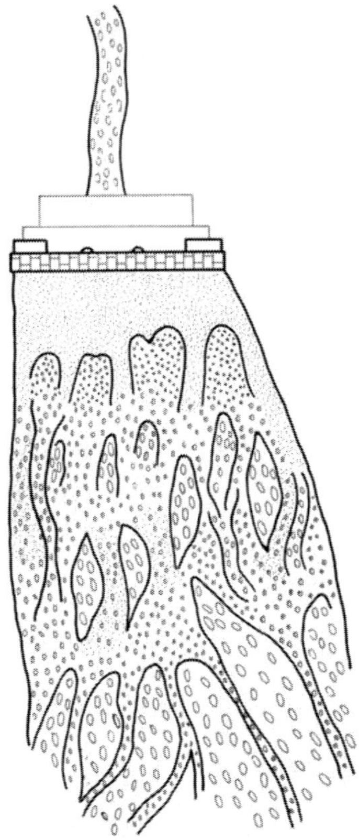

Figure 5. Idealized spatial organization of sediments upstream of a check dam in a braided river. Source: García-Ruiz and Puigdefábregas (1985).

Figure 6. Debris flows in the middle Ijuez River check dam, where coarse sediment overlaps finer sediment.

Figure 7. The Estarrún River downstream a check dam. The river is incising the alluvial plain, and the vegetation colonized progressively the coarse sediments.

Figure 8. Incision and armour layer downstream the lower check dam in the Ijuez River.

A good example of channel evolution after the construction of a checkdam is shown in Figure 9, representing the main characteristics of the Aurín River channel downstream the dam. In 1956 a typical braided morphology prevailed, with a wide alluvial plain where several channels wandered. Most of the alluvial plain was composed of recent, coarse and heterometric sediments. Plant cover was in an incipient stage, dominated by annual herbaceous communities, a sign of immaturity and instability of the system. Some older deposits (remnants of non-frequently flooded bars) had a denser woody vegetation, and a small lateral terrace was occupied by cultivated meadows.

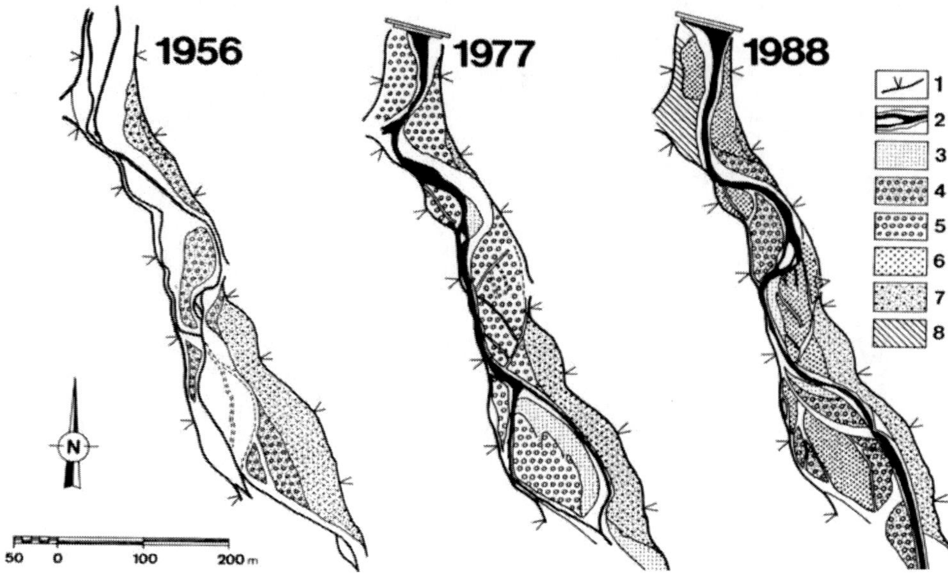

Figure 9. Channel morphology before and after the construction of a check dam in the Aurín River. 1: Limits of the original alluvial plain. 2: Main river channel. 3: Recent deposits without plant cover. 4: Recent deposits with woody, open vegetation, 5: Bars with woody, relatively dense vegetation. 6: Bars with dense shrub cover. 7: Older terrace with a grazing meadow. 8: Extremely human-disturbed area. Source: Martínez-Castroviejo *et al.* (1990).

In 1977, eleven years after the construction of the dam, the channel pattern had changed remarkably. A new, confined channel was developed up to 500 m downstream the dam. As a consequence, the braided pattern disappeared and the old instable deposits of the alluvial plain appeared slightly perched over the incised channel, favouring a plant colonization dominated by willows. The presence of new sediments (bare gravels) only appeared when incision ceased, that is, 500 m downstream the dam. Most of the secondary channels had been abandoned and the main, incised channel migrated laterally, with an incipient meandering pattern that eroded the terrace occupied by the meadow. The retreat of the terrace is about 15 m, that is, a retreat rate of 1.4 m per year.

In 1988 the channel was still changing, and the effects of the dam were progressively advancing downstream. In the studied stretch there were no signs of braiding and the incised channel continued 850 m from the dam. At both sides of the channel there were several deposits slightly perched, stabilized levels with different plant colonization. The incipient meandering of 1977 had migrated downstream and the curve was more exaggerated. The activity of the main channel caused a new retreat in the old meadow, with a rate reduced to 0.5 m per year.

The gravel size distribution downstream the check dam also illustrates the changes produced in the channel. Figure 10 shows the sampling points and the sediment size frequency. Table 3 includes other characteristics of the samples. The most significant result was the rapid decrease in the gravel size, in such a manner that gravels were very coarse immediately downstream the dam and decreased toward the extreme of the studied stretch.

This is expressed by the average data, D_{90}, and the frequency histograms, though there are some anomalies within this general trend. The average gravel size was 203.3 mm at point 1 and 83.6 mm at point 11.

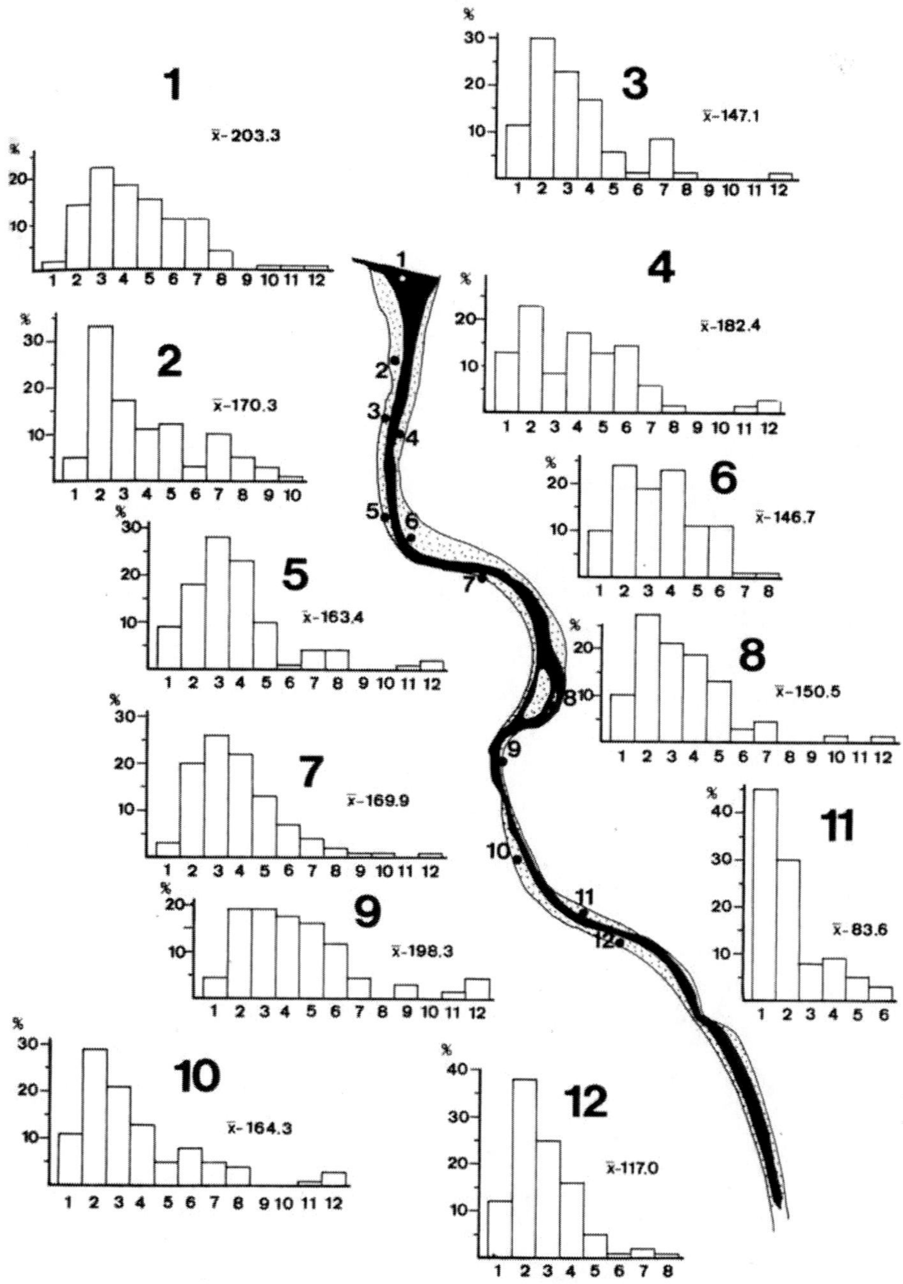

Figure 10. Location of the sampling points of gravel size distribution downstream of the check dam in the Aurín River, and the corresponding frequency histograms. The classes of the histograms range from 0-50 mm (class 1) to more than 600 mm (class 12). Source: Martínez-Castroviejo *et al.* (1990).

Table 3. Characteristics of the sampling points downstream of the check dam in the Aurín River

Sample point	Distance from the dam	Slope	Average grain size (mm)	D90 (mm)
1	5	0.05	203.3	334
2	30	0.025	170.3	350
3	60	0.01	147.1	321
4	80	0.02	182.4	315
5	150	0.018	163.4	322
6	190	0.015	146.7	275
7	240	0.01	169.9	277
8	335	0.005	150.5	268
9	410	0.04	198.3	332
10	480	0.01	164.3	334
11	570	0.01	83.6	200
12	630	0.025	117	193

In the same way, the histograms were progressively narrower downstream because the group of coarser gravels disappeared. Anomalies in this decreasing trend appeared at point 4 (due to the fall of coarse gravels from old deposits) and at points 9 and 10, due to a local gradient breaking caused by the foundations of an old bridge.

The evolution of the average grain size was measured in 70 points in the Ijuez River from the headwater to the mouth. Figure 11 shows that there was a remarkable and expected decreasing trend, though there were also a number of anomalies caused by the arrival of sediment from tributary ravines. Nevertheless, the highest anomalies, both over and beneath the adjustment curve, were caused by each one of the check dams along the Ijuez River. Immediately downstream each dam a strong increase in the average sediment size occurred, related to the development of a surface armour layer. The average sediment size underwent a sudden decrease upstream each check dam (Martínez-Castroviejo and García-Ruiz, 1990).

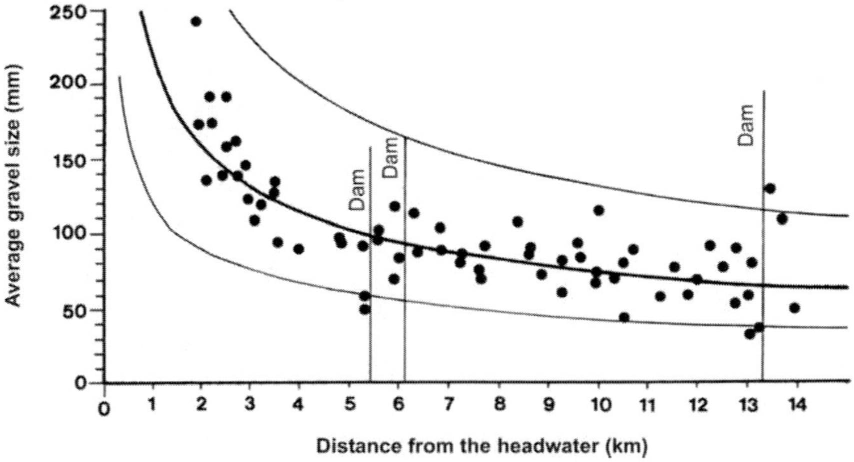

Figure 11. Evolution of the average gravel size from the headwater to the mouth of the Ijuez River. The disturbances caused by the construction of check dams can be observed. The curves delimits the 95 per cent confidence interval. Source: Martínez-Castroviejo and García-Ruiz (1990).

The geomorphic effects of the check-dams was also observed in different ravines close to Biescas, in the Gállego Valley, with a steep gradient in the lowest stretch (almost 20%), before arriving to the alluvial fans. The torrential behaviour of the ravines was seemingly corrected by the construction of a number of check dams in the Arás, Escuer, Orós and Oliván ravines (Figure 12), as well as with the reforestation of a part of the basin. The most important consequence was the development of an entrenched stretch in the alluvial fans, due to a deficit in the sediment supply (Figure 13).

Figure 12. Check dam in a steep, torrential ravine, the Arás ravine, Gállego River basin.

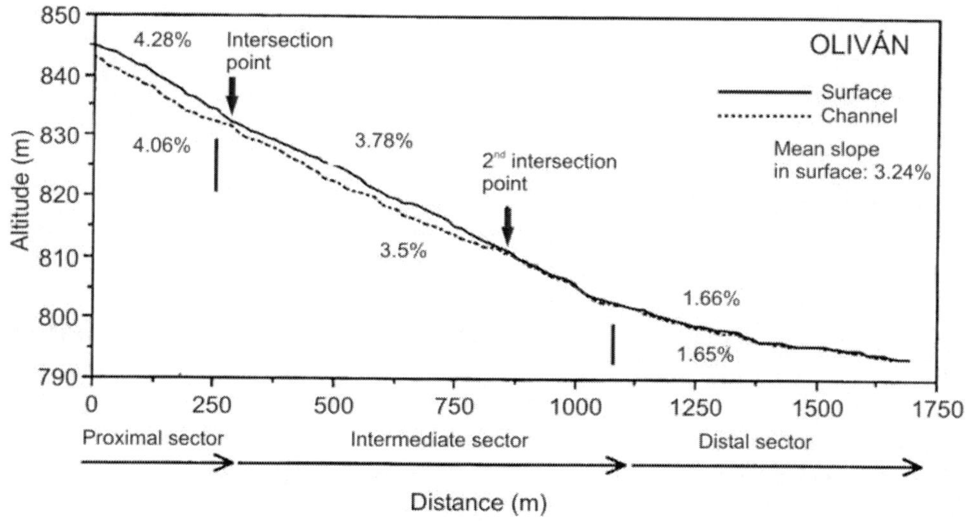

Figure 13. Longitudinal profile of the Oliván alluvial fan, showing both the surface and the channel profile. Source: Gómez-Villar (1996).

The entrenchment reached 5 m in the proximal sector of the alluvial fans (Gómez-Villar, 1996), eroding not only the fan sediments but also the flysch bedrock. Sedimentation begun in the intersection point, close to the distal area of the fan and even allowed the transfer of sediment to the main river. This contributed to the stabilization of the alluvial fans, since new sediments were only deposited in the lowest part of the fans, and the entrenchment impeded channel wandering.

The capacity for trapping sediments was studied after the Biescas campsite disaster on August 7th 1996 in the Arás ravine. An extreme hydrological event related to a very intense rainstorm (about 200 mm in 1 hour and 15 minutes) (White *et al.*, 1997) caused the sudden destruction of 38 check dams located immediately upstream the beginning of the alluvial fan (García-Ruiz *et al.*, 1996). This allowed to coarsely estimate the sediment accumulated by the check dams since their construction in 1940-1945 as 90,000-100,000 tons, that is between 1,800 and 2,000 tons per year. Taking into account that the basin has an area of 18.8 km^2, a sediment accumulation rate of about 95-106 t km^{-2} yr^{-1} was estimated. This rate should be much higher for the time during which the check dams were active (probably up to the middle of the fifties), that is, when they had some capacity to accumulate sediments before being completely infilled. This phenomenon demonstrated (i) the high capacity to retain sediments and to change, at least temporary, the dynamics of very active and torrential rivers, and (ii) the fragility of these structures during extreme hydrological events, when extraordinary destabilization of the channel may cause the destruction of the dams and the motion of all the sediments accumulated during decades, in a process that has been described in other environments (Castillo *et al.*, 2007; Conesa-García and García-Lorenzo, 2007).

CONCLUSION

The construction of check dams in braided rivers introduces important changes in the fluvial dynamics, because they represent a sudden decrease in the coarse sediment transport and induces erosion downstream the check dams. Most authors (i.e., Gregory and Park, 1974; Knighton, 1978; Petts, 1979; Castillo *et al.*, 2007) described the occurrence of a clear incision downstream the check dams, leaving perched the old deposits of the braided alluvial plain. In the case of the rivers studied in the Flysch Sector of the Spanish Pyrenees, the incision in a confined channel represented (1) a process of stabilization of the alluvial channel which reached at least 500 m downstream the dam; (ii) the development of a straight channel that tended to meandering; (iii) a local increase of the erosion risk in the river bank due to the evolution of meanders; (iv) the development of an armoured layer immediately downstream the dam; (v) the organization of coarse sediments following a progressive increase of the smaller gravels with the distance from the dam; and (vi) the colonization of the old alluvial plain by new plant communities adapted to the lowering of the water table.

Upstream the dam, sedimentation was encouraged by the new base level forced by the dam, and sediments also showed a clear spatial organization, with different environments according to the distance from the dam, the gradient and the arrival of debris flows during extreme events.

In very torrential ravines, the sediment storage by the check dams can change the morphology of alluvial fans (entrenchment in the proximal sector), shrinkage of the most active sectors and plant colonization (Gómez-Villar and García-Ruiz, 2000), though in any case this can be considered as a sign of definitive stabilization, as the Arás ravine disaster demonstrated (White *et al.*, 1997).

ACKNOWLEDGMENTS

Support for this research was provided by the projects PROBASE (CGL2006-11619/HID, Consolider) financed by the Spanish Commission of Science and Technology, and PI032/08, financed by the Aragón Regional Government. The authors also acknowledge support from the Program of Research Groups of the Aragón Regional Government, and from RESEL (the Spanish Ministry of the Environment). The contribution of L.C. Alatorre has been possible thanks to a scholarship granted by The National Council for Science and Technology of Mexico (CONACYT).

REFERENCES

Arnáez-Vadillo, J. and Gómez-Villar, A. (1990). Efectos de presas de corrección en la evolución geomorfológica de un torrente de montaña: la morfología del canal. In *I Reunión Nacional de Geomorfología*, Teruel: Sociedad Española de Geomorfología.

Beguería, S., López-Moreno, J.I., Lorente, A., Seeger, M. and García-Ruiz, J.M. (2003). Assessing the effects of climate oscillations and land-use changes on streamflow in the Central Spanish Pyrenees. Ambio, 32 (4), 283-286.

Beguería, S., López Moreno, J.I., Gómez Villar, A., Rubio, V., Lana-Renault, N. and García Ruiz, J.M. (2006). Fluvial adjustments to soil erosion and plant cover changes in the Central Spanish Pyrenees. *Geografiska Annaler*, *88 A* (3), 177-186.

Castillo, V.M., Mosch, W.M., Conesa-García, C., Barberá, G.G., Navarro-Cano, J.A. and López-Bermúdez, F. (2007). Effectiveness and geomorphological impacts of check dams for soil erosion control in a semiarid Mediterranean catchment: El Cárcavo (Murcia, Spain). *Catena, 70,* 416-427.

Conesa-García, C. and García-Lorenzo, R. (2007). *Erosión y diques de retención en la cuenca mediterránea*. Murcia: Fundación Instituto Euromediterráneo del Agua.

García-Ruiz, J.M. and Puigdefábregas, J. (1982). Formas de erosión en el flysch eoceno surpirenaico. *Cuadernos de Investigación Geográfica, 8*, 85-128.

García-Ruiz, J.M. and Puigdefábregas, J. (1985). Efectos de la construcción de pequeñas presas en cauces anastomosados del Pirineo Central. *Cuadernos de Investigación Geográfica, 11*, 91-102.

García-Ruiz, J.M., White, S.M., Martí, C., Valero, B., Errea, M.P. and Gómez-Villar, A. (1996). *La catástrofe del barranco de Arás (Biescas, Pirineo Aragonés) y su contexto espacio-temporal*. Zaragoza: Instituto Pirenaico de Ecología.

García-Ruiz, J.M., Arnáez, J., White, S., Lorente, A. and Beguería, S. (2000). Uncertainty assessment in the prediction of extreme rainfall events: an example from the Central Spanish Pyrenees. *Hydrological Processes, 14*, 887-898.

Gómez-Villar, A. (1996). *Conos aluviales en pequeñas cuencas de montaña*. Logroño: Geoforma Ediciones.

Gómez-Villar, A. and García-Ruiz, J.M. (2000). Surface sediment characteristics and present dynamics in alluvial fans of the Central Spanish Pyrenees. *Geomorphology*, *34* (2), 127-144.

Gregory, K.J. and Park, C.C. (1974). Adjustment of river channel capacity downstream from a reservoir. *Water Resources Research*, *10*, 870-873.

Knighton, A.D. (1978). Reservoir influence on the distribution of mean grain size along a river profile: A case study from north-east Derbyshire. *East Midland Geographer*, *7*, 19-31.

Leopold, L.B. (1994). *A view of the river*. Cambridge: Harvard University Press.

Leopold, L.B. and Wolman, M.G. (1957). River channel patterns: braided, meandering and straight. *U.S. Geological Survey Professional Paper*, 282 pp.

Leopold, L.B., Wolman, M.G. and Miller, J.P. (1964). *Fluvial processes in Geomorphology*. San Francisco: Freeman.

López-Moreno, J.I., Beguería, S. and García-Ruiz, J.M. (2006). Trends in high flows in the Central Spanish Pyrenees: response to climatic factors or to land-use change? Hydrological Sciences Journal, 51 (6), 1039-1050.

Lorente, A., García-Ruiz, J.M., Beguería, S. and Arnáez, J. (2002). Factors explaining the spatial distribution of hillslope debris flows. A case study in the Flysch Sector of the Central Spanish Pyrenees. *Mountain Research and Development, 22* (1), 32-39.

Martínez-Castroviejo, R. and García-Ruiz, J.M. (1990). Coladas de piedras (debris flows) y dinámica fluvial en ríos torrenciales del Pirineo Central: el caso del río Ijuez. *Cuadernos de Investigación Geográfica*, *16*, 55-72.

Martínez-Castroviejo, R., Inbar, M., Gómez-Villar, A. and García-Ruiz, J.M. (1990). Cambios en el cauce aguas abajo de una presa de retención de sedimentos. In *I Reunión Nacional de Geomorfología*, Teruel: Sociedad Española de Geomorfología.

Ortigosa, L., García-Ruiz, J.M. and Gil, E. (1990). Land reclamation by reforestation in the Central Pyrenees. *Mountain Research and Development, 10* (3), 281-288.

Petts, G.E. (1979). Complex response of river channel morphology subsequent to reservoir construction. *Progress in Physical Geography, 3* (3), 329-362.

Richards, K. (1982). *Rivers. Forms and process in alluvial channels*. London: Methuen.

Romero-Díaz, M.A., (coord.) (2007). *Los diques de corrección hidrológica. Cuenca del río Quípar (Sureste de España)*. Murcia: Universidad de Murcia.

Schumm, S.A. (1977). *The fluvial system*. New York: John Wiley.

Schumm, S.A. (2005). *River variability and complexity*. Cambridge: Cambridge University Pres.

Serrano-Muela, M.P., Lana-Renault, N., Nadal-Romero, E., Regüés, D., Latron, J., Martí-Bono, C. and García-Ruiz, J.M. (2008). Forests and their hydrological effects in Mediterranean mountains. *Mountain Research and Development, 28* (3-4), 279-285.

White, S., García-Ruiz, J.M., Martí, C., Valero, B., Errea, M.P. and Gómez-Villar, A. (1997). The 1996 Biescas campsite disaster in the Central Spanish Pyrenees, and its temporal and spatial context. *Hydrological Processes, 11,* 1797-1812.

Wolman, M.G. (1954). A method of sampling coarse river-bed material. *Transactions of the American Geophysical Union, 35*, 951-956.

Commentaries on the Book

"This book will be of immense value to those who need a broad guide to Check-Dams, Morphological adjustment and Erosion control in torrential streams" - Prof. Eng. Sergio Fattorelli; Full professor of "Water Resources Planning and Management", University of Padova, Italy.

"This book brings up to date a constant issue in torrential streams, especially those in mountain areas: the need to know and regulate them" - Prof. Juan Ángel Mintegui Aguirre; Professor at the E.T.S. I. de Montes. Universidad Politécnica de Madrid, Spain.

"The Authors have epitomised and integrated in an excellent way the various aspects of the topics concerning erosion problems in torrential streams" - Prof. Geol. Mario Panizza; Full professor in "Applied Geomorphology", University of Modena and Reggio Emilia, Italy.

"This is a valuable synthesis of perspectives on the implementation and effects of check dams and is a very useful contribution to increased understanding of their use and consequences in environmental management" - Prof. Janet Hooke, Professor of Physical Geography, University of Liverpool, UK.

LIST OF CONTRIBUTORS

Dr. Luis C. Alatorre - Instituto Pirenaico de Ecología, CSIC, Campus de Aula Dei, Apartado 13.034, 50080-Zaragoza, Spain.

Dr. Simon Beecham - Centre for Water Management and Reuse (CWMR), University of South Australia; Mawson Lakes Campus, Room P2-39 , Mawson Lakes Sa 5095, Australia

Dr. Santiago Beguería - Estación Experimental de Aula Dei, CSIC, Campus de Aula Dei, Apartado 13.034, 50080-Zaragoza, Spain

Dr. Anne Chin - Department of Geography, University of Oregon, Eugene, OR 97403-1251, U.S.A.

Dr. Francesco Comiti - School of Science and Technology, Free University of Bolzano, via Sernesi 1, Bolzano, Italy

Dr. Carmelo Conesa-García - Department of Physical Geography, University of Murcia, Campus de la Merced s/n. 30001 Murcia, Spain

Dr. Vincenzo D'Agostino - Department of Land and Agroforest Environments, Università di Padova, Viale dell'Università 16, 35020 - Legnaro (Padova), Italy

Dr. William E. Doolittle - University of Texas, Austin, USA

Dr. Nada Dragović - Faculty of Forestry, University of Belgrade, Kneza Višeslava 1, 11030 Belgrade, Serbia

Dr. Shinji Egashira - NEWJEC Inc, 3-20, Honjyo-Higashi, 2-chome, Kita-ku, Osaka 531-0074, Japan

Dr. Vito Ferro - Dipartimento di Ingegneria e Tecnologie Agro-Forestali, University of Palermo, Viale delle Scienze 13, 90128 Palermo, Italy

Mr. Rafael García-Lorenzo - Autonomous Community of Murcia Region. Environmental Integration and Information Service. C/ Catedrático Eugenio Ubeda, s/n. Murcia, Spain

Dr. José M. García-Ruiz - Instituto Pirenaico de Ecología, CSIC, Campus de Aula Dei, Apartado 13.034, 50080-Zaragoza, Spain.

Dr. Amelia Gómez-Villar - Department of Geography and Geology, University of León, Campus de Vegazana, , 24071-León, Spain

Dr. Jeffrey P. Haltiner - Phillip Williams & Associates, Ltd., 550 Kearny Street, Suite 900, San Francisco, CA 94108, U.S.A.

Dr. Jian LIU - College of Civil Engineering, Shenzhen University, Nanshan District, Shenzhen 518060, China

Dr. Ali Morad Hassanli - Shiraz University, Shiraz, Iran & University of South Australia, School of Natural and Built Environments; Center for Water Management and Reuse (CWMR) Room H3-24; Mawson Lakes Sa 5095, Australia

Dr. G. Mathias Kondolf - Department of Landscape Architecture and Environmental Planning, University of California, Berkeley, CA 94720, U.S.A.

Dr. Stanimir Kostadinov - Faculty of Forestry, University of Belgrade, Kneza Višeslava 1, 11030 Belgrade, Serbia

Dr. Mario Aristide Lenzi - Department of Land and Agroforest Environments, Università di Padova, Viale dell'Università 16, 35020 - Legnaro (Padova), Italy

Dr. Luca Mao - Department of Geography, University of Hull, Cottingham Road, Hull, United Kingdom

Dr. J.-P. Malet - IPGS, CNRS UMR 7516, University of Strasbourg. 5 rue René Descartes, 67084 Strasbourg Cedex, France

Mr. Shuhei Minami - NEWJEC Inc., 3-20, Honjyo-Higashi 2-chome, Kita-ku, Osaka, 531-0074, Japan

Dr. Stefano Pagliara - Department of Civil Engineering, University of Pisa, Via Gabba 22. 56122 Pisa, Italy

Dr. Michele Palermo - Department of Civil Engineering, University of Pisa, Via Gabba 22. 56122 Pisa, ITALY

Dr. Alexandre Remaître - IPGS, CNRS UMR 7516, University of Strasbourg. 5 rue René Descartes, 67084 Strasbourg Cedex, France

Dr. Rabindra Osti - International Centre for Water Hazard & Risk Management, Public Work Research Institute, Minamihara 1-6, Tsukuba 305-8516, Japan

Dr. Guang-Qian Wang - Department of Hydraulics and Hydropower Engineering, Tsinghua University, Beijing, 100084, P. R. China

Dr. Ellen Wohl - Department of Geosciences, Colorado State University, Ft. Collins, CO 80523-1482, U.S.A

Dr. Xiang-Zhou Xu - School of Civil and Hydraulic Engineering, Dalian University of Technology, Dalian 116024, P. R. China

Dr. Yi Xu - School of Civil and Hydraulic Engineering, Dalian University of Technology, Dalian 116024, P. R. China

Dr. Hong-Wu Zhang - Department of Hydraulics and Hydropower Engineering, Tsinghua University, Beijing, 100084, P. R. China

Dr. Ou-Yang Zhang - Bureau of Hydrology, Changjiang River Water Resources Commission, Wuhan, 430010, P. R. China

INDEX

A

B

C

D

E

F

G

H

I

J

K

L

O

P

R

S

T

U

V

W

Y

Z